Power Production: What are the Risks?

Two liquefied gas storage containers ablaze after a series of explosions at a storage depot on the outskirts of Mexico City. The disaster, in November 1984, left at least 500 dead and many injured. Photograph © Associated Press Ltd.

Power Production:
What are the Risks?

J H Fremlin

Emeritus Professor of Applied Radioactivity,
University of Birmingham

Adam Hilger, Bristol and New York

British Library Cataloguing in Publication Data

Fremlin, J. H. (John Heaver)
  Power production: what are the risks?. – 2nd ed.
  1. Great Britain. Energy industries. Hazards
  I. Title
  363.1'1962'042'0941

  ISBN 0-85274-132-4
  ISBN 0-85274-133-2 pbk

US Library of Congress Cataloging-in-Publication Data

Fremlin, J. H.
  Power production: what are the risks?/J. H. Fremlin. – 2nd ed.
    p. 24cm.
  Includes bibliographical references.
  ISBN 0-85274-132-4. – ISBN 0-85274-133-2 (pbk.)
  1. Energy industries–Health aspects–Great Britain. 2. Energy industries–Great Britain–Safety measures. I. Title.
  HD7269.E472G744   1989
  363.1'19621042'0941–dc20   89-19780

First published 1985
Second edition 1989

Consultant Editor: **Dr R H Taylor**, CEGB

Published under the Adam Hilger imprint by IOP Publishing Ltd
Techno House, Redcliffe Way, Bristol BS1 6NX, England
335 East 45th Street, New York, NY 10017-3483, USA

Typeset by Mathematical Composition Setters Ltd, Salisbury, Wiltshire
Printed in Great Britain by J W Arrowsmith Ltd, Bristol

Contents

Appendices

Preface

Since I first wrote this book a great deal of new information has been published on subjects as widely apart as windmills, radon, smoking risks and the disposal of nuclear wastes, among others, and I have tried to bring this book up-to-date. In particular, the disaster at Chernobyl happened, so a new section has been incorporated in Chapter 9.

There has also been an increase in public concern on subjects such as acid rain and the increase of carbon dioxide in the atmosphere, but I have not had to make much change in my original version of these dangers. I have always wished to preserve the environment and to conserve our natural resources; not only coal and oil but also rain forests and the varied fauna and flora of the world. We have a dilemma here: we can try to make the world more comfortable and safe for ourselves and our descendants, but the natural world could be conserved better if there were fewer of us and we did not multiply so fast.

Industry in general has increased the efficiency of use of energy. This has however led to much greater use of electrical power, and the need for this is rising. In the thirty years 1958 to 1988 the average electricity use by householders went up from 1700 kWh to 4000 kWh per year. The use of renewable sources is increasing, but will not during the next couple of generations be able to cope with the increasing demand.

With all these factors in mind I agree with the 70% of the British public found by a recent poll to believe that more nuclear power will be needed in the future; and that we have to plan future needs *now*.

I make no apology for using rems rather than the inconveniently large sieverts, although the ambiguity of the contraction Sv is unlikely to cause confusion. Rads and rems are used in America and in the USSR, and disintegrations per second can be more easily remembered than becquerels.

J H Fremlin
1989

Preface to the First Edition

It is an extraordinary feature of the long and healthy lives that we now enjoy that we are worrying more about the dangers of the abundant energy supplies that have made these advantages possible than we do about the enormously greater and potentially controllable hazards such as those arising from the ordinary things that we eat and drink and breathe. This is not due to public stupidity, but to ignorance fed with well-meant misinformation by ill-informed media avid for the frightening and the horrible.

In this book I have attempted to provide a factual explanation and discussion of the actual risks arising from each of our important present and prospective sources of energy. Unfortunately, to do this needs a lot of numbers and some arithmetic. Arsenic and strychnine in large quantities are dangerous poisons. In small quantities they are useful tonics. To answer the vital question 'how much is dangerous?' or 'how dangerous is it?' you need figures. I have tried to write in such a way that you can skip the figures if you wish but can still get the message. But if you want to understand in depth, the facts and figures are there.

The most misleading part of the book is the appearance of my name alone on the cover. My wife Reinet will not count herself as joint author; but in clarifying my complicated sentences, in simplifying my explanations in the light of her 19 years of physics teaching, in pointing out omissions and in producing felicitous ways of putting things, not to speak of doing much of the arithmetic, she has done more than many joint authors do. Unlike most joint authors she also typed the entire manuscript, the labour of which will be appreciated by those who know my handwriting, and drew all of the line diagrams.

I am grateful to many others who have contributed; firstly to Mrs Eileen Shinn who made a fair copy for the publishers against the organised and persistent delaying tactics of successive word-processing machines; to Mrs Margaret Baggott who traced the illustrations to a high professional standard; to non-scientific friends who made suggestions on style; and to all of my friends in the University of Birmingham and elsewhere who wittingly or unwittingly contributed to the data I have used.

John Fremlin
September 1984

Part 1

Introduction to Risks

Chapter 1

Introduction

My objectives in writing this book are twofold. The first arises from my concern about the destruction of the world's non-renewable resources, in particular its resources of easily available energy, consequent upon the explosion in consumption in the technologically developed countries and the uncontrolled growth of population in the less developed. A huge increase in the use of fossil fuel is needed by the Third World if it is even to approach the standards of living, education and health now regarded as a minimum in the rich countries. For this to be possible, the developed countries must use their technology to develop alternatives to the use of fossil fuels. Accordingly, I would like to see the most rapid extension of wind, wave and solar power that people can be persuaded to pay for, together with that of nuclear power.

My second objective is to reduce the exaggerated fear of nuclear power which is hindering its development. With this aim I shall make a detailed examination of the hazards characteristic of each of our possible energy sources.

I am often attacked for being biased towards nuclear power, and accused of being an enthusiast for its use. This is not so. I am biased in favour of facts and against the unnecessary use of fossil fuels, especially oil. I am driven by these views to support the building of more nuclear power stations, only because nuclear power plus the renewable alternatives can conserve fossil fuel more rapidly than can the renewable alternatives alone.

RISKS OF THE PAST

The first real humans were for a long time hunters and gatherers of wild food supplies. Their chief dangers were from accidents, predators and famine and from fighting between different groups, but during this time two other hazards appeared.

Fire

The first new hazard arose from mastering the first source of energy other than human muscle—fire. We do not know when this was achieved; there must have been many attempts over a long period before useful fires could be kept going indefinitely, and a longer period still before they could be started by friction rather than by taking advantage of a grass or brush fire started by lightning.

From the beginning fire must have been dangerous, and initially it would have been of only marginal value as a source of warmth in the tropical and sub-tropical regions in which early man lived. It would have had a real value in warning off predators, for cooking excessively hard or otherwise indigestible foods, and for hardening wooden spear points; but if a committee of elders had had to be assembled to decide on permission for its use and to satisfy those afraid of its ability to burn both forests and savannah in the dry season—in fact to destroy their whole known world—northern Europe might still be inhabited only in the summer by visiting bands of hunters. Either there were no committees, or they were not obeyed, and fire opened the cold-winter lands to human occupation.

For all its essential advantages fire was genuinely dangerous and even now has not been made safe, although over the last 30 years the move away from open fires in the home has reduced the risks. At the beginning of this century 2500 lives were lost per year in fires in Britain, and even now there are still about 700 fire-caused deaths per year.

Disease

The second hazard, which grew to be significant rather than appearing for the first time, was disease. Serious disease is rare among thin and scattered populations; any disease or parasite needs a critical density and mass of susceptible hosts to spread successfully. With increasing populations in long-settled areas the persistent debilitating diseases such as the milder forms of malaria could have developed; but parasites of slow-growing creatures such as man, spread sparsely as hunter–gatherers have to be, cannot afford to kill their hosts, or even to make it too difficult for them to earn their livings. Accordingly, it is unlikely that disease was at this stage a serious danger.

The first really important rise in the risk of disease came with the large-scale development of agriculture, between 10 000 and 20 000 years ago. This made it possible to feed populations large enough and dense enough to reach the critical conditions for air- or waterborne epidemic diseases. Throughout the historical period disease has killed farmore than have been killed by war or famine. The influenza epidemic of 1919 killed more people worldwide than did the fighting over the whole period of the 1914–18 war. Indeed, in spite of the unpleasantness of wars and the very considerable numbers that on occasion they are able to kill, they have not been the main limiters of population since records began, and it is unlikely that without war the world population would have grown appreciably faster than it has. In the absence of wars, and of the destruction of food supplies they cause, populations would have more rapidly reached the critical density for the next epidemic but would have been reduced by it to much the same absolute level as before.

In view of this, the differing risks due to the increasing variety of energy sources were unimportant to the mass of people, important as they could be to individuals. Nevertheless, the early sources of energy were far more dangerous per unit of output than the energy sources of today, and those who believe that we could reduce the dangers facing us by returning to a simpler life are making a big mistake. Happier we—some of us—might be. Safer we would not be.

EARLY ENERGY SOURCES

The first source of mechanical energy other than human muscle was animal muscle. To maintain the continuous gigawattage (one gigawatt (GW) is a million kilowatts (kW)) of mechanical work that can be produced by the electricity supplied by a modern power station, would have required 5 or 10 million oxen working in relays. The fuel (grass and hay or other vegetation) to feed them would have required perhaps 10 000 square miles of good land to produce. Oxen are not dangerous animals, though their fathers are, but their droppings must have provided an important additional source for the flies which carried the bacteria responsible for killing diseases such as infantile diarrhoea. Power stations do not breed flies.

The more powerful horse was little used until the twelfth century because harness suitable for oxen was liable to choke a horse if it pulled hard. The invention of the horse collar raised the animal's efficiency as much as Watt's later introduction of a condenser increased the efficiency of the steam engine. Horses then gave a much larger output than oxen. The dangers arising from the fall-out of waste were very similar, though perhaps more serious owing to their more frequent presence in the cities as well as in the countryside.

In many instances—for example in the pyramid-building period in Egypt—slave power was used as well as animal power. A slave provided less mechanical output than an ox, had to be fed on more expensive foods, and might take nearly 20 years to produce and work up to full output whereas a modern power station takes 8 to 10. Two sexes were needed in nearly equal numbers to keep up the supply, and a 30% availability over 24 hours and round the year would have been good. Hence 150 million or more would have been needed to keep going a continuous gigawatt of mechanical work. Against this, they had built-in control systems of much greater versatility than had the oxen and could be set to work on far more complex tasks.

In ordinary conditions slaves were not very dangerous; but excessive use of large numbers, without proper repair and maintenance, could lead to revolts in which not only were a lot a of slave-owners killed but many of the female owners might get raped; a hazard for which not even the nuclear component of our modern supplies can be blamed. While the risk per decade of such uprisings was small, the real contribution of slaves to the death rate has rarely been recognised and at the time it was not recognised at all. This contribution arose from pollution of the environment by their toxic wastes. Disease, as we have seen, was then the main cause of death, and the contribution of slaves to the size and density of population must inevitably have induced more frequent epidemics and have been responsible for an important fraction of the total death risk of the time, for slaves and slave-owners alike.

THE CONQUEST OF EPIDEMIC DISEASE

In Britain the death rate from disease remained high up to the early part of the nineteenth century, both from periodical epidemics of such acute conditions as cholera and smallpox and from endemic diseases such as tuberculosis and fly-borne infantile diarrhoea.

As the wealth based on manufacturing industry expanded, the concomitant improvement in food supplies led to improved resistance to endemic diseases among the poor. The reporting by Jenner around 1800 of vaccination against smallpox using lymph from cowpox sufferers, together with the discovery reported by Snow in 1849 that cholera was spread by contaminated water, made it possible to avert the periodic disastrous epidemics of these two diseases. The Public Health Act 1848 established 400 local boards of health, concerned with the supply of pure water and with disposal of sewage. A major contribution was made by sanitary engineers. By the 1880s all of the big towns of Britain had efficient sewerage systems to replace the repellent system of throwing the night's excretions into the street. Well into last century the cry of 'gardyloo' (perhaps from *gardez l'eau*?) could be heard in the streets of Edinburgh, warning the passers-by that slops were about to descend from upper windows.

Airborne infectious bacterial diseases persisted for some time, but were defeated by the development in the last 50 years of first the sulphonamides and then penicillin and other antibiotics, which allowed the curing of bacterial diseases by positive action. Effective vaccines were developed against poliomyelitis and the killing virus diseases of childhood such as diphtheria, whooping cough and measles. In 1979 nobody in Britain died of diphtheria or acute poliomyelitis; 17 (all under 10) died of measles and 7 of whooping cough. The number of deaths from infectious diseases in Britain has now fallen to only about 4 per 1000 of total deaths, mostly among the over 60s. The death rate in youth or middle age fell to unprecedentedly low levels, and the hitherto neglected risks arising from the large increase in energy production and consumption, that had made all these improvements in health possible, began for the first time to arouse real concern.

WIND, WATER AND STEAM POWER

The first non-living source of mechanical power was the wind, used for propelling ships and, by mediaeval times, being capable of providing tens or hundreds of kilowatts for this purpose in suitable conditions. It was used for pumping water for irrigation and for grinding corn in Britain until the early years of this century. Simple water-mills were first used around 100 BC, and by 1086 the Domesday Book recorded nearly 6000 water-mills in England, mostly used for grinding corn. These were intermittently used, and the average output over the year for each was less than a kilowatt.

Windmills probably killed more construction workers than members of the public (even if they were quixotically attacked by short-sighted persons). Water-mills, however, could be of direct danger to people swept over or under water-wheels, and the mills had to have millponds for storage of water in slack periods. One death by drowning, mostly of children and drunks, per 200 years per mill would seem a reasonable guess of the risk. Taking an optimistic estimate of 1 kW as the round the year output, a million mills would be needed for a gigawatt output, killing some 5000 people per year. Air pollution by raw-coal burning in our big cities 100 years ago may have reached this rate of killing, but nothing we have today in the industrialised countries even approaches it.

By far the largest increases in availability of mechanical energy arose of course from the use of steam produced by burning coal, followed by the development of electricity for power transmission and of internal combustion systems based on the burning of liquid fuels.

The risks involved in the use of these will be discussed in detail in the following chapters.

Chapter 2

The Risks of Modern Life in Britain

In order to see clearly the importance of the dangers resulting from the use of the main sources of energy it is necessary to discuss as a background the dangers to life and health from all causes. It will become apparent that the production and use of power are responsible for only a tiny part of our death and disability rates. Furthermore, the far smaller risks we face today as compared with those prevalent before the development of modern energy sources, stem to a large extent from this very development, which has liberated an unprecedented proportion of human time and effort from the simple drudgery of the earlier ways of keeping us warm and fed.

Throughout this book I shall be discussing only the dangers in the richer countries such as those of Europe and North America, and most of my examples will be concerned with Britain. Paradoxically, though understandably, in the rich countries, in which the major traditional hazards of epidemic diseases and tuberculosis have been abolished, a great deal of money and effort is available to reduce the remaining risks† that we can still hope to control, even though doing so may add only a few years to our mentally and physically active lives. It is therefore of practical value to know their correct order of practical importance.

The first requirement is qualitative; there are many kinds of danger and these will be rated differently by different kinds of people. Thus a teenage boy will often accept a quite significant risk of killing himself in some completely unnecessary activity to prove his courage to himself. A young mother will not accept such a risk.

In the following chapters I shall discuss only the magnitudes of risk to life or health, together with the longer-term risk of mutation, and not the risks of such things as losing your job or failing an examination or being late for an appointment, although in real life people often worry

---

† Although the words are often used synonymously, it is desirable to distinguish the meaning of risk from that of hazard. An unfenced deep hole in the ground is a hazard; the risk is the probability that somebody will fall into it multiplied by the seriousness of the consequences. If these are measured from 0 for no effect to 1 for death, the numerical risk of death is simply the numerical probability that death will occur.

more, and more often, about these mundane matters than they do about the more serious but less probable calamities.

UNEQUAL ATTITUDES TO DEATH OR INJURY

Even when dealing only with death, illness and injury, it is quite inadequate simply to give figures for the number of cases per 100 000 per year that arise from different causes. Different causes affect people of different ages differently. Taking all cancers together for example, there are very few deaths before the age of 30, but the number per 100 000 almost trebles each decade thereafter until around 70, and appears to fall only among the over 80s when perhaps those more susceptible or those exposed to specific important carcinogens have already died, and other causes of death are increasing even faster. Accidental deaths on the other hand are high both for the very young and the very old, and among British males show a marked rise in the late teens, leading to a peak in the early 20s, contributed to very heavily by the popularity of high-powered motorcycles and other hazards which appeal less to older people. No such peak in death rate appears among girls. During the last few million years of our evolution there were no motorcycles, but it is likely that a community would have gained as the result of aggressive and dangerous exploration outside their normal territory by expendable young males, owing to the valuable experience gained by the survivors.

The importance as well as the probability of death or injury varies with age. The proportion of deaths in the long run is of course always the same: one person, one death. Nothing that power production can do or cannot do will make us immortal, and what is important is therefore not *whether* one dies, and only to limited extent *how* one dies, but *when* one dies. The interesting quantity is how much life one gets before dying, and hence the seriousness of any particular death risk depends on how much more life you could have expected if you hadn't died of what you did die of. The death of a 20 year old woman with an expectation of life of 55 years seems obviously worse than the death of an 80 year old man with an expectation of life of five years. In terms of the lost contribution to society, the disproportion is of course greater still.

THE CAUSES OF DEATH IN MODERN BRITAIN

The chief present causes of death in Britain, which do not differ greatly from those in other rich countries, are shown in table 2.1, together with the average loss of expectation of life. All the major causes of death have been included in this table, covering 99% of total deaths in the year 1986.

In addition I have included some subsections of particular causes, where they are of special interest.

If we look at the ages at which people have died of a particular cause we usually get a graph like that in figure 2.1, which has been plotted for all acute myocardial infarctions (a common form of heart disease). This shows a change with age which is very similar to that for the other major causes of death. The greatest number of deaths occurs in the age range 70–74; for other causes the most dangerous age range is given in the fourth column of the table.

Table 2.1 A selection of causes of death in England and Wales in 1986.

Cause of death	Number of deaths	% of total deaths	Rate per million per year	Age of most deaths	Average loss of life expectancy (yr)
Neoplasms (cancers) —*all sites*	M: 73 717 F: 67 084	M:25.6 F:23	M:3021 F:2610	70–74	12.9
Trachea, bronchus and lung†	M: 25 235 F: 10 022	M: 8.8 F: 3.4	M:1034 F: 390	70–74	12.0
Leukaemia— all kinds	M + F: 3572	M + F: 0.6	M + F: 71	75–79	17.4
Diseases of the circulatory system	M:136 414 F:142 335	M:47.4 F:48.0	M:5591 F:5544	75–79	9.3
Acute myocardial infarction	M: 58 711 F: 43 957	M:20.4 F:15	M:2406 F:1712	75–79	10.5
Diseases of the respiratory system	M: 32 999 F: 30 053	M:11.5 F:10.2	M:1352 F:1171	80–84	8.1
Pneumonia and influenza	M + F: 28 211	M + F: 4.8	M + F: 563	85–89	5.9
Injury and poisoning	M: 11 271 F: 7483	M: 3.9 F: 2.6	M: 462 F: 291	80–84	26.6
Road accidents (excluding pedestrian deaths)	M: 2397 F: 768	M + F: 0.54	M + F:63.2	15–24	41.4
Accidental falls	M + F: 3732	M + F: 0.64	M + F:74.5	80–84	9.9

† For comparison, total deaths in the USA are 136 000 per year.

The frequency graph which shows the greatest difference in shape is that for deaths from road accidents. This is given in figure 2.2 and separates pedestrians who are rarely the active agents in road accidents. There is a very striking peak at age 15–24.

From the figures of deaths at different ages it is possible to calculate the average 'loss of life-expectancy'. A child of 10 in Britain expects on average another 63 years of life, whereas a grandparent of 75 expects on average another 10 years. The final column of table 2.1 shows the average 'loss of life-expectancy'. The figure for road accidents is strikingly different from all others. The greater the figure in the last column, the more young people are affected. (If road accidents are subtracted from the general 'injury and poisoning' figures, the average loss of life-expectancy in this section is reduced from 26.6 years to 23.6 years).

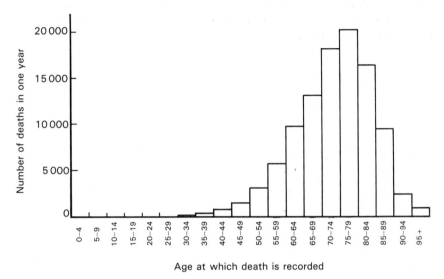

Figure 2.1 Deaths from acute myocardial infarction in 1986.

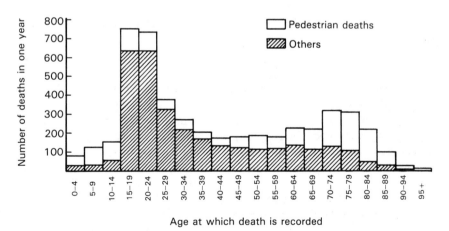

Figure 2.2 Deaths due to motor traffic accidents in 1986.

It will be seen that in terms of loss of life-expectancy the biggest single cause of death, namely diseases of the circulatory system (mainly heart disease), is much less serious than accidental death on the roads. The loss of life-expectancy from all neoplasms (cancers) put together is rather greater, in part as a result of the fact that leukaemia is a significant cause of death among children under 20 (about 20 per million per year).

Table 2.1 represents an enormous change from earlier periods. From 1780 to 1800 the death rate in Britain from tuberculosis alone was 10 000 per million per year of the total population, almost the same as the *total* death rate of today (to which tuberculosis adds less than 20 per million per year). The total death rate itself has dropped by a factor of two in the last 100 years, corresponding to a near doubling of expectation of life at birth and, although the death rates above 75 years of age have altered little, the probability that a child of 1 will die before 50 has decreased between 5 and 10 times.

These quite moderate general figures do not give an adequate impression of the importance of the changes. If no girls between 1 and 25 died at all, it would make only about 0.75% change from the present value in the number of survivors. We can say that since the proportion of female deaths before the end of child-bearing age is small (about 1.5%) future British population figures will, if we avoid war, depend almost entirely on numbers of births and on emigration since no possible improvements in health services can make any significant difference. There are other interesting demographic effects. For example, we can judge with considerable accuracy in 1989 how many applicants qualified on present standards could be trying to enter university in the year 2006.

WHY YOUNG PEOPLE ARE CONCERNED WITH VERY SMALL NEW RISKS

Of more relevance to this book is the fact that over the opinion-forming years between 15 and 30 the average risk of death is little more than one in 10 000 per year. People in this period of life have never bothered about known risks of death at that level, and indeed in the past have been little concerned about much larger risks. It may be of particular importance too that the residual risk of death over this period is largely due to accidents. There is of course a sense in which any accident could be said by the victim to have been unavoidable; if it were avoidable he would have avoided it. But this applies only to the moment of the accident. If you skid uncontrollably at 85 mph on a 750 cc motorcycle and fail to avoid oncoming traffic it is true that the accident was unavoidable from the moment of the skid. But you do not have to ride a 750 cc motorcycle or anything else if you do not think it worth the risk of accident. You do not have to climb mountains in bad weather with inadequate clothing and no emergency food or equipment. The whole of the excess of the five-times greater death rate in traffic accidents among boys than that among girls in the same age range must result from this kind of accepted risk.

Altogether, the probability of death from causes over which the individual has no influence at all between 15 and 25 can be only a few in 100 000 per year. It is not unreasonable therefore that young people should grow up with a feeling that even the tiniest extra external risk is unacceptable—after all, a new extra risk of death of one in a million per year, about which no previous generation could possibly worry, adds several per cent to the pre-existing uncontrollable risks of the present generation. Naturally, young people do not do this kind of calculation; they merely grow up without ever knowing personally anyone in their age range who has been killed by any uncontrollable cause, but read about such deaths in newspapers which publish them because they are rare.

There is another factor of significance. At these levels of risk, quantitative discussions mean nothing to a very large number of people, including nearly all of those in their formative years. I do not suppose that very many will think that a one in a million chance must be a lot greater than one in a thousand chance. However I am quite sure that for a lot of people there is very little difference between 10 000 and a million or 1000 million—all of these represent large quantities beyond real comprehension—and that someone who is perfectly prepared to neglect a one in a thousand chance of something going wrong in a not-too-vital project may find it quite unacceptable to regard a one in a million chance as negligible compared with a one in a thousand chance.

To a mathematician it may seem odd that so many fail to appreciate the significance of large numbers. In fact, it is the mathematicians that are odd. Our senses and brains have evolved to *detect* small changes or differences in our environment, not to *measure* large ones. In comparing the weights of two similar objects we can with practice detect changes of about 10%, such as the difference between 1 g and 1.1 g, or between 1 kg and 1.1 kg, but the extra 0.1 g that we can detect when added to 1 g cannot be detected when added to 1 kg. We can detect a change of less than 10% in brightness between adjacent subjects in a picture whether we are looking at it in bright sunlight or in moderate lamplight, which may give a change in the total intensity of the light reaching our eyes of several thousand times. With dark-adapted eyes we can distinguish dark objects from their less dark background at 1000 times lower intensity still. Our hearing can discriminate fractional variations over an even larger range in intensity. We do not however *feel* that the brighter lights and the louder sounds that we can accept without distress are a million times as intense as the poorer lights and the fainter sounds by means of which we still get similar information.

When it comes to counting, we can distinguish intuitively and immediately the number of irregularly arranged objects up to four or five, but beyond that we have to count. It is said, I do not know with what truth, that if five men with guns approach a flock of rooks feeding in a field and four men go away leaving one hidden, the rooks which left the field on the men's approach will not return, recognising that not all have gone; but if six or more men leave one behind the rooks will return without realising that not all have gone. If this is true, our direct appreciation of small numbers is pretty primitive, and depends little on intelligence.

The really basic numerical difference is that between one and two. It is easy to recognise that one plank is twice as long as another, that a bottle is half full, or that this class of children has about twice the number of that class. If we multiply ten twos together (2^{10}) we get 1024, very close to 1000, and multiplying this by a further 2^{10} gives just over a million. It may not be a coincidence that unless you have a quite unusual flair for arithmetic the 'feel' of the word 'million' is much more like ten times the 'feel' of the word 'thousand' than of a thousand times a thousand.

Some writers purposely use this difficulty of understanding large figures to give an exaggerated impression, by saying for example 'fifty people will die in Britain in the next 10 years as a result of such-and-such'. The 50 deaths are real and sound serious; the fact that this means only one in 10 million per year has no impact on most of the readers. It could be made to do so simply by adding 'and about six million will die from other causes in the same time'. Most people will feel, correctly, that six million is a lot more than 50.

NUMERICAL PREDICTION OF RISK

Since I shall be showing in this book that no single lethal risk to the public arising from any source of energy is as great as one in 10 000 per year, it is inevitable that the quantitative arguments based on numerical evaluations of risk will be found unconvincing by the vast majority whose lives do not require them to deal extensively with large numbers. It is accordingly essential to give practical comparisons from several fields, and I shall also try to indicate the way in which those who do not readily think numerically do in fact evaluate risks.

There follow three examples of known risks with which the risks arising from power generation can be compared.

A risk which is at present uncontrollable by most of the potential victims is that of dying from acute heart disease. Death from acute myocardial infarction may occur at any age, and little can be done to reduce the danger, but the probability increases rapidly as you get older.

When you are 22 there is one in a million chance that you will die of this within the next two years. You will probably not have heard of any one you have met, of your own age, who has done so. At 32 there is a one in a million chance that you will die of it in the next month (you are still most unlikely to have met a case) and by the time you are 72 there is a one in a million chance that you do so in the next five hours. This means that you have the far larger chance of one in a thousand of dying of it within the next 7 months, and by the time you have reached this age you will probably have known of several people who have died of it before they reached 72 and it will be a matter of genuine interest.

These figures are based on records of causes of death among the total population (some 50 million) of England and Wales (Mortality Statistics 1986) and are therefore the result of direct experience. If a risk is very small the accuracy obtainable from experience will be poor since very few cases can have been observed and random chance may become important. The average number of people who die each year as the result of wasp stings is between one and two; representing an average risk of about one in 30 million per year. 30 people were killed directly in England and Wales between 1949 and 1969 (Spradbury 1973). In many years there will be no such deaths and occasionally there could be 10 or more. A warm spring with no late frosts will favour the establishment of wasps' nests, and a sunny late summer may tempt more people out into waspy territory for wasp-attracting picnics. This is not only a small risk, but an avoidable one. If you stay in a large city it can be made almost zero; if you keep a good look out for wasps' nests, examine your sandwiches and beer before you take a mouthful, and know how to do a tracheotomy if you do get stung in the throat and start to choke, you can be nearly as safe.

In both of these examples the magnitude—or minitude—of the risk can be sufficiently closely estimated from past experience. The examples also give an indication of the way in which people evaluate traditional risks, without any thought whatever of calculation but getting much the same answer about which risks matter and which do not, by social experience. It is sometimes quite possible however to make a quantitative estimate when there has been no record at all of anyone being killed by the risk considered.

For example, although there is an ancient Chinese record of the destruction by a meteorite of several chariots and 10 men in 616 BC, there are no records of anyone being killed in Britain by being struck on the head by a meteorite, and there have probably been no such deaths in Britain in its entire period of human occupation. Nevertheless, any meteorite of mass greater than 100 g could be relied upon to kill anyone whose head it did strike; and at least some tens or hundreds of such meteorites per century must strike somewhere in the 230 000 square

kilometres of Great Britain. If we take the vulnerable area of an average
head to be 200 square centimetres, we can work out the probability that
a particular head will be struck in the coming year as somewhere near
one in a million million.

Since I do not know accurately the frequency of fall of meteorites of
this size or larger, and have not allowed for the fact that most people
spend most of their time under the protection of a roof, this estimate
could easily be many times too small or too large. If it mattered,
however, there exist enough data on meteorite falls over the areas in
which they would be sure to be observed to be more accurate than this;
and before large numbers of tourist vehicles begin to explore the surface
of the moon, unprotected by an atmosphere, it might be as well to have
some more accurate figures to encourage them.

PREDICTION OF THE RISKS OF MULTIPLE DEATHS

All of the risks that I have discussed so far have been concerned with
hazards to individuals. More difficult and sometimes more important is
the estimation of the risk of dangers which affect large numbers
together. Epidemic diseases and famines—which never occur unless
they can affect large numbers—are, with the possible exception of
influenza, no longer a danger to the rich countries. There are however
several natural hazards which, though rare, are capable of killing very
large numbers of people at once. Earthquakes, volcanoes, tidal waves,
floods and hurricanes are well known examples. Less frequent and
hence less often discussed are large meteorites or comets, and less well
known are solar flares.

In 1908 a fortunately remote part of Siberia was struck by what is now
believed to have been the head of a small comet, and trees were blown
down and scorched to a distance of nearly 20 miles. If it had hit the
centre of New York, Tokyo or London, then five or ten million people
might have been killed. It has been suggested that the extinction of the
dinosaurs followed the impact of a meteorite or small asteroid, perhaps
10 km in diameter, which could have produced a dust cloud which
reduced the sunlight to a level too low to maintain normal plant growth
for weeks or even months over the whole surface of the earth.

Figure 2.3 shows the observed probabilities for a number of catas-
trophes (Fryer and Griffiths 1979). 10^6 is a million, so a probability of
10^{-6} means one chance in a million that a meteorite killing 1000 people
will fall in the next year. In all the kinds of serious natural catastrophe
shown in figure 2.3 the numbers of people likely to be killed increase
faster than the probability of the catastrophe declines. Thus an earthquake occurring once in 10 years may kill 100 000 people somewhere in

the world, an average of 10 000 people a year, while a smaller earthquake killing 1000 people will occur on average every year. Accordingly it is the largest events that are responsible for the largest number of deaths per century.

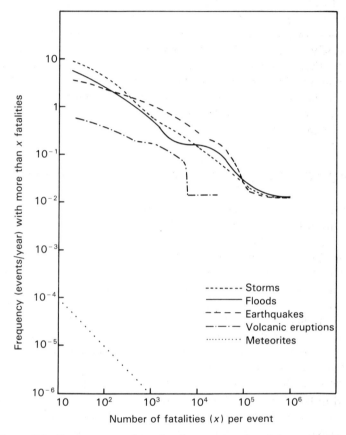

Figure 2.3 Frequency of natural events involving fatalities (worldwide).

The worldwide dangers of death from storm, flood and earthquake are surprisingly similar, and each is at least 10 000 times greater than the danger from meteorites; collectively they give rise to nearly 100 000 deaths per year. Averaged over the world population this would represent a risk of death to each individual of about one in 100 000 per year. This would not be negligible in the countries with a lower general death rate, and corresponds in Britain to about 500 deaths a year. However a worldwide average, quite suitable for consideration of the far smaller risk from meteorites, is quite unsuitable for the terrestrial risks. In Britain the risks to individuals from any of the three hazards mentioned are far below the world average. We do have earthquakes, but such small ones that I have not found any record of anyone ever being killed. Storms and floods do occasionally kill people, lightning being responsible for around a couple of deaths a year, but must between them kill fewer than 10 a year if we neglect the accidents at sea which may be caused or contributed to by storms.

Large earthquakes, floods and hurricanes occur only in known limited areas of the world, giving local risks as far above the world average as those in Britain are below it.

If one makes a quite plausible guess that San Francisco and Los Angeles are likely to have on average 100 000 deaths from earthquakes every 100 years, the risk to an individual must be approaching one in a thousand per year. It is unlikely that such a risk, which is well understood by the inhabitants of both cities, would be acceptable if it arose from any kind of human activity. Evidently the reliably sunny climate, protected from serious overheating by the neighbouring Pacific Ocean, is seen by the inhabitants as a direct reward which makes the risk worthwhile—putting it into the same class of high but acceptable risks as are presented by driving motor vehicles or by climbing mountains.

In practice, the risk of deaths from meteorites, which with very low probabilities could kill very large numbers indeed, could be eliminated by a quite moderate development of our present technology. An errant asteroid on a collision course with the earth could be split into pieces passing harmlessly on either side by a team of engineers with a few hundred hydrogen bombs, while still several hundred million miles away (a thousand times as far as the moon); or could be equipped with a nuclear powered rocket drive to change its velocity by the necessary tiny proportion to make it cross the earth's orbit a few minutes before or after the earth reaches the point concerned. We are likely to have plenty of time to work out effective procedures, because an asteroid whose orbit has been changed by Mars or Jupiter to cross the orbit of earth has a chance of only about one in 60 000 of finding the earth at the crossing point the first time it crosses, so that we should be able to study its orbit for many years before we have to do anything about it. There are several small asteroids which already come quite close—in astronomical terms—to the earth's orbit and whose movements are carefully recorded.

ACCEPTABLE RISKS

While the great majority of non-technical people are put off by large numbers, and indeed have little enthusiasm for any numbers at all, they are often better than many scientists in seeing qualitative differences.

Some extremely well qualified scientists have expressed surprise at the fact that people fail to worry effectively about the frighteningly visible 5000 deaths per year on the roads, but worry a lot about tiny numbers of deaths, which as we shall see may indeed occur but for which there is no observational evidence at all, due to levels of radiation well below the natural background.

On the first issue it is clear to most teenagers and to every cigarette smoker that it is worthwhile and sensible to take a bit of a risk to do something that you yourself enjoy, but that it is *not* worthwhile to take even a small risk so that somebody else should make some money. If no one gained any pleasure or feeling of relaxation and well-being out of smoking, one could not get many people to smoke by pointing out that for every death five to ten people would earn their livings for a year in the tobacco industry, and that the manufacturers would be able to sell £75 000 worth of cigarettes. Actually, many regular smokers may fail to recognise that the Government's pusillanimous euphemism 'smoking can seriously damage your health' really means that in Britain 35 000 of them die every year from lung cancer and at least twice as many die of heart or other conditions as a result of smoking (Royal College of Physicians Report 1983). It is to be hoped that the present varied and more explicit warnings from the Health Departments' Chief Medical Officers will have more effect.

Driving a car or a motorcycle certainly involves a risk of death and a larger risk of serious injury. Most motorists feel that the chief danger does not arise from their own driving but from that of the careless and selfish so and sos with whom they share the roads, but practically all are aware that they are taking *some* risk. They are however convinced that the freedom to go where you like when you like is more than worth it. With access to and understanding of the figures I am convinced of this myself.

One can therefore usefully employ the driving of 50 miles in a car, or the smoking of $1\frac{1}{2}$ cigarettes, as an indicator of the *magnitude* of a one in a million additional risk, but neither of these has any relevance to the *acceptability* of a death risk of one in a million from some quite different cause giving you no personal reward.

Complications arise when large-scale operations give a clear reward, by increasing their safety, reducing their costs or otherwise, to some people but not to others, while causing risks to some who have no such gain. Where everyone, or at least an effective majority, understands both factors, there is no difficulty. This is just the kind of problem to which democracy can give the fairest answer; do what the majority of those who care either way wish to be done—preferably with an unwritten but humane proviso that large risks should not be forced on others, and that the bigger the risk the bigger should be the majority who gain more than they lose over those who only lose.

It is often unnoticed that most of our civilisation is founded on a series of very unequal majority decisions of this type, shown very strikingly by the case of North Sea oil. Nearly everyone in Britain gains by the exploitation of the oil, with no increase of risk over that of buying oil abroad. A small number of divers and skilled operatives, however, have to take really serious risks of death or permanent injury. In my first edition I said 'This problem is solved to everyones's satisfaction by the paying of sufficiently high wages for an adequate number of competent people to feel that the gains ... are more than worth the risk.' Accidents in which numbers of workers have been killed in helicopter crashes have not materially altered the situation, but the 165 deaths in the Piper Alpha fire, and the less serious accident shortly after, have changed the equation; and it is certain that extensive and visible improvements in safety precautions as well as increased wages will be required for the continuation of production of North Sea oil. The problem facing the power industries is not that people will not work in them but that there is at present an inadequate realisation of the absolute dependence of our very lives in our highly technical civilisation on large supplies of power, and a widespread and serious disagreement on the magnitude of the risks.

In deciding what kinds of power we should use, whatever the total quantity required should be, there are many factors to be considered. Cost, availability and extent of resources and effects on the environment must all be taken into account; but the risks to human life and health must always be important and may be overriding. It is hoped that this book will help to reduce the range of disagreement on these risks.

STRONG DRINK

One more specific risk is worth discussing as it illustrates some important new points. This is the risk arising from excessive drinking of alcohol. The direct death risk from this is quite small in Britain; two or three per million per year from the directly toxic effects, with little difference between the sexes, and more than a dozen per million per year (twice as many males as females) dying from liver conditions recognised as produced by alcohol, plus a few deaths in fights which were probably so produced. This is certainly a good deal better than a century ago, but the deaths caused indirectly in alcohol-assisted accidents, especially on the roads, must add up to nearly another 50 per million per year, many of the deaths being those of people who had not themselves been drinking.

A quite unusual characteristic of alcohol in the form of fermented drinks is that in moderate quantities it can be positively valuable to health, for example beer can be an important source of vitamin B for people on otherwise B-deficient diets. During our evolution, when fermenting fallen fruit could be a useful food item, we developed the ability to metabolise alcohol, if absorbed at not too great a rate, as a source of energy, as well as to eliminate its toxic effects on the nervous system. When tired by a physical activity, a glass of beer may not be better than a cup of well-sugared tea or coffee but it is certainly better than nothing, and in hot conditions can be much more quickly refreshing.

Alcohol's most important killing effects are more indirect and far less obvious than its ability to increase the risk of accidents. They arise from the capacity of alcohol to increase the effectiveness of other agents. This type of interaction is known as a synergistic effect. Figure 2.4 shows the effect of drinking on the death rate from cancer of the oesophagus, for light and heavy smokers of cigarettes (Tuyns *et al* 1977 as quoted in Doll and Peto 1981).

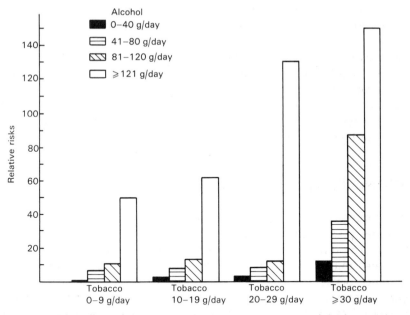

Figure 2.4 The effect of smoking and drinking on cancers of the oesophagus.

The annual death rate in Britain from cancer of the oesophagus is about 4000, and the number of deaths of those who neither smoke nor drink, although unrecorded by Tuyns and co-workers (perhaps there were none in the French population they studied) must evidently be very small indeed. The multiplicative rather than the additive effect of tobacco and alcohol consumption is very clear. Over the range considered, the increase of cancer rates with tobacco consumption for a low and constant alcohol intake is 7.8 times; and the increase of the cancer rate with alcohol consumption up to 100 g a day for a low and constant tobacco consumption is 11.7 times. With both together it is 87 times higher. Examination of figure 2.4 in detail will show that this particular synergistic effect is not as simple as mere multiplication; the proportional increases produced by alcohol are much greater at the lower smoking levels and those produced by smoking are greater at the lower alcohol levels.

A similar synergistic relation exists between smoking and asbestos in the initiation of lung cancer. The incidence of lung cancer among asbestos workers who are also heavy smokers is some 40 times that among asbestos workers who do not smoke at all. Indeed it is unlikely that the carcinogenic (cancer-producing) properties of asbestos would yet have been discovered if all the workers in the industry had happened to be non-smokers.

It will be important to remember that synergistic effects may well occur between the carcinogens derived from fossil fuels, from ionising radiation and from other sources—or even with substances which by themselves are not carcinogens at all, which may well include alcohol itself. Another way of expressing this is to consider alcohol or tobacco primarily as *promoters* of cancers initiated by other causes, such as air pollution or diet. This might help to explain why in Japan cigarette smokers have little more lung cancer than have non-smokers.

PUBLIC PERCEPTION OF RISKS

In early primitive conditions in a stable environment there were no risk analysts, but the public perception of risk was in close agreement with the actual probabilities of death or injury. By the time you had reached the ripe old age of 40 you would have seen a lot of deaths from predators, accident or disease. For the first two you would know from experience the best ways of minimising the risks, and word of mouth tradition would have warned you of the more serious of the rarer dangers such as flood or forest fire. The unavoidable risks of hunting would be made as small as collective experience could make them without impairing too much the success of the hunt, and the residual risks would be accepted without resentment and even eagerly for the public credit which would accrue from success in the more dangerous undertakings. Nobody would make numerical estimates of risk, but at least the relative magnitudes of the commoner dangers would be well appreciated and the perceived importance of each would lie fairly close to the actual value. I do not need time travel or clairvoyant capacity to know this. Where dangers are real and large, natural selection would have rapidly eliminated the groups who got their risk-avoidance priorities seriously wrong. If you keep the children away from playing in the stream because of crocodiles when there *aren't* any crocodiles, but don't stop them playing in the woods instead because of leopards when there *are* some leopards, there will soon be a shortage of children.

Nowadays the situation is entirely different. It really makes no practical difference what men or women in their early twenties believe about risk, apart from the voluntarily accepted risks of non-essential but exciting leisure activities. They can be as afraid as they like about imaginary dangers without risk to themselves, and even the unavoidable real risks that they do run are so tiny that it really doesn't much matter whether they know them or not.

The result of this is that there no longer needs to be any relation between the actual magnitudes of risks and the subjectively perceived magnitudes. Consequently there no longer *is* much relation. Indeed, the situation may be even worse than having no relation at all. If you have no personal knowledge of anyone in your age range who has ever died of anything, you can get your information only from reports in the media. 'Man bites dog'—that's news; it's the rare and spectacular or horrifying kinds of death that get reported, and inevitably it will be the rare and memorable ways of dying that come to seem important.

In 1979 a group of psychologists at Decision Research, Oregon, published a table showing how various groups of people rated the relative risks of 30 different activities or technologies (Slovic *et al* 1979). The results are shown in table 2.2. The second column shows the order in which the risks were put by a group of 15 people professionally involved in risk assessment from different regions of the USA, and the figures in the last three columns show the orders in which the three other groups placed the same set of hazards.

Where the figures are readily available, I have added in the first column the number of deaths in one year in England and Wales. The deaths attributed to x-rays and to nuclear power have not been observed but have been inferred from the radiation doses to which the operators and the public have been subjected, using the relation of deaths to dose due to the International Commission on Radiological Protection. It must be remembered that the gain to the patients from x-rays far exceeds the risk. Alcohol must indirectly have been responsible for far more deaths than those directly recorded due to its toxicity. Furthermore, the two deaths attributed to the nuclear power industry are those for which the radiation exposures in 1979 were *expected* to be responsible. The basis of the expectations will be explained in Chapter 4 and owing to the long latent period of cancer the two deaths are quite unlikely yet to have occurred.

Table 2.2 Ordering of perceived risk for 30 activities and technologies (USA).

	British deaths in 1979	15 experts	40 members of League of Women Voters	30 college students	25 Active Club members
Motor vehicles	5800	1	2	5	3
Smoking	25 000†	2	4	3	4
Alcohol	700	3	6	7	5
Handguns		4	3	2	1
Surgery	155	5	10	11	9
Motorcycles	1000	6	5	6	2
X-rays	(200)‡	7	22	17	24
Pesticides	3	8	9	4	15
Electric power	25	9	18	19	19
Swimming	110	10	19	13	17
Contraceptives		11	20	9	22
Private flying		12	7	15	11
Large construction		13	12	14	13
Food preservatives		14	25	12	28
Bicycles	51	15	16	24	14
Commercial aviation		16	17	16	18
Police work		17	8	8	7
Fire fighting		18	11	10	6
Railways	115	19	24	23	20
Nuclear power	(2)‡	20	1	1	8
Food colouring		21	26	20	30
Home appliances		22	29	27	27
Hunting		23	13	18	10
Antibiotics	1	24	28	21	26
Vaccination	1	25	30	21	29
Spray cans		26	14	13	23
College football		27	23	26	21
Power mowers		28	27	28	25
Mountain climbing		29	25	22	12
Skiing		30	21	25	16

† Estimated for lung cancers only (Mortality Statistics 1981 and Doll and Peto 1981).

‡ Estimated from radiation risks (see Chapter 4 for assumptions made).

The order adopted by the experts in the USA, putting the often dangerous sport of mountain climbing near the bottom, makes it clear that this group was concerned with the total number of people killed in the USA, where only a few people spend much time in serious climbing. Some of the differences between their list and those of the other groups may have resulted from confusion in the minds of some of the respondents between the risk definition adopted by the experts and the alternative interpretation, concerned with the risk to individuals directly involved.

There is evidence that people think what they would *like* to think, independently of any evidence they may have had. For example, the college students who enjoy swimming put this last, while the experts put it tenth. The really striking difference from the view of the 'experts' is the top position given by two of the groups to nuclear power. The 'experts' were not, of course, experts on nuclear power, but neither are those other two groups. Whatever the actual risks, which I shall be explaining at some length in later chapters, it is clear that these groups can have got their views only at secondhand from the media, which are not themselves overburdened with experts on nuclear power. As we shall see, deaths due to radiation as well as to 'ordinary' accidents are to be expected as a result of the operations of the nuclear industry, but it is very doubtful whether anyone except the risk analysts could have known enough about these to take them usefully into account.

Without wishing to run ahead of the evidence by discussing the question of which group is right, I do want to stress the importance of the very big differences of view that appear in the table. Somebody's perception of risk must be seriously wrong to an extent which would have been impossible in palaeolithic times.

It is worth noting that results of the survey, published in April 1979, could hardly have been affected by the accident at Three Mile Island, which occurred on 8 March 1979 and which was nowhere mentioned in the text of the paper.

My wife and I subsequently (four years after that accident) conducted a similar study in Britain, on final year university students and student nurses in Birmingham and on sixth-form science students in Monmouth. The results are shown in table 2.3. The respondents were explicitly asked to give the order which they believed to correspond to the total number of deaths per year in Britain. It is interesting to note that the opinions of future nurses and school sixth-formers differed little from those of university students, and that the opinions of students of different subjects differed little between themselves. The most serious and least serious risks were well recognised by all groups.

Like the American respondents, they gave a low rating to medical x-rays, no doubt for the same reason—that perception of the number of deaths resulting 20–30 years after irradiation is overwhelmed by knowledge of the very much larger number of lives which are saved from much more imminent hazards.

Table 2.3 Ordering of perceived risk for 22 activities and technologies (Britain).

Order from annual deaths as given in next column	Deaths in one year†	6th form science group (36)	Civil engineers (26)	Physicists (26)	Social scientists (24)	Nurse (53)
					Order from median vote by each group	
1 Lung cancer	34 000	2	2	3	2=	2
2 Motor vehicles	5200	1	1	1	1	1
3 Accidental falls	4400	10	7	8=	12	9=
4 Skin cancer	1200	15	13=	17	9=	13=
5 Alcohol‡	1000	4=	10	6=	5	6
6 Influenza	730	16=	17	15=	19=	19=
7= Fire in private dwellings	650	3	3	2	2=	3
7= Accidental poisoning in the home	630	13=	11	13	9=	9=
9= Drowning	360	7	8	8=	13=	12
9= Homicide and injury by assault	350	6	5	6=	6	4=
11 Pedal cycles	280	8	9	5	7=	13=
12= Medical x-rays§	200	21	20=	20	21=	22
12= Appendicitis	190	18	20=	18	18	17=
14 Accidents due to machinery in industry	130	4=	4	4	4	4=
15 Railway accidents	95	13=	12	14	7=	13=
16 Drug dependence‡	85	9	13=	10	13=	7=
17 Electric shock	60	11	15	11=	13=	9=
18 Air transport accidents	50	16=	18	19	17	16
19= Gas explosions	11	12	6	11=	9=	7=
19= Abortion (including spontaneous)	9	19	16	15=	19=	19=
21= Lightning	3	22	22	21=	21=	21
21= Nuclear power§	2	20	19	21=	16	17=

† Figures from Mortality Statistics, 1979 and 1981.
‡ Directly attributed in death certificates. An extra number should probably be added f
other deaths due to these conditions.
§ Number calculated from the effects of radiation—see Chapter 4.

A curious feature is the fact that every group put the rating of accidents due to industrial machinery in exactly the same place, fourth, which is far higher than is justified by the actual figures. This suggests that machinery is regarded as the main cause of industrial accidents, whereas far more of such accidents are caused by fires or falls, or to having something dropped on one from a height. The low rating given to influenza is no doubt because, except in major epidemics, the deaths occur mainly among the elderly; young people describe most feverish colds as 'flu, and regard it as a trivial complaint.

Unlike the American students no group gave a high place to nuclear power, although one or two individuals in two groups did so.

In a democracy increasingly based on complex technologies, and in which risks to people are regarded as important, a reasonably accurate perception of which are the hazards carrying the largest risks matters a good deal. In this respect our sample of a well-educated rising generation in Britain appears to have a clear advantage, perhaps as a result of being less swayed by the media's concentration on the unusual and frightening rather than on the common dangers.

The importance of the differences between perception and reality was well summarised by the apocryphal Josh Billings 'The trouble with people is not that they don't know, but that they know so much that ain't so.'

It is my hope that the later chapters in this book will help to bring perceptions of risk in the energy field closer to the real risks than they have been over the last few years.

Chapter 3

Risks of Cancer and Mutation—the Background

No mention has yet been made of any significant number of deaths arising specifically from the power-producing industries. This is because exceedingly few individual deaths can be attributed with certainty directly to the industries. There are specific hazards to coal miners, uranium miners, and to the divers and platform operators engaged in the extraction of oil. These will be discussed in detail later.

Apart from these, the risk of death to workers in the power industry differs little from the risk in any other industry. Death rates from a series of industries in Britain are shown in table 3.1. Building construction is the most dangerous field in which to work, and although the figures for the power industries differ from each other none of them differ greatly from the average of British industries.

Of more interest to most of us are the risks to the general public, but death certificates in Britain do not record the conditions, the environment or the specific industry initiating the condition which caused death. Lorries carrying coal, petrol or nuclear fuel occasionally kill people on the roads, and gas leaks cause explosions. Faulty wiring causes fires and deaths from electric shock; but it would be inappropriate to blame the coal industry for 80% of the deaths from faulty wiring because it supplies the fuel for 80% of our electricity production!

The difficulty lies in the fact that most of the deaths among the public arising from the main power industries—coal, oil, gas and nuclear—are indistinguishable from far larger numbers of similar deaths due to other causes. This is especially true of deaths caused by cancer, which arouse the greatest amount of fear. In this chapter I shall therefore consider the general risks of dying of cancer before examining in detail the responsibilities of the power industries for producing cancers.

As was said in the previous chapter, cancer is rare below the age of 30, but above this the incidence of new cases, and the death rate, rises at an increasing rate until the age of 80. Above this age the death rate falls off somewhat, suggesting that people may vary in their susceptibility to cancer, and that by 85 or 90 most of those susceptible have already succumbed.

The other two major killers in the better-off countries, heart disease and diseases of the respiratory system, also increase with age in much the same way, except that rates do not fall off at advanced ages. However the long drawn out and painful character of death from cancer, together with the low probability of cure, has led to much greater popular fear of this than of heart and lung diseases, in spite of the fact that the combined death rate from these two is three times that from cancer. All are of course far less important in the poorer tropical countries in which epidemic and parasitic diseases are still killing many of the young. In fact, one can make a good guess at the health status of a country from the gross cancer death rate alone: the higher the cancer rate the healthier the country.

Table 3.1 Number killed per year in Britain in various occupations and accidents. (Data from *H and S Statistics* 1989 and *Mortality Statistics* 1985).

Industry	Number of deaths
On the farm (1985)	81
Metal goods, engineering and vehicle industries (86–87)	36
Oil and gas extraction (1986–87)	9
Chemical and allied industries (1986–87)	29
Food, drink and tobacco manufacturing (1986–87)	29
Rail transport (1985)	41†
Accidents in the home	4478

† Excluding trespassers and suicides

The gross death rate is the ratio of number of deaths per year to the number of the whole population. For comparison between populations of different sizes one will usually give the number of deaths per 100 000 per year. For many purposes this is less useful than age-specific rates, such as the number of deaths per year per 100 000 between 70 and 74, or age- and sex-specific rates such as the number of deaths per year from breast cancer per 100 000 women between 50 and 60.

CANCERS IN BRITAIN

My aim in this section is not only to provide some background information based on the things that we know, but also to warn the reader that there are more, and more important, things that we do not know.

There are about 140 000 deaths from cancer every year in Britain, nearly half of them associated with cigarettes (*Hansard* 1988). The incidence of cancer varies considerably from place to place; for example in the Birmingham area the gross rate averaged over all ages and over three consecutive years, 1968–70, was 312 per year per 100 000, while the rate in the Manchester area was 326 and in the SE and SW Metropolitan areas was 368 per year per 100 000 (Registrar General 1975). There would of course be some random chance variations in all of these figures, but if the risk of cancer had been identical in the three areas, the chance variations would have been only a few tenths of 1%. The risk to the average inhabitant of London must therefore be greater than that to the average inhabitant of Birmingham. We do not know why.

If we look at individual types of cancer, the differences from place to place are very much greater still. For example in Worksop there were 18 deaths from liver cancer between 1958 and 1971 in a population of 36 000 while in Sheffield—only 15 miles away—there were just over 3 deaths per 36 000 in the same period (Hill *et al* 1973). The numbers of deaths in Worksop are not large, but this difference could hardly be due to chance. The most obvious relevant difference between the two towns is their nitrate consumption. Up to at least 1953 the concentration in the Worksop water supply was higher than in any other borough in the UK; 10 parts per million (ppm) against 5 ppm for Sheffield. Nitrates can be reduced to nitrites in the anaerobic conditions in the human gut, and the bacteria normally living in the gut can convert nitrites to N-nitroso compounds, which are powerful carcinogens (cancer producers). Soon after 1953 the Worksop water supply was changed to one with a low nitrate content, but owing to the long latent period for cancers it will be some time before the effects of this show themselves. Nitrites themselves are used extensively as food preservatives, in ham for example, without as yet having produced recognisable increases in cancer, and such high liver cancer rates as at Worksop have not been reported (or, perhaps, looked for) from other areas where nitrate concentrations are high. Furthermore, a high consumption of vegetables, which have a large nitrate content, seems to *reduce* cancer rates.

If we compare British cancer rates with those in well-recorded areas of other countries we find differences on a far larger scale. Many places show much lower rates than we have for most of the important cancers (IARC 1982).

If the British stomach cancer rate were as low as that of Utah in the USA, of lung cancer and large bowel cancer as low as that of the Bombay and Poona areas of India, and of breast cancer as low as that of Japan, we should have 35 000 fewer deaths from cancer each year. These four cancers are responsible for less than half of our total cancer death rate, and it is probable that British excesses could be found among many of the remaining 60%. (England has however a lower rate than most other countries in cancer of the oesophagus and some other cancers.)

The differences are not due simply to different racial susceptibilities. Second generation pure Japanese in Hawaii have the same pattern of cancers as do the Americans of European origin there, although the patterns were widely different in the areas from which the Hawaiian Japanese came (Doll and Peto 1981, p. 1201).

An important difficulty facing investigators of the causative factors of cancer is that there is a long delay between the initiating cause and the observable effect. For leukaemia (blood cancer) the delay may be five years or less, with a maximum incidence after about eight years†. Leukaemia is relatively rare and for most other cancers the average delay will be 20 to 30 years or even more. Most cancers of the lung arise from cigarette smoking, which usually begins between the ages of 10 and 20, while diagnosis of active lung cancer rarely occurs before 40 and has a maximum incidence around 60. This does not, however, prove that the average delay is 40 years or more. Smoking has a cumulative effect and it is found that the risk to a previously regular smoker who gives it up begins to fall, from the level that it would otherwise have reached, after 10 years or less.

These complexities have two important effects. The effect of medical importance is that with such large differences, for unknown reasons, in cancer death rates, and with such long delays between the causative action and the expression of the effect, we have little hope of detecting minor causes by statistical study of deaths. The effect of political importance is that even the wildest hypothesis concerning the causes of cancer can be given apparent support by careful selection from perfectly genuine data. The dental profession can be grateful for the fact that the water supply of Worksop had not been fluoridated. If it had been, the large excess of the rate of liver cancers over that of (then) unfluoridated Sheffield would inevitably have led to impassioned attacks by the anti-fluoride organisations, since many of their members would have been completely convinced that a causal connection between fluorides and cancer had been incontrovertibly proved.

Similarly the nuclear industry can be happy that it had not built a nuclear power station near Worksop.

HOW HONEST STATISTICS CAN BE MADE TO TELL DAMNED LIES

If one takes any statistically significant difference as evidence of a causal connection, a careful search will enable one to find the most extraordinary relations. An example of what can be done if you try is shown in figure 3.1. I am much indebted to my wife for demonstrating an

† The geographical distribution of child leukaemias will be discussed in Chapter 11.

impressive degree of correlation between the incidence of lung cancer
and the alphabetical position of the first letter of the name of the hospital
region concerned. The vertical bars in the figure indicate the variation
that should be attributed to random chance, and are calculated on the
assumption that the chance of dying of this cancer is independent of
everything except this alphabetic relation. If there were no dependence
on the letter of the alphabet, the points should all lie on or close to the
horizontal dotted line representing the average value over the whole
country. The probability that the results actually observed should appear
by chance would be less than one in 1000 million. Evidently the results
are not due to chance.

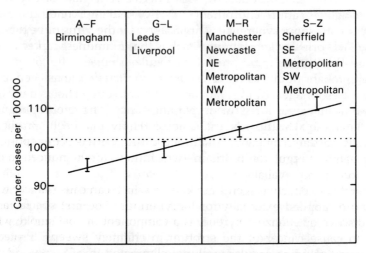

Figure 3.1 Graph demonstrating a surprising relationship between
the alphabet and incidence of lung cancer.

Clearly the real risks in the Metropolitan areas must genuinely have
been considerably greater than in Birmingham, although, as stated
above for the total cancer rates, we don't know why. It must be
remembered that cancers in 1968–70 depended on conditions 20 years or
more before.

The really important technique used in this piece of trickery was to
find a plausible excuse for leaving out the points which did not agree
with the chosen hypothesis. If you start with a random distribution of
points you can *always* produce evidence for *any* hypothesis by omitting
a few of the points which most strongly disagree with the hypothesis.
What we did was to omit five regions on the plausible-sounding excuse
that they were 'less-industrialised'.

To come down to earth, the lesson of this exercise is that any particular
genuine causal relation of cancer death rates to real factors will have to

be large if it is not to be obscured by other real variations for which the causes are unknown.

CHEMICAL CAUSES OF CANCER

As we saw in the last chapter, several causes of cancer have been reliably identified where the effects are sufficiently large. Besides the effects of cigarette smoking and the largely synergistic effects of alcohol and asbestos, the only chemical agent which has been shown positively by direct observation to cause cancers in human beings when swallowed is aflatoxin. This is a product of the fungus *Aspergillus flavus* which contaminates peanuts and other foods in hot damp conditions. This, perhaps with some synergistic amplification by the virus of hepatitis B, leads to the primary liver cancer which is the commonest cancer type in large areas of Africa.

Most probably, but less certainly, bracken (*Pteridium*) has been found to cause oesophageal cancer. Japanese who eat young shoots daily have three times as great a risk of developing cancer of the oesophagus as have Japanese who do not eat it at all (Hirayama 1979). Bracken is generally not eaten by people in Europe but cancer has been reported in sheep eating it regularly in Britain—which they can be induced to do if other food is not available.

Several very effective agents are known which can cause cancer when inhaled or applied externally for long periods. The first specific agent ever discovered, benzo-a-pyrene, is a component of coal smoke which used to cause cancer of the scrotum in chimney sweeps. Protective measures, including regular washing, eliminated this.

Lubricants for metal-cutting machinery and a number of other chemicals used in industry have similar effects and also require and receive protective measures. Weakly carcinogenic chemicals used in several different large industries could cause large numbers of cancers without being recognised.

After allowing for all the facts so far mentioned, we are left with 100 000 deaths per year from cancer in Britain for which we do not know the cause. Of these, around 80 000 result from types of cancer which vary sufficiently from place to place, even when proper allowance is made for different age distributions, to give very strong evidence of dependence on the environment, including food eaten, water drunk and air breathed. This does not necessarily mean that the remaining 20 000 are *not* dependent on the environment—a carcinogen which occurred in all three of tea, coffee and beer and in several common foods would not produce appreciable variations from place to place. Benzo-a-pyrene does in fact occur in many foods (see table 3.2), but at levels thought to be negligible. What we can say is that at least 80 000 cancer deaths per year are unlikely to be due to something happening spontaneously inside the victims independently of external factors.

Table 3.2 Amount of benzo-a-pyrene in foodstuffs. Concentrations in micrograms/kg. Data from Royal Commission on Environmental Pollution (1981).

Food	Concentrations of benzo-a-pyrene
Cooked meats and sausage	0.17 – 0.63
Cooked bacon	1.6 – 4.2
Smoked ham	0.02 – 14.6
Cooked fish	0.9
Smoked fish	0.3 – 60
Flour and bread	0.1 – 4.1
Vegetable oils and fats	0.4 – 36
Cabbage	12.8 – 24.5
Spinach	7.4

Our inability to relate any large number of deaths to particular carcinogens in the environment does not mean that we know of no such compounds occurring in significant amounts. Benzo-a-pyrene and a number of other chemical carcinogens can be detected in city air, and some quantitative information about their effects is available.

There are some occupations, such as the production of coal gas or working with hot pitch or asphalt, that have entailed the inhalation of sufficient quantities of airborne carcinogens to produce an unambiguous and measurable increase in lung cancers (Pike *et al* 1975 and Hammond *et al* 1976 quoted in a more readily available publication by Doll and Peto 1981). Not all of the carcinogenic compounds involved have been identified or can be easily measured. Accordingly, Pike and colleagues measured the concentration of the known compound benzo-a-pyrene and used the results as indicators of total carcinogenic activity. If we assume that it forms a roughly constant fraction of the total carcinogens wherever it occurs, we can estimate the probable risks on the basis of measurement of this single compound.

COLLECTIVE DOSE OF CARCINOGENS

Making the common assumption that the risk of producing cancer is proportional to dose, together with the known increase of incidence of lung cancer among men who had breathed large known concentrations of benzo-a-pyrene for known periods, Pike *et al* deduced that breathing a steady level of 10^{-3} micrograms (μg) of benzo-a-pyrene per m^3 (cubic metre) of air would lead to an extra four cases per million per year among men between 40 and 74 years of age with average British smoking habits. Since about 80% of all lung cancer deaths occur in this age range, we can take five per million per year as the total effect on the whole male population.

A reasonable inhalation rate for a man doing light work for six hours a day would be 5000 m^3 per year. Then the total annual dose inhaled from $10^{-3}\mu$g per m^3 would be 5 μg. Since this would be expected to lead to five deaths from lung cancer per million, we can say that if a million men inhale 1 μg of benzo-a-pyrene per year, on the average one per year will die.

The total inhaled by the whole million can be described as a collective dose of one million man-micrograms, and this collective dose on the average will be responsible for one cancer death. This will be true whether a million men each inhale 1 μg, or 10 000 men each inhale 100 μg; if the product of number of men and average dose is one million there will be on average one death.

The concept of collective dose will be dealt with more fully in connection with radiation-induced cancers in the next chapter.

Lung cancer is much less common among women than among men, although with the increase in cigarette smoking they are catching up. It is not known whether women have a correspondingly smaller lung cancer rate from the inhalation of benzo-a-pyrene. It would be playing safe to assume that there will be the same risk to women as to men, and that the collective dose of one million man-micrograms per death applies to both sexes.

The air in some cities has concentrations of benzo-a-pyrene sufficient to give an annual dose of 100 μg or more. This could clearly be responsible for large numbers of deaths, and will be discussed in more detail in Chapter 10, after the risks from absorption of radioactive materials have also been described.

IDENTIFICATION OF CARCINOGENS

Animal experiments have identified large numbers of widely distributed chemical compounds which are carcinogenic—to the animals—in minute quantities. Naturally we cannot confirm these results on humans, and it is not obvious that a substance that will induce malignant tumours in a mouse inside a year will also induce tumours in people 20 or 30 years later. This is quite reasonably inferred, however, from the fact that substances such as benzo-a-pyrene that are known to induce cancers in man after long induction periods do induce them in mice in very much shorter times—in much the same fraction of the normal expectation of life of the mouse as in ourselves. While there are few, if any, of the compounds known to be carcinogenic to humans that are harmless to mice, the relative effectiveness often differs considerably.

Preliminary tests can use creatures much less closely related to humans than are mice. It has been known for a long time that most carcinogens are also mutagens—substances that produce inheritable mutations—and vice versa. The Ames test for potential carcinogens consists of examining their ability to produce mutations in bacteria. Large numbers of bacteria (*Salmonella*), of a strain whose ability to make for themselves the essential compound histidine has already been lost in a previous mutation, are exposed to the test substance and then placed in a nutrient medium lacking histidine. If a back-mutation has restored the capacity to synthesise the missing compound, colonies of bacteria visible to the naked eye will appear in the course of a few days. This proof of the mutagenicity of the substance tested then indicates a high probability that it is also a carcinogen.

The Ames test is not 100% reliable; it gives a few per cent of false positives—substances mutagenic to bacteria but not carcinogenic to small mammals—and, more seriously, a similar proportion of false negatives in which no bacterial mutations occur with substances known to be carcinogenic to mammals. Nevertheless it is a useful preliminary test, and may be regarded as adequate where a substance is used only in small quantities with little risk of absorption by humans (Weissburger and Williams 1981). The Ames test can now be expanded to a battery of tests, including a similar test on cultures of mammalian cells in the laboratory, and tests of ability to cause visible breaks or abnormalities in chromosomes. Positive results in every test give nearly 100% certainty of carcinogenicity in humans. There is still a proportion of false negatives, which should be reduced as the work progresses.

As a result of these various tests, it is found that there are many carcinogens present in our current environment, including even appreciable amounts in traditionally prepared foods which we have eaten for centuries. Apart from the N-nitroso compounds already mentioned as derivable from the nitrites used in curing ham (which themselves were identified by experiments on mice) traditionally-smoked fish and bacon carry considerable quantities of carcinogens, including the benzo-a-pyrene shown in table 3.2. (If it were as efficiently absorbed when swallowed as when inhaled a considerable fraction of our 80 000 environmental cancers could be accounted for by benzo-a-pyrene alone in our ordinary diet.)

Although little has been done to control the carcinogens in traditional foods, considerable efforts are now being made to prevent their being added to food, drink, cosmetics or toilet preparations, and manufacturers of new additives are required to make extensive—and very expensive—tests for carcinogenicity.

Quite apart from cancer producers already present, even frying food in fat can produce benzo-a-pyrene and other carcinogens when overheated, especially when the fat is used repeatedly. Good cooks do not usually overheat food much, but even normal cooking procedures at 100–200 °C may produce small amounts of carcinogens. Most of the benzo-a-pyrene that we inhale in urban areas in Europe and the USA is produced by the burning of coal and oil—largely in vehicle exhausts.

To many people the relation between chemical carcinogens and mutagens may come as a surprise—although the fact that ionising radiation can produce both cancers and mutations has been known for a long time. It has quite important implications. The number of serious mutations produced in a population exposed to a given level of radiation dose is less well known than is the number of cancers produced, but the evidence seems to show that the numbers are similar. It would be wise to assume that whatever the causes of the 140 000 cancer deaths each year in Britain, a comparable number of serious mutations are also being annually produced. (There could be many more minor ones such as changes in the genes for eye colour, which would not be noticed and would be of no importance.)

A detailed discussion of a possible way in which damage by chemicals or radiation to the nucleus of a single cell can initiate a cancer or a mutation is given in Appendix 1.

Chapter 4

Radiation and Cancer

Myself when young did eagerly frequent
Doctor and saint and heard great argument
About it and about, but evermore
Came out through the same door as in I went.

Omar Khayyam

We have irrefutable evidence from medical therapy, from the effects of over exposure to x-rays of radiobiologists and radiographers, and from Hiroshima and Nagasaki, that large doses of radiation, if they do not kill outright, can cause a wide range of different forms of cancer.

Those people close to the A-bombs dropped on the Japanese cities who were not killed outright by blast or fire died in a few months of damage to the gut or of anaemia due to damage to the bone marrow where the red blood cells are formed. The men also suffered from a temporary sterility, which usually cleared up within a year if they survived. People further from the explosions, up to a couple of kilometres away, would be ill for weeks or months from these conditions but would recover.

A few years later an unusually high proportion of these survivors, together with others who had shown no early effects, began to die of leukaemia, and later still an unusual proportion began to die of other types of cancer. We can reasonably infer that the much lower but far more widely distributed radiation doses to which the public are exposed from natural sources and by the activities of the nuclear industry may also be causing cancers, even if there are not enough of these to be recognised among the very large numbers of irregularly distributed cancers due to other causes that were mentioned in Chapter 3.

In order to estimate how many people are likely to die of cancer induced by radiation, it is clearly necessary to know how much radiation each person has received and just how large is the probability of dying of cancer that follows exposure to any particular dose. In simpler terms, if you want to know how many will die or what is the risk to the individual, you must know how much radiation was received and how dangerous it was. This, unfortunately for people who dislike both of them, involves some understanding of radioactivity and a lot of arithmetic.

ATOMS AND RADIOACTIVITY

Every one of the atoms of which we and the rest of the visible universe are built consists of a positively charged central core or nucleus which accounts for more than 99.9% of the mass of the atom, surrounded by a number of negatively charged electrons (starting with one for the simplest atom, that of hydrogen). In chemical interactions some of the outer electrons of an atom can be lent to or shared with other atoms, but the nucleus remains unchanged. The nucleus itself is composed of two kinds of particles of almost equal mass: the positively charged proton and the neutral neutron. Carbon-12 for example has 6 protons and 6 neutrons, and the neutral carbon atom will have 6 electrons. Other 'isotopes' of carbon will always have 6 protons, but will have differing numbers of neutrons—carbon-14 has 8 neutrons. All the isotopes of a particular element will have the same number of protons in the nucleus and will have the same chemical and biological properties, but some may be stable and others radioactive.

A radioactive atom is one in which the nucleus will spontaneously change itself, and we call this process radioactive decay. When they decay, radioactive atoms may give off any of three different things†. In alpha decay, which occurs only in a few heavy elements such as uranium, radium and plutonium, the nucleus breaks into two very unequal parts, one of which is always a helium nucleus. This, usually called an alpha particle, is thrown off at a speed around 15 000 kilometres per second—a twentieth of the speed of light. It is brought to rest by a few centimetres of air or by less than 0.1 mm of water or organic material. This means that alpha particles will not even penetrate the thicker parts of the skin, and accordingly you could sit for hours on a sheet of a pure alpha-emitting radioactive material without harm, with nothing more than a thin sheet of polythene between you and it—to protect the material from corrosion by sweat, not to protect yourself from the radiation from the material.

At the same time, since the alpha particle dissipates the whole of its very large kinetic energy in so short a distance, it can do an enormous amount of local damage if it is emitted inside the body as a result of swallowing or inhaling an alpha-emitting atom. Practically every one of the hundred thousand or so molecules through which it passes (in its 0.1 mm path) would be disrupted.

---

† Neglecting spontaneous fission, which is too rare to have any direct health significance.

In beta decay, which is the mode of decay of the great majority of radioactive elements, an electron is emitted from the nucleus, moving with a speed that may be 90% of the speed of light. This may be able to pass through as much as a few metres of air or a millimetre or two of skin or flesh. Although they move much faster, electrons are nearly 10 000 times lighter than alpha particles, and the total kinetic energy they carry is usually several times less. The damage they do is therefore thinly spread out over their much longer track.

Many of the alpha or beta emitters also emit the third type of radiation, gamma rays. These consist of radiation of the same nature as light or x-rays (electromagnetic radiation). They usually have a penetrating capacity equivalent to that of hard x-rays; that is to say they can pass clean through the human body or through hundreds of metres of air. Perhaps surprisingly, the gamma rays that pass right through the body are entirely harmless, but a proportion, larger for the lower energies, is stopped, and in stopping ejects a fast electron with exactly the same properties as the beta-particle electrons. The effect of any *external* source of gamma rays is therefore just the same as the effect of any *internal* source of beta particles, distributed over the region within which the gamma rays are stopped.

HALF-LIFE

A very important characteristic of radioactive atoms is the time they take to decay. Some radioactive atoms are likely to decay in less than a millionth of a second, others may last for thousands of millions of years. Each type of radioactive atom has a well defined probability of decay in the next second, which is usually recorded in terms of half-life; this is the average time taken for half of any group of such atoms to decay. Thus, if we start with a million atoms of strontium-90, after 28.1 years half a million of them will have decayed to yttrium-90, each giving off a beta particle as they do so. Unlike human beings, the half million atoms which have failed to decay will not have aged in any way; it will take a further 28.1 years for half of *them* to decay, leaving a quarter of a million and another 28.1 years for half of these to decay, and so on. On this basis, some of the original atoms will last for a long time. It will take about 20 half-lives to reduce the original number a million times, about 40 half-lives to reduce it a million million times and so on.

While the half-life must be known to forecast the future of a piece of a radioactive element, the thing that matters immediately is how radioactive it is. People are often frightened by the fact that an element has a long half-life. This can obviously matter in the future, but the longer a half-life is the longer its atoms last and the fewer can be breaking down per second now. Thus a milligram of thorium-232, with a half-life of 16 000 million years, is emitting only two alpha particles a second, presenting no significant hazard at all; one could swallow the lot with little risk. On the other hand a milligram of polonium-210 with a half-life of 138 days would be emitting 200 000 million alpha particles per second, and would be capable of killing 100 000 people if properly shared out among them. But it would not be dangerous at all 10 years later.

Quantities of radioactivity are in fact measured in terms of the numbers of atoms decaying per second. Two different units, the becquerel and the curie, are currently in use. One becquerel (Bq) is defined as one decay per second, which is too small for use in most applications. The old unit, the curie (Ci), is defined as 3.7×10^{10} decays a second (the decay rate of 1 g of radium) and is the unit that I shall usually use in later chapters.

The atom into which a radioactive atom decays may be stable (non-radioactive). For example carbon-14 decays into nitrogen-14, which is stable and makes up the greater part of our atmosphere. Or it may be radioactive—the yttrium-90 into which strontium-90 decays is itself radioactive, with a half-life of 64 hours, emitting another beta particle and changing into zirconium-90 which is stable. Some of the heavy radioactive elements go through a long series of radioactive changes before reaching stability and some important examples are given in Appendix 2.

MEASUREMENT OF RADIATION DOSE

The physical magnitude of a dose of radiation to either people or materials is measured in terms of the energy absorbed per kilogram, and can be stated in rads. One rad represents 0.01 joule (J) of radiation energy absorbed per kilogram. A unit convenient for large doses, and recommended in the SI system of units, is the gray (Gy) equal to 100 rad; but this is little used in the medical field and is inconveniently large in the field of health physics.

The biological effects of doses of a rad of x-rays, gamma rays or beta particles are very similar. They vary a bit with the kinetic energy of the electrons which do the actual damage, though the difference is not enough to be generally worth allowing for, but a dose of 1 rad of alpha particles may do 20 times as much damage as do any of the others. This factor of 20 is known as the quality factor. Its value varies somewhat with the effect produced—death, mutation, cancer initiation, etc—but this variation again is small compared with the difference between the alpha particles and the other types of radiation mentioned, and will be neglected. The quality factor for beta particles or gamma rays is unity.

Since we often want to consider the combined effect of mixed doses of different radiations it is convenient to use as a unit the product of the rad and the quality factor. This is called the rem†.

Then 1 rem of alpha particles has just the same effect as 1 rem of x-rays, gamma rays or beta particles. For the remainder of this book I shall quote most doses in rems. There is however one further practical point to be mentioned. A dose of 1 rem may be incurred by the whole body, meaning that every kilogram in the body has received the equivalent of 0.01 J of gamma rays, or may be incurred by only part of the body. For example in a chest x-ray the few kilograms of the chest and its contents will each receive 0.01 J of x-rays for a partial-body dose of 1 rem. A given whole-body dose is clearly more serious than the same dose given only to an arm or a single organ such as the liver. A whole-body dose of 300 rem given in a short period leaving no time for repair will have about an even chance of killing a human being after a few weeks or months of serious illness if untreated, while the same dose to only a hand or foot would lead only to some local and (with luck) temporary inflammation, well known to early workers in the form of x-ray dermatitis. (A 300 rem whole-body dose of x- or gamma rays to a 70 kg man means altogether 210 J absorbed. 300 rem to 500 g of a hand means only 1.5 J absorbed.) Rats show a lower sensitivity, having an even chance of death after a short-period dose of about 700 rem.

Doses of hundreds of rems to the whole body, which could produce the so-called prompt effects of anaemia and damage to the gut or local inflammation, could occur in peacetime only as a result of a major accident in the military or civilian nuclear industry. The likelihood of such accidents will be dealt with in Chapter 9. In this chapter I shall be concerned with the effects of the much lower doses likely to be encountered in the absence of such accidents. The whole-body doses to be expected are then of the order of rems or millirems per year, but may be incurred by large numbers of people, so that even a small risk per person may be important. At such doses cancers, appearing years later, and mutations are the only risks known to exist.

As I showed in the previous chapter, we cannot hope to detect the small effects from small doses directly, so we have to deduce these small effects from the observed effects of much larger doses. Probably the largest source of data could have been the death rate from leukaemia and other cancers among radiologists and radiographers two or three decades ago. Unfortunately their exposures to radiation, though often large, were neither accurately known nor recorded, since in the relevant few decades earlier still the danger had not been appreciated.

† The -em termination was chosen to represent 'equivalent-man'. 100 rem = 1 sievert.

INTERNATIONAL COMMISSION ON RADIOLOGICAL PROTECTION

An international body, the International Commission on Radiological Protection (ICRP) was set up in 1928 to consider quantitatively the data available on injuries to workers with x-rays and radioactive substances, and to make recommendations on appropriate limitations of radiation dose to various categories of people. Until about 1950 it was commonly assumed that ionising radiations did not cause cancer unless they were intense enough to cause obvious damage to the irradiated tissue (Doll and Peto 1981 p. 1254). Accordingly it was assumed that for radiation, as for acutely toxic chemicals, there was a safe dose below which no long-term effects need be expected. This dose was described as the 'threshold' dose. This dose was initially set at the equivalent of 50 rems per year, well below the level at which burns might appear on the hands. As late as 1950, 0.1 rem per day was considered permissible. As records of the deaths of radiologists who had entered the profession before 1921 mounted up, this 'safety' level was successively lowered until it was finally abandoned altogether, although the existence of such a threshold at some low level cannot be disproved.

Studies of the cancer rates among radiologists (Smith and Doll 1981) showed that those who had entered the profession before 1921 had a 75% higher death rate from cancer than medical practitioners, while those radiologists who had entered later showed no significant difference from medical practitioners. The change is likely to have been due to the protective measures that came into widespread use after 1921. It may also be relevant that the radiologists who were already established in 1914 would have been working under emergency conditions on war casualties with no time or opportunity to bother about protection.

Although no dose figures were available, the results obtained were extremely useful in destroying a myth (Doll 1981). To quote Doll:

> Within months of the report that small mammals had their lives shortened by a single sub-lethal dose of radiation or by chronic exposure at comparatively high dose rates, it was believed throughout the scientific world that whole-body doses of one roentgen (approximately one rem) would reduce the expectation of life by between one and five days ... a remarkable illustration of what Trotter called 'the mysterious viability of the false'.... The results of the study of British radiologists ... showed that, whatever harm ionizing radiations might do, the total effect was so small that even doctors who had specialised in radiology from its earliest days suffered no higher mortality than either doctors in general or all men of the same socio-economic class.

THE INFORMATION AVAILABLE ON CANCER PRODUCTION

The most important data now available have been derived from three sources: medical therapy, the nuclear bombs dropped on Japan, and animal experiments. In all of these cases, numbers of cancers recognisably larger than normal could be observed only for large doses. The largest and most reliably known doses were delivered in the course of medical therapy, in particular to patients being treated for a progressive condition of the spine known as ankylosing spondylitis, for which one of the standard treatments consisted of large and regular irradiations by x-rays. Since these were spread over many years, leaving time for repair of injury to healthy tissue between exposures, local doses of 1500 rems or more could be accumulated without any important prompt effects. After such doses the numbers of patients who developed leukaemia, a relatively rare disease, were many times the numbers expected to develop it without the x-ray treatment. The chance of developing leukaemia was found to be roughly proportional to total dose down to a few tens of rems, consistent with a linear increase in risk from zero dose upwards. This would mean that even the smallest dose of radiation would give some—but of course correspondingly small—risk.

Similar results for leukaemia were obtained from a systematic follow-up study of the survivors from the bombing of Hiroshima and Nagasaki. The survivors in Japan received a whole-body irradiation, while in the spondylitis patients only 35–40% of the active bone marrow of the body was irradiated. Since the bone marrow, in which the affected blood cells are formed, must be the region concerned, the spondylitis patients had a smaller risk per rem than did the Japanese survivors. Allowing for this by scaling down the dose recorded for the spondylitis cases by 2.7 times to give the equivalent dose to the whole body, we find that the number of leukaemias among the spondylitis cases is about half that which would have been expected from the number observed among the bomb survivors. It seems likely that this is due to the fact that doses to the spondylitis patients, though larger, were spread over a considerable period, while the total dose to the bomb victims was practically instantaneous.

It was originally believed that the Japanese victims were irradiated by neutrons as well as by gamma rays; but test studies of the radiation from similar bombs and allowance for the loss of neutrons due to collisions with the nuclei of hydrogen in the moist air over Japan have now shown that the doses received at Hiroshima and Nagasaki were due almost entirely to gamma rays, and thus were less than originally believed.

A massive psychological trauma was induced both in Japan and in Western countries by the effects of the nuclear bombs, and of the unprecedented and lingering deaths due to radiation during the following months. This has not unnaturally led to an exaggerated expectation in most people's minds of the risks of cancer and mutation. It may therefore come as a surprise to many that of the 6035 people who received doses of 50 rems or more—nearly 500 times what we normally receive spread over a year from natural sources—only 192 extra, above the normal cancer rate for this number of people, had died of cancer, including leukaemia, by 1982. (About 40% of the extra deaths were from leukaemia.) The variation of cancer death rate with radiation dose is shown in table 4.1. A striking feature is that for doses between 0.5 and 5 rem there are actually 108 fewer deaths than would have been expected if no dose at all had been received. The statistical uncertainties however are large, and the confidence level is no more than 90% that the radiation dose was responsible for any reduction at all in the number of deaths.

Table 4.1 Cancer deaths, including leukaemia, from 1950 to 1982, of 91 231 survivors of the bombs, within the cities of Hiroshima and Nagasaki (*Life Span Study Report* 1986). The dose figures have been corrected to allow for the most recent estimation of the doses actually received in Japan.

Dose range (rad)	0	0.5–5	5–25	25–50	50–100	100–150	150–200	200+
Number of persons	37 173	28 855	14 943	4224	3128	1381	639	887
Total cancer deaths in study group	2438	1815	1095	326	270	150	63	113
Predicted for same number of person-years for zero dose	2438	1923	992.6	281.4	209.8	94.0	42.8	57.4
Observed minus predicted	0	− 108 ± 60	102.4 ± 40	44.6 ± 19	60.2 ± 17	56.0 ± 12	20.2 ± 8	55.6 ± 11
Observed minus predicted above and below 50 rad		39 ± 76				192 ± 25		

It is also certain that further cancer deaths due to the bombing are still to come as the remaining group of survivors gets older, which will more than compensate for the high death rates among survivors already elderly at the time of their exposure to radiation.

On the basis of all the data both from Japan and from medical therapy, the International Commission on Radiological Protection (ICRP) estimated that there was an extra risk of dying eventually of cancer of about 1 in 10 000 per rem of whole-body dose received (ICRP 1977). The figure of 1 in 10 000 is an average for an average range of ages in a population; clearly a dose of 1 rem to a man of 85, who is unlikely to live for another

10 years, is rather unimportant, while a similar dose to people in their twenties would give time for any type of cancer to appear before they died from other causes.

The ICRP 1977 publication gives the maximum cancer rates to be expected for specific organs for standard radiation doses, given at the least desirable age. For example, sections 35 and 36 give the following maximum rates, all per million people per rem of exposure: for leukaemia 20 (for exposures below the ages of 50 for men and 55 for women); for lung cancers 20; for bone cells 5; for thyroid 5; for breast cancer 50 for women and 0 for men, giving a population average of 25; and for all other organs 50. Detailed figures are given for the expected variation of leukaemia rate with age. These fractional organ-figures add up to 125 per million per rem, which at first seems inconsistent with the figure of 1 per 10 000 per rem given above for whole-body irradiation. The apparent inconsistency is due to the fact that the 1 per 10 000 (or 100 per million) is averaged over all ages, while the separate organ risks are given for doses incurred at the age at which the effect will be greatest. The difference is small.

Two comments are desirable. The first is that the ICRP figure takes no account of dose *rate*; the effect of 1 rem delivered in a second is assumed to be the same as the effect of 1 rem delivered over 10 years. There is much evidence that the effects may be smaller for a given dose at lower dose rates. The second comment is that for large doses delivered quickly as at Hiroshima the figure of 1 per 10 000 per rem will certainly be too small, both because of the new dosimetry and because more cancers than expected are still appearing among the Japanese survivors after over 40 years. A figure of 1.65 deaths per 10 000 per rem is being suggested. On the other hand the quality factor of 20 for alpha particles may be too large; in some conditions it seems to be nearer 10. These comments thus suggest opposing effects, and for simplicity I shall throughout this book continue to use the figure of one death per 10 000 per whole-body rem. If a different figure than 1 per 10 000 is agreed upon in the future for low doses, the figures that I give can easily be corrected.

Collective dose

The ICRP formulation of risk as 1 in 10 000 per rem implies that risk is proportional to dose; 10 rems would give a risk of 1 in 1000, and a millirem would give a risk of 1 in 10 million. If a million people receive 1 rem each we should expect 100 deaths. We should equally expect 100 deaths if 10 000 people had 100 rems each. In both cases the product of the number of people multiplied by the number of rems is one million. We could describe this situation by saying that in each case there was a *collective dose* of 1 million man-rems. Thus for every 10 000 man-rems the ICRP formula would predict on average one death from cancer.

In a recent (1987) statement the ICRP indicates that a preliminary reassessment of the Japanese survivors has raised the total cancer risk estimate for the exposed population by a total factor of the order of 2. This information, in the view of the ICRP, is not considered sufficient to warrant a change in dose limits for occupational exposure (NRPB GS9, Nov. 1987). It must be remembered that there is no direct evidence that low doses of radiation at low dose rates are responsible for any deaths at all; and as a reminder of this I shall usually describe the deaths resulting from the ICRP formula as hypothetical deaths.

BEIR COMMITTEE ESTIMATES

Evidence has been cited by the American Committee on the Biological Effects of Ionising Radiation (BEIR) in their report (BEIR III 1980) to show that the ratio of radiation-induced cancer incidence to dose increases with age, as does the normal cancer incidence. This Committee also considers, in the part of its report supported by the large majority, that the simple linear model employed by the ICRP is unlikely, and that at low levels of beta or gamma irradiation some combination of linear and quadratic effects is the most probable assumption †. Figure 4.1 shows the difference at low doses between the effects which would be expected on the linear, quadratic, and linear plus quadratic hypotheses if these all predicted the same effect at 10 rems, which is the lowest figure for which there is any evidence of the numbers of cancers produced.

Besides preferring the linear plus quadratic hypothesis, the BEIR Committee makes the forthright statement (paragraph 5 of the Summary and Conclusions) that 'The Committee does not know whether dose rates of gamma or x-rays of about 100 mrad per year (equal to 100 mrem per year for those radiations) are detrimental to man.' For reasons detailed below, I am in complete accord with this, but would stress that it may not apply to low dose rates of alpha particles, which probably do some permanent damage even at the lowest doses.

For higher doses, the Committee does not regard as useful a simple single figure of cancer risk corresponding to the ICRP figure of 1 in

---

† A linear effect means that the cancer rate is simply proportional to the dose of radiation, so that doubling the dose doubles the risk of death. A quadratic effect means that the cancer rate increases with the square of the dose, so that twice the dose gives four times the risk (and half the dose gives a quarter the risk) or ten times the dose gives a hundred times the risk. A linear plus quadratic effect would lie between the two, the linear component being most important at very low values and the quadratic component at high values of dose.

10 000 per rem. Instead, they propose an estimate of risk proportional to the rate of cancer production by other causes. They suggest (BEIR III p. 3) for a single whole-body dose of 10 rems a risk between 0.5% and 1.4% of the 'naturally' occurring cancer mortality, or 3% to 8% for continuous lifetime exposure to 1 rem per year. In view of the forthright statement quoted above, these percentages should presumably be lower than in proportion for doses below 10 rems; at 100 mrem which is one hundredth of this for example, where the Committee is uncertain of any deleterious effect, the range should presumably be from 0% to less than 0.014% of the cancer mortality from other causes. At any dose level, the cancer rate will be less in young populations than in old. Over the world these 10 rem figures would bracket the ICRP estimate, which would simply be 10 deaths per 10 000 people, but for the elderly British population, with a cancer death rate of 22%, application of the BEIR estimate would lead to between 11 and 30 deaths per 10 000.

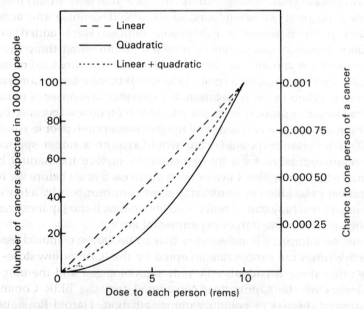

Figure 4.1 The effect of different assumptions on the chance of imitating a cancer, starting with a 1 in 1000 chance at a dose of 10 rems.

The majority report of the Committee is seriously criticised by four of its members, who have given specific reasons for disagreement. The longest and most detailed criticism is due to the Chairman of the Committee himself, Professor Edward P Radford, an epidemiologist

(BEIR III pp. 229–49). He expresses doubts about the relevance of the quadratic relations found in a number of animal studies, and points out that the abandonment of a purely linear relation had not been properly discussed or agreed by the Committee. He also stresses the fact that it had been agreed that the figure given for cancer risk should depend on the rates of cancer *incidence* rather than of cancer *mortality*. This change was based on the consideration by the subcommittee concerned that any radiation-induced cancer produces a major psychological, social and (in the USA) economic cost to the individual affected, whether or not the cancer is ultimately the cause of death. This makes a large difference in the figure for female breast cancer, and an even larger difference in the case of thyroid cancer which is usually curable if diagnosed in time.

I agree with Professor Radford about the importance of the effects of incidence as distinct from mortality, but I disagree with both him and the Committee about the desirability of using the obviously larger (by around 30%) figures for incidence rather than those for mortality in assessing risks. Many people express the view that they would prefer to be killed outright to being blinded. It would confuse the accident statistics badly however if the people blinded were added to the mortality statistics. Death may or may not be the worst thing that can happen, but it is certainly qualitatively different from any kind of injury.

Finally, Professor Radford reminds us that if there were an abnormally susceptible group in the population, the effects of low doses of radiation could be larger in proportion to the effects at high doses—because by the time high doses were reached only the less susceptible people would be left. This is certainly possible, but would require a rather specialised level of susceptibility for a lot of people to survive the natural background (which contributes five or six rems to each of us before we reach 40) and yet to be killed in statistically significant numbers by a few extra rems before reaching the 10 rems or so above which the Japanese results look linear within the (large) experimental error.

Professor Radford's conclusion is that to be safe we should assume some five times the cancer rate accepted by the ICRP at low doses.

The other three dissidents, the only radiobiologist and the only two radiologists on the Committee, considered that the BEIR Committee *overestimated* the risk of beta or gamma radiation. Harold Rossi, of the Biological Research Laboratory, Columbia University, expressed the view that '...the most plausible estimate of the cancer risk from low-LET † radiation is lower than any of the ones given in BEIR III.' I am in agreement with this. See below.

Edward Webster, of the Division of Radiological Sciences, Massachusetts General Hospital, is more explicit. He stated (BEIR III

---

† Low linear energy transfer, e.g. beta particles or gamma rays.

p. 263) 'The average incidence risk given by Table V-14 (BEIR III) is 18 cases per million per year per rad, which is about *13 times greater* than the 1.4 fatal cancers deduced from the Japanese study.' Dr Mays supported this and was particularly concerned that the risk coefficient derived by the Committee exceeded by a factor of about 13 that derived directly from the A-bomb life span mortality data.

With this wide range of variation in the experts' opinions derived from the same set of facts, the BEIR Committee is undoubtedly correct to accept an uncertainty of a factor of nearly three. This is less misleading than the ICRP presentation of a single take it or leave it figure, with qualifications that will be read by only a tiny proportion of those who will accept and use the figure given. I certainly accept too the approach that regards the total effect of a specific small dose rate of radiation as likely to be a small specific percentage increase of the cancer rate due to other causes, rather than being a definite addition to this rate independent of all else. This would be analogous to a small synergistic effect of the kind described in Chapter 2. On the other hand, I am not confident that either the Committee or its critics have adequately allowed for the effect of the enormous dose *rates* both in Japan and in medical therapy, which may impair the cellular repair processes which undoubtedly take place at low dose rates.

An important corollary of the existence of a large quadratic component would be the devaluation of the concept of the collective dose. On the basis of the ICRP linear assumption, 10 rems to each of 1000 people would lead to one cancer death, and 0.1 rem to each of 100 000 people would also lead to one death. On a quadratic hypothesis 0.1 rem to 100 000 people would lead to only a 1% chance of a single death.

If we do assume simple linearity of risk with dose independent of dose rate, as the ICRP recommends, we can calculate the cancer risk from any dose whatever, although it is clearly not useful to calculate the probability of dying of cancer in about a decade if the dose given is large enough to kill directly in a few months.

The same results can be given in two different ways. To the individual the more interesting way is to say that a whole-body dose of 1 rem will add one extra chance in 10 000 to the 24% chance of dying of cancer before you die of anything else which is currently typical in Britain, so that your total chance is now 24.01% instead of 24%. This method has the disadvantage that the figure given applies in reality only to someone around or a little above the average age of the population, since it has been derived from the effects on people of mixed ages at Hiroshima, while ankylosing spondylitis attacks young to middle aged adults with a smaller life expectancy than the young people of Hiroshima. Elderly people will have smaller changes to their total risk. The other approach is appropriate to the consideration of the effects of continuous exposure on large groups or whole populations. Then for a continuous exposure applied to a whole population one could say that there would be one hypothetical cancer per year per 10 000 per rem, or 100 such cancers per year per million per rem.

Figure 4.2 gives a kind of ready reckoner from which the number of deaths to be expected can be determined from the size of a population and the average dose received.

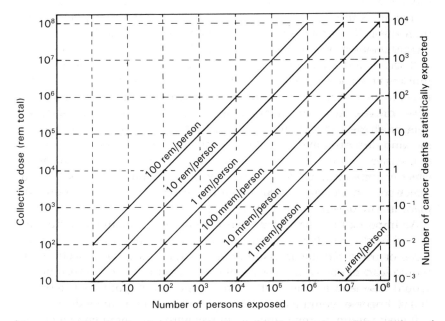

Figure 4.2 Graphs for the determination of collective dose and number of hypothetical deaths from cancer to be expected for a wide range of doses and populations, on ICRP figures. A figure of 10^{-3}, for example, on the right-hand axis means that there is 1 chance in 1000 that one such death will result.

The dose rate considerations and the evidence against linearity emphasise the suggestion that at low doses and at dose rate levels below 1 rem per year, even the ICRP estimate of deaths is too simple and may be too high. In the absence of complete certainty it is however wise to play safe by acting on too high an estimate of risk rather than on one too low.

LATER EVIDENCE FROM JAPAN

Now, nearly 45 years after the bombing of Hiroshima, the leukaemia rate among survivors which, unlike the other cancers, showed a rise as little as two years after the bombing, is down to the level expected for unexposed persons; but the rates of the longer-delayed solid cancers, some of which began to show a rise only after 20–30 years, are still considerably raised. This was not apparent from the records of the spondylitis patients, relatively few of whom had survived for as much as 30 years after the beginning of their period of irradiation at the time the data were analysed.

An interesting but unexpected finding of the study of Hiroshima casualties is that, in spite of the observed increase in cancer rate, the number of the survivors in the study group who were still alive after 30 years was over 600 greater than would have been expected from the controls. It was suggested by Professor Rotblat that the survivors, who must have faced serious injuries from blast and fire, were tougher than the non-survivors. This seems quite reasonable, although the lives of survivors from chlorine attacks in 1917–18, and of surviving prisoners of war in Japanese camps, were shorter than the average. More thorough medical supervision of the survivors may have been even more important.

Some important studies of the effects on human beings of large alpha-particle doses at moderate dose rates have been reported by Professor C W Mays (Mays 1980, 1982).

According to the first study 800 patients in Germany were given a series of injections of the 3.6 day alpha-emitter radium-224 as treatment for ankylosing spondylitis or tuberculosis. Radium is concentrated in the surfaces of bones—which is why it was used—and very large doses were given over periods of up to four years. In a small number of cases this later led to a death rate of 40% from bone sarcoma (cancer).

The average rate worked out at between five and eight bone sarcomas per million skeleton-rems, in good agreement with the ICRP expectation of five per million.

The second study was concerned with the liver cancers induced by the injection of colloidal thorium dioxide into an artery or vein to give contrast in an x-ray. About 65% of the thorium—a very long-lived alpha emitter—is deposited in the liver which was then subjected to a dose of a few hundred millirem a day for several decades, and led to an average of 15 deaths from liver cancer per million man-rems of alpha radiation. This again is consistent with the ICRP figure of 50 per million per rem to 'other organs' of which the liver would be an important one.

RISK OF MUTATION BY RADIATION

Evidence soon appeared suggesting that the bombing of Hiroshima might lead to the production of children with serious genetic defects.

Many irradiated pregnant women lost their babies soon after the bombing; and others, irradiated at an earlier stage of pregnancy, produced in due course full-term babies with major physical defects. It was realised that these defects were congenital rather than genetic, i.e. that the foetus was gravely injured in the womb, like the thalidomide children, without there necessarily being in either case any inheritable contribution, although newspaper reports often failed to distinguish the two kinds of defect. However, this led to a thorough investigation of the babies born to irradiated survivors more than nine months after the bombing, so that the effects observed could arise only from damage producing inheritable defects in the DNA† of either father or mother.

The investigation consisted in organising the midwives of Hiroshima to record every defect in every baby born of irradiated women among the survivors. The initial report of the midwives was horrific. The numbers of defects reported were several times larger than had been recorded anywhere before. The story goes that members of the Japanese Women's Institutes, which had been active in organising the investigation, were anxious for immediate publication of these results but that the President of the Japanese National Association of WIs insisted that a control study should be made, before publication, in Osaka, a coastal industrial city similar to Hiroshima. Being a determined woman, she got her way. The quite unexpected result was that the proportion of defects found in babies born in Osaka was indistinguishable from the proportion in Hiroshima.

The reason for the large numbers was that every small defect, even such a minor blemish as a hammer toe, that would always have been disregarded in the past, was now being reported. Few of us are free from minor defects at that level.

The identical result for Osaka was most surprising at the time, but it has been borne out by the results obtained over many years since the first study (Royal Commission on Environmental Pollution 1976). Neither among the 78 000 children born to the irradiated survivors and so far examined, nor among those born to sufferers from spondylitis, have any increases in hereditary defects been observed (although a small decrease, of doubtful statistical significance, in the proportion of males might be interpreted as due to sex-linked mutations lethal at an early stage of development *in utero* (Grosch and Hopwood 1979)). This does not necessarily mean that no mutations were produced. The latest UNSCEAR Report (1982) estimates that a population uniformly exposed to 1 rem of beta or gamma radiation per generation is likely to produce 20 cases of dominant hereditary disease and 60 recessives per million births. Fifty times this number of recessive mutations added to the

† See Appendix 1.

natural pool of perhaps a few million such recessives carried by the
300 000 significantly irradiated survivors would hardly have been detectable, and the mutagenic effects of radiation must be of much less
importance than the large number of new mutations produced each year
by chemical carcinogens.

Unfortunately, the early results from Hiroshima leaked out before the
control study had been done, and although the story of conspicuous
numbers of disastrous defects—all unspecified defects are disastrous—is
completely untrue, it is still very widely accepted as an incontrovertible
experimental result.

It is a sad fact that an exciting myth can never be overtaken and
destroyed by a dull truth, and I don't doubt that the myth will still be
conscientiously passed on by honest and intelligent but uninformed
believers for a very long time to come. But I hope that the number of
believers will gradually fall.

LOW DOSE RATES AND CONTINUOUS EXPOSURES

I shall now return to the risk of producing cancers, about the existence of
which there is no doubt whatever.

Many people feel that the ICRP had already erred seriously on the side
of optimism, and should, even before the reappraisal of dose risks, have
reduced its recommendations for acceptable doses. I believe that the
ICRP is more likely to have been pessimistic in the past, owing to its
neglect of the effect of dose rates. All living things have been exposed to
low-level radiation since life began, and the more complex creatures
such as ourselves could not have evolved without very effective repair
mechanisms to deal with damage to the enormously complex assembly
of a million or more genes in the nucleus of every one of our cells.

With a natural background of 100 millirems (0.1 rem) of beta-plus-
gamma radiation, the nucleus of each cell in our bodies, including the
cells of the reproductive system, will on the average suffer the passage
of a fast electron about once every five years. Each electron during its
passage will cause hundreds of ionisations, i.e. will displace several
hundred electrons, the sharing of which holds the molecules together,
and will give rise to comparable numbers of highly reactive groups of
atoms known as free radicals, which are equally capable of disrupting
the molecules which control the operations of the cell. Between them
these must cause many hundreds of disruptions of gene structure—and
a gene changed in any way represents a mutation. In the absence of
efficient repair systems many hundreds of mutations would be added
every generation to every individual child. Fortunately, many previously damaged cells will die without descendants.

The repair system must have evolved also to cope with the natural chemical carcinogens and mutagens which, with a lot of artificial assistance, now produce many more cancers than does radiation. The human species would be deteriorating pretty rapidly if we could not repair DNA damage at all.

Since repair takes at most an hour or so, its efficiency in coping with radiation should not be seriously reduced until the frequency of electron passages through a cell rises to several an hour, even though the repair system has evolved to cope with less than one a year due to radiation—plus a steady small flow of chemical injuries happening one at a time. Remembering the safety factor which appears in most biological systems, 10 to 100 passages of an electron through the nucleus per hour might be needed to overload the repair mechanism. This would correspond to dose rates of 5 to 50 rems per hour. It is reasonable (Fremlin 1980) to suppose that it could be overloaded by 10 to a 100 or more rems in a fraction of a second as at Hiroshima, or by 200 or 300 rems in a few minutes as in medical therapy.

It is worthwhile to compare the radiation risks with the risks of other ways of getting cancer. Thus the risk of dying of cancer in Britain from real but unknown causes is 100 000 in 50 million, or 20 in 10 000, per year. The official figure for the risk due to 1 rem whole-body is 2 in 10 000, so that the unknown causes are equivalent to 10 rems per year. Thus a 1 rem dose is equivalent in cancer risk to the inhalation of 200 micrograms (μg) of benzo-a-pyrene or to living for 18 days in an average part of Britain eating an average British diet—a risk that we can hardly avoid.

Another hazard, with which radiation hazards are often compared, is that of smoking cigarettes. Updating the figures given in the Flowers Report (1976), 1 rem is equivalent in risk to the smoking of 300 cigarettes—about one a day for a year. This of course is purely a comparison of the sizes of the risks. It has no relevance to the relative acceptabilities.

NATURAL RADIATION BACKGROUND

As has already been pointed out, the entire biosphere has always been radioactive. The mildly radioactive element potassium is an essential constituent of every living cell, and even the viruses can multiply only inside a living cell. Radioactive carbon-14, formed by bombardment of nitrogen with neutrons produced by cosmic rays in the upper atmosphere, forms a part of every living thing including viruses. The oceans, all soils and all rocks contain uranium, thorium and 10 or 12 other heavy radioactive elements, many of which appear in small traces in our food supplies and become permanently established in our bodies. Several of these elements emit gamma rays which reach every point and the interior of every building on the surface of the earth. The radioactive gas radon diffuses out of the walls of our houses and the ground beneath them into the air which we breathe, and several mesons per second produced by cosmic rays at the top of the atmosphere pass clean through the body of each of us, leaving a line a foot or more long of ionised atoms and disrupted molecules along their track. Table 4.2 shows the contribution of different sources to the average radiation background in Britain.

Table 4.2 Estimates of average background radiation doses (Clarke and Southwood 1989).

Source	Type of radiation	Annual dose (mrem)	
		UK 1988	US 1987
Cosmic rays	Mesons and gamma	25	27
Ingested natural radionuclides	Alpha, beta, gamma	30	39
External gamma rays	Gamma	35	28
Radon from earth materials	Alpha	120	200
Thoron from earth materials	Alpha	10	
Fall-out (including Chernobyl)	Beta and gamma	1	0.6
Occupational	Beta and gamma	0.5	0.9
Nuclear industry	Beta and gamma	0.1	0.05
Miscellaneous	Alpha, beta, gamma	1	5–13

The total dose to each of us from all of these sources may vary over the range of 130 to 250 millirems a year in Britain, with an average of perhaps 180 millirems per year. This means that the 50 million Britons receive a collective dose of around nine million man-rems, corresponding to an ICRP estimate of 900 cancer deaths each year. Fall-out from the bomb tests of the 1950s and 1960s is still delivering between three and four millirems annually to each of us, corresponding to a further 15 or 20 cancer deaths a year.

The only artificial sources giving a dose of radiation comparable to the natural background are the x-ray machines used in medical diagnosis and therapy. These give an average dose to each member of the British population of about 30 millirems each year. If this were distributed uniformly over the population it would correspond to 150 hypothetical deaths from cancer annually. In fact, much of the dose is given in treatment to people who have not long to live anyway and the number of lives saved by medical x-rays is of course many times greater than this.

A number of attempts have been made to observe directly the effects of the natural background for comparison with theory. There are several parts of the world, notably in Brazil and on the thorium-phosphate-containing monazite sands of Kerala in southwest India, where indigenous populations have lived for many generations with a natural background dose of a rem or more per year. With a lifetime dose of 100 rems or so one could expect an extra 1% of the population to die of cancer. Several surveys have been made, which in total show no significant increase over the rate for similar populations in less radioactive areas. Unfortunately many of the peoples concerned have simple standards of life and little education, and their vital statistics are unreliable. Where no records of births exist, no reliable data on age distribution are possible. In such circumstances the ages claimed by individuals, who will certainly suspect the investigators of having ulterior motives for the questions, will depend a great deal on what the said individuals believe these motives to be. If they fear a military call-up, very few men will turn out to be of military age; if there were to be a subsidised training scheme for youths there will be a lot of youths, and so forth. Since general cancer rates roughly treble for every decade of life between 30 and 60, an error of even a few years in the average age can make a lot of difference to the cancer rate to be expected in the absence of radiation.

Diagnoses of cancer may also be unreliable. (The ancient Greeks called gangrene cancer, and this was also the case in England in the Middle Ages.) An investigation of the proportion of defective infants born, in the hope of detecting the effects of radiation on mutation rates, would be even more difficult. The bearing of a defective infant may well be a shameful event likely to be hidden by the mother if she can manage it.

THE ARGONNE STUDY

A very large-scale study of cancer rates in the USA was made by Frigerio et al (1973) at the Argonne National Laboratory in USA to see whether the effects of differences in natural radiation background in the different states could be shown to have any effect. Age-adjusted cancer statistics for the entire white population of the USA over the 18 years from 1950 to 1967 were compared in each state with the average natural background for each state. The total collective dose was over 20 million man-rems per year, which on ICRP expectations should be leading to at least 2000 cancers per year.

The actual results are shown in figure 4.3. Not only does this fail to show a rise in cancer rates with radiation dose; it actually shows a reduction. Application of probability theory shows that there is a probability of less than one in 100 000 that this result could be due purely to chance.

If it had happened that the cancer rates in the seven high-background states had been higher than in the other states, it would have been regarded with certainty as showing an effect of radiation some 10 to 20 times the value expected by the ICRP. Taken at its face value, the actual result must demonstrate with exactly the same degree of certainty that radiation is good for you, reducing the chance of contracting a cancer for a US citizen by about 0.2% per rem. This is not impossible, as will be seen below when the results of animal experiments are being discussed; but in my view it is more likely that the non-radiation cancers are, for reasons nothing to do with radiation, more prevalent in the low-background states.

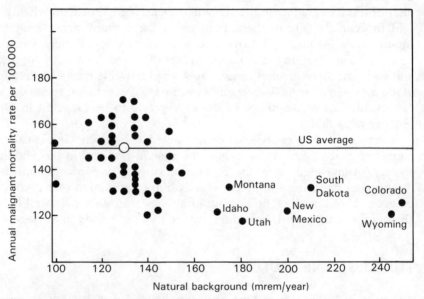

Figure 4.3 Malignant mortality rates for the US white population from 1950 to 1967 by state and natural background.

The seven low-cancer high-background states are all more than 1000 metres above sea level; Wyoming and Colorado, the two highest-background states, being almost entirely above 2000 metres. It is this that leads to their large natural background; more cosmic-ray-produced particles can penetrate the thinner atmosphere. So the first and most obvious proposition is that more people of the high-cancer low-background states live in cities, with greater air pollution. However, this does not at first seem to be so—the proportion of city dwellers in the low-cancer states is much the same as in the high-cancer states. However, the cities in the industrial east are much closer together and must receive a good deal of pollution from each other, while the high-altitude states not only have fewer cities more widely spaced, but must also have significantly stronger average wind velocities to blow their pollution away. The air pollution that matters most will be close to the ground, so that car and diesel exhausts—which must be breathed regularly by commuters—may be particularly important. A good hill breeze, morning and afternoon, might help a lot†.

A smaller study than that at the Argonne Laboratory, but one using far less mobile populations, was carried out in China between 1972 and 1975 (HBRRG 1980). A detailed comparison between two populations each of about 70 000 persons, all of whom had lived in their areas for 2 to 16 or more generations, and whose natural backgrounds differed by about 200 mrem per year, showed a barely significant increase in chromosome anomalies, usually clinically unimportant, but about half the cancer rate, in the higher background area.

ANIMAL EXPERIMENTS ON CANCER PRODUCTION

Many kinds of animals have been subjected to external irradiation by x-rays, gamma rays and neutrons, and to internal irradiation by inhaled or swallowed radioactive materials. The delay between exposure to radiation and the development of a tumour seems to be roughly proportional to the animal's normal expectation of life which suggests some more than coincidental relation between the factors controlling cancer and those controlling normal aging.

The fact that mice develop cancer 30 or more times as quickly as we do is of enormous practical value, but does not necessarily mean a higher susceptibility to radiation damage. Different mammals differ by less than a factor of two in their LD50 (the radiation dose following which 50% of animals die from damage to the cells of the gut and the blood-forming systems) independently of size. This is to be expected, since we are all built of similar cells doing similar jobs; the insects, built in different ways, have LD50 values in the region of 10 000 rems against our 200, and many simpler organisms have higher levels still. It seems likely that we can also expect similar mutation rates to those of other mammals, which is of even more importance, since neither from Hiroshima nor from anywhere else have we any reliable evidence of mutations in humans caused by radiation. In the absence of animal experiments we

† Another factor is that a large part of the population of Utah, and smaller parts in neighbouring states, are Mormons who neither smoke nor drink.

should have had no warning even of the *possibility* that they could be produced by radiation.

A possible example of genetic damage, though not of mutation, produced by radiation has been reported by Kochupillai *et al* (1976) from Kerala, southwest India. Here there was a marked increase in Down's syndrome (mongolism) in a study population in a natural background of 1.5 to 3 rems per year over that in a similar population, in a background of 0.1 rem per year, 12 cases being found in the first and none in the second. However, the number per 1000 in the study population was 0.93, only 14% above that in India as a whole, and the abnormality seems to lie in the zero value for the control population.

RADIATION HORMESIS

Hormesis is a biological term defined as 'the stimulus given to any organism by non-toxic concentrations of toxic substances'. Well known examples are for example the fact that arsenic and strychnine, lethal in quite moderate quantities, are useful tonics in very small quantities. In rather larger quantities arsenic improves the complexion and physical stamina.

The word hormesis may also be stretched to include the fact that a quantity of a poison that would be lethal when all taken at once may be harmless or helpful if taken over a period. Thus a hundred aspirins washed down by half a litre of whiskey would be reliably lethal, but one aspirin and five millilitres of whiskey taken each day for a hundred days would be harmless. This is well known to be true also for ultraviolet radiation. In a long series of short and mild exposures ultraviolet on our skin can be valuable, or indeed necessary, to enable us to produce the essential vitamin D for ourselves when there is a shortage of vitamin D in our diet. In long and intense exposure it can do prompt and serious damage, including the production of skin cancers—a small proportion of which are lethal.

If instead of learning about ionising radiation on the way down from Hiroshima we had learnt about it on the way up from the natural background in which all life has evolved, we should have assumed this to be good and would have taken radiation hormesis for granted when the unexpected and no doubt initially disbelieved dangers of large doses were discovered.

Quite certainly, without our radioactive background human beings would never have existed. A considerable percentage of the countless billions of mutations responsible for our evolution from the first living cells must have been due to this background; without it some fairly advanced form of life might have developed by now, but it certainly couldn't be us.

The evolution of the DNA repair system which alone made possible the existence of any long-lived multicellular life must have been driven by our natural background; and these repair systems are now helping to reverse the damage to DNA done both by radiation and by the numerous novel and dangerous chemicals with which civilization is now faced.

The fact that radiation was responsible for our development in the past is not proof that it is doing us more good than harm at present; and while the idea of radiation hormesis would probably have been taken for granted if Hiroshima and Nagasaki had not been bombed, they *were* bombed, and as a result a large proportion of the world population has been brought up to believe only that radiation is dangerous. Indeed radiation can be dangerous, and it is needful to present and to repeat the evidence for radiation hormesis when in a happier world it would have seemed nonsensical to deny its existence, and needful only to learn the conditions in which it was important.

It is widely believed that the effects of the bombs on Japan have proved that even low levels of radiation do increase cancer rates. This is not so. Here it is worth looking again at table 4.1 (page 45) showing the excess deaths for all cancers up to 1982 among survivors of the bombs. Leukaemias develop relatively quickly and few if any further cases due to the bombing are likely to appear. Further 'hard' cancers are to be expected however, but there is no reason to expect a statistically significant increase in deaths at low doses. In any case, the point at issue is not that the table gives strong support to the idea of radiation hormesis but that, in spite of popular belief, it quite definitely provides no evidence whatever against the existence of radiation hormesis.

Positive evidence *for* radiation hormesis is provided by a number of animal experiments. Experiments on small and short-lived animals have given some most important qualitative results. For example, the significance of dose *rate* has been shown quite unequivocally. As stated earlier in this chapter, rats, which like us die in a few days after a dose of less than 1000 rems delivered in a few minutes, will stand 16 rems a day for a year or more without 'prompt' effects (Lamerton *et al* 1960), although there is a significant reduction in life expectancy, largely but not wholly due to the cancers which result with a probability comparable with that characteristic of humans. This is consistent with the ability of people treated for ankylosing spondylitis to stand total doses of 1500 rems or more over a few years without serious prompt injury.

The unexpected effect is the *increase* in life span which has been observed in several species of animal with moderate radiation doses at moderate dose rates delivered discontinuously over long periods. This does not seem to be due to a reduction in cancer rates, but to the deferment of death (and thus a reduction in death rate) due to a number of other common causes.

For example, figure 4.4 shows the median age at death of male Sprague–Dawley rats, which received regular doses of radiation for a year. Starting with young adults at four months of age, the dose was spread uniformly over 16 hours in each 24, leaving 8 hours dose-free (apart from the natural 100 mrem per year) giving time for repair (Carlson and Jackson 1959). Each group contained 22 rats. Half of the control animals living at 26 °C with no irradiation were dead after 445 days, but half of those given 0.8 rem a day were dead only after 585 days. No zero-irradiation controls were kept at 28 °C and 35 °C, but at both temperatures the maximum life span occurred at about 2.5 rems per day, a total dose of over 900 rems. If the largest increase in longevity found in these experiments applied to humans, people now dying at 75, if similarly irradiated, could expect to live to 120. If the effect were linear

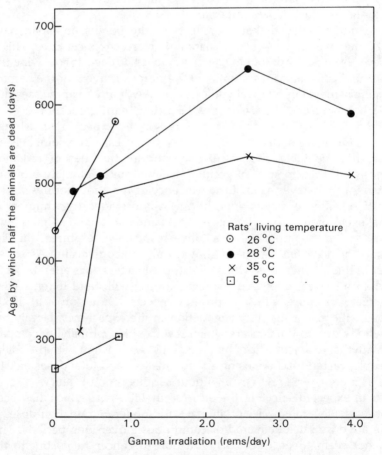

Figure 4.4   Median age of irradiated rats at time of death.

at low doses the gain in life expectancy would be about 18 days per rem per person, and 10 000 man-rems should increase collective life expectancy by 500 man-years.

Similar results were obtained by Lorenz (1950) for mice, which showed an increase of median age from 23 months for zero irradiation to 27 months with 0.11 rem spread over 8 hours each day for the entire period, a total of 90 rems. Extrapolated to humans, this would give us only about 15 years greater expectation of life; guinea pigs did less well still, with 0.11 rems giving a two month increase in median life over a control value of 46 months with a total dose of 180 rems.

Figure 4.5 shows the effects of continuous irradiation of mice by neutrons (assuming a quality factor of 10) and by gamma rays (Spiers 1968). Again, this shows an increase of survival time for 1 rem per week of gamma rays, but not for neutrons.

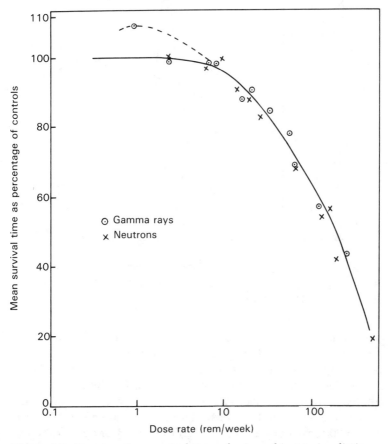

Figure 4.5 Comparative survival time of mice subject to irradiation.

Further support for the existence of radiation hormesis is given by a different type of experiment, the results of which are shown in figure 4.6 (Hoffsten and Dixon 1974). Here the effectiveness of antibody formation in the immune systems of mice was determined by measuring the number of micrograms of a foreign hormone that were bound by 1 millilitre of blood plasma after six weeks of irradiation at a series of dose rates. The standard errors are considerable, but there is a very clear maximum at around 170 rems a week, or 1 rem an hour.

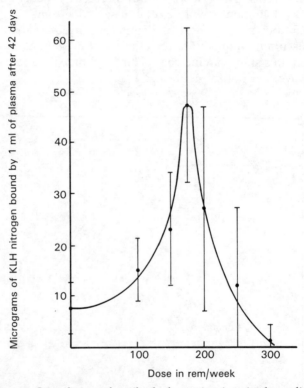

Figure 4.6 Stimulation of antibody formation in mice by radiation (Hoffsten and Dixon 1974).

A gain from low-level irradiation is not confined to mammals, as has been shown by an irradiation experiment made in Canada on the Chinook salmon. As with other species of salmon, the eggs are laid in a stream well away from the sea, and the young fish, when they are big enough, leave their birthplace and migrate to the sea. Two, three or four years later they return to their birthplace to breed themselves. Doses of from 0.5 to 20 rems per day were given to batches of eggs from shortly after fertilisation until the young fish hatched and had started feeding. A series of batches, each of 100 000 or more young fish, were released in each experiment. Abnormalities in the young fish were increased at all dose rates, but the number of adult fish returning to breed was not decreased. No damage to the fish stock was observed over several generations. The relevance of the experiment to the present discussion is that the groups given low levels of irradiation returned in greater numbers, and produced a greater number of viable eggs, than did the unirradiated control fish (Pentreath 1980).

These results and a number of other similar ones have received extraordinarily little attention, perhaps because they seem unreasonable. This is natural; we have a great capacity for perfectly genuine failure to remember data which do not fit with our established knowledge. I was myself jolted into remembering them only after seeing a couple of examples of increased longevity in mice produced by small doses of radiation, shown without any comment by either author, in a single session of a conference in Vienna in 1980, when these apparent anomalies appeared in slides illustrating papers on the damaging effects of larger doses. And the reason I remember is not just because I saw the same anomaly twice in three hours, but because a possible reason for it suddenly occurred to me—that the well protected animals in laboratories ordinarily received very little challenge by viruses or bacteria to their immune systems and that perhaps the irradiation gave their immune systems some kind of stimulus which, like exercise of any other faculty, made them more effective in later life. Whereupon, having now found the means to make the facts fit into my preexisting pattern of knowledge, I have been able to remember them ever since. (The fact that my hypothesis improved my memory is no proof that my hypothesis is right. It is however well known that animals reared completely free of germs of any kind are abnormally easily killed by minor infections.)

There remained, of course, a chance that some irrelevant coincidental effect had reduced the expectation of life of the control animals in the two experiments described in Vienna. The probability of this did not seem large, and seemed still smaller two hours later at the same conference when I proposed my hypothesis to Dr Stanley Cohn, the Director of the medical physics laboratory at Brookhaven. He avoided comment on my ideas about the immune system, but remarked that he had obtained the same result in a mouse-irradiation experiment when he was a research student, but had never dared to publish it!

I have since found that Carlson, Scheyer and Jackson (1957) had suggested that the mild injury of irradiation might be beneficial by stimulating renovation processes. This, like the experiments themselves, seems to have had little attention.

While the attention paid to these results has not been large, it has not been negligible. The question was discussed by Lindop (1965) in the course of an article on the reduction of life span by radiation. The main object of the article was to report experiments which showed that the life-shortening effect of a dose of 100 rems of gamma radiation was less when the dose was delivered at lower dose rates. The lowest dose rate given was 10 rems per minute, which corresponds to an electron traversing each cell nucleus every three seconds, and cannot be regarded as low in terms of the time left for repair. There was a linear relation between the total dose, over a range of up to 800 rems, and the life-shortening effect, over a range from 0 to 50 weeks. In a discussion of the life-lengthening results of Lorenz, Carlson and Jackson described above, Professor Lindop pointed out that very small changes in the conditions of the laboratory animals can produce large changes in life span, and that the control groups of animals might have been accidentally given less favourable conditions.

This is clearly possible, but requires that quite a large number of independent experimenters have failed to give the essential equal treatment to their controls and to their experimental groups. The same doubt could of course be applied to the experiments in which the control animals had longer lives than the irradiated ones. Indeed, better conditions for the controls would be more easily missed because they gave the results expected; and the suggestion does not apply to those of the Carlson and Jackson results shown in figure 4.4 in which all groups were given some irradiation. It is difficult also to see how it could apply to the large-scale studies of salmon reported by Pentreath.

The increases in life span are not incompatible with the ICRP or the BEIR expectations that even the lowest doses will add some risk of cancer. The actual development of cancer depends not only on its initiation in a single cell, but also on its later development and on the failure of the body to recognise and destroy it. It seems probable that a large majority of cells which have been given the potential of initiating a cancer are recognised and destroyed or are in an environment unsuitable for their multiplication. Increased irradiation in early life must increase the number of cancer initiations, but if at the same time it improves the chance that the cells concerned should be recognised and destroyed by the immune system, the number of deaths even from cancer itself could be reduced.

No information was given by Lorenz of the causes of death of his animals, but Carlson and Jackson recorded visible or palpable tumours in their rats, although these were not stated to have been the cause of death. There were 10 tumours among the 44 animals in the two groups receiving close to four rems a day, three tumours in the 44 animals receiving the lowest dose of 0.3 rems a day, and none in the 22 control animals. At least part of the increase of tumours among the irradiated animals must have been due to the non-radiation-induced increases of cancer rate with age that are familiar in human beings.

The assumed possibility that a threat of some kind to our immune system can sometimes make the system more efficient in other ways has been proved by Kneale *et al* (1986), who showed that the risk of childhood cancers was reduced in children who had received inoculations for any of eight infections: smallpox, diphtheria, tetanus, whooping cough, measles, German measles, poliomyelitis and tuberculosis (BCG). The numbers were in some cases small, and the statistical value correspondingly small, but altogether there were nearly 12 000 matched pairs and the statistics were completely convincing. The average risk of childhood cancer for each individual was reduced by 9%, and for immunisation after 10 years old by 24%. The authors themselves concluded that the results could be a sign that simulated infections have immune system effects which impede or prevent further development of a cancer *in situ*, and that an unrecognised effect of recent immunisation programmes has been a reduced frequency of solid tumour deaths in young adolescents.

Newborn babies have no effective immune system of their own, relying for some time upon the defences derived from their mothers; and under modern hygienic conditions it may be some time before they receive any stimulation by harmless microorganisms to prepare them for dangerous ones. And our natural radiation background, to which we have been exposed throughout our entire evolution, may still be playing an important if not essential part. There are not many permanent features of the environment that living creatures have not found ways of exploiting.

Until recently I have supported the view that radiation hormesis will occur only after low-LET interactions. Dr Bernard Cohen has however recently reported a comparison of the risks of lung cancer with radon exposure in 411 counties from all states of the USA. He found a *negative* correlation between the average radon concentrations and the average lung cancer rates, i.e. where the indoor concentrations were high the lung cancer rates were low. This looks on the face of it to be solid evidence for a large degree of radiation hormesis following alpha-particle damage. This is pretty convincing, but it is not impossible that there could have been an alternative explanation of Cohen's results, as was the case for the similar results of the Argonne study. (See page 58.)

There is an interesting corollary to radiation hormesis. This is that the trouble and expense of keeping down the radiation doses received by the public from the nuclear industry are counter-productive, and that the reductions recommended by the NRPB must be *preventing* the increase of life span that could be expected as a result of a moderate increase of annual dose. I do indeed think that this is highly probable, but until we have more direct evidence for the occurrence of radiation hormesis among humans, as well as among all the other organisms that have been studied, it would not be wise to do anything about it. The gains are likely to be too small to be worth the torrent of opposition to be expected from practically all of those who would almost certainly live longer with a larger annual dose.

SUMMARY

I have tried to put into perspective the risks of cancer and mutation that arise from exposure to ionising radiation. The additional risks due to artificial sources are discussed, and the reasons for using the figure of one extra death per 10 000 man-rems for low-level exposures are explained.

The possibility that in some circumstances exposure to low levels of dose at low dose rates might lengthen the expectation of life is discussed.

Part 2

Power Production, Conservation and Needs

Chapter 5

Energy Supplies—'Renewable' Resources

No energy is more expensive than no energy.

Homi Bhabha

DEFINITIONS

Before going on to discuss the great variety of ways in which we can extract energy from the world around us, it is necessary to explain the terms I am going to use. Two things are important: the amount of energy we use and the rate at which we use it. To start with the second, when you switch on a one kilowatt electric fire you begin immediately to use electrical energy at a *rate* of one kilowatt (kW). If you switch if off again at once you won't have used much energy and it will hardly show on your meter. If you leave the fire on for an hour, using electricity at the rate of 1 kW for the whole hour, you will have used up a kilowatt hour (kWh) of electrical energy. This is the unit for which you are charged, and which will cost you about six pence.

All forms of energy are measured in the same units, so that in running your fire you are turning electrical energy into heat energy and the kWh of electricity that you have used will have produced a kWh of heat. If you use the kilowatt of electricity to drive a perfectly efficient 1 kW electric motor, the motor will be doing mechanical work at a rate of 1 kW. This would do quite a lot of mechanical work in an hour. It takes a tough and healthy man to do physical work all day at a rate of 100 watts (0.1 kW), so that he would need 10 hours of actual hard labour to produce the kWh that you get from the mains for sixpence. One kWh of petrol combustion energy would drive a Mini Metro about 10 miles, the efficiency of the engine being 25–30%.

Modern power stations have to produce electricity at a very high rate to supply a town of even moderate size; the rate of production of heat to warm as few as 60 houses in winter may be 1000 kW, which is called a megawatt (MW). If the houses were warmed equally well by oil or gas, these would need just the same 1 MW of heat from the chemical energy of gas or oil burning in oxygen.

The output of a large electric power station may be a million kilowatts—which is called a gigawatt (GW)—and the total amount of electricity produced in a year may be put in thousands of gigawatt hours or in gigawatt years (GWy). A gigawatt year of electricity (GWye) will be worth £200 or £300 million.

VARIETY OF ENERGY SOURCES

For the purposes of discussion, the kinds of energy available to us are often divided into the renewable sources, such as wind, wave and solar power, and the non-renewable sources such as coal, oil and natural gas, stored up aeons ago. The distinction is important, but it is not always easy to decide into which class a specific source should be put. Direct use of sunlight for heating or electricity production obviously comes in the renewable class; the sun will last a long time. Oil and natural gas are obviously in the exhaustible class; they are unlikely ever actually to run out, but extraction will undoubtedly become too expensive to use them simply for burning. Oil and natural gas could reach this stage in a few decades, though they will probably last more than the 20 years or so that is often suggested. Coal should last a few hundred years.

On the other hand it could be a mistake to put into the renewable class the growing of fuel crops. The sunshine, water and nutrient elements are indeed available for an unlimited time, but with the current growth of world population the large land areas required will be available only for a very limited period. This will be discussed further in the section on current biological sources of energy.

Nuclear energy is usually placed in the exhaustible class, depending as it does on uranium from the limited number of ores rich enough for our current methods of extraction at current costs. The million million tonnes of uranium distributed at 2–4 grams per tonne through the top 2 km of the world's land surfaces could, if employed in breeder reactors, last the whole present world population 100 million years at the rate of energy expenditure per head now characteristic of Europe and the USA. This thinly distributed uranium will however never be useful for thermal reactors because it would need more energy to extract than it could produce when extracted. A good fraction of the 4000 million tonnes of uranium in the ocean, however, could be extracted at a cost of less than 10 times the current cost of uranium (Davies *et al* 1964, Curran and Curran 1979, Anderson 1986). These estimates are highly speculative, but should be achievable by the thirties of the next century if needed. This may later be acceptable even for thermal reactors, and for breeder reactors would add only a few per cent to the cost of the electricity that they produce. This would last 100 000 years for the present world population. Since the human race cannot last for as much as 1000 years on this planet at the present rate of multiplication, and must for survival limit its numbers to match the resources available, nuclear power would be better regarded for practical purposes as being in the renewable class. Figure 5.1 shows the mix of UK sources of primary energy in 1987, as well as in the US.

The different sources will now be examined in some detail, with particular emphasis on the new sources that might soon become available.

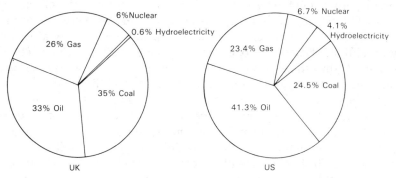

Figure 5.1 UK and US sources of primary energy (1987).

DIRECT USE OF SOLAR ENERGY

The power density (P_s) of solar radiation reaching the earth's orbit each day is 1.37 kilowatts per square metre. The total amount of this intercepted by the earth is then $\pi r^2 P_s$ where r is the radius of the earth. This works out as 170 million GW, which is more than we are likely to use for some time, while the average power per square metre absorbed round the year is then $\frac{1}{4}P_s$, since the total surface area of the globe is $4\pi r^2$.

There is a popular misconception that the light and the heat reaching us from the sun are separate kinds of radiation, the latter being invisible (i.e. infrared). Most of the heat radiated by an electric fire running at 800–900 °C is indeed in the infrared, but the main energy derived from the sun is radiated at 5800 °C in the visible light range. The energy of the light, like that of the infrared, is converted to heat with 100% efficiency when it is absorbed.

Near the equator at midday and in the absence of clouds or dust, about 1 kW/m² reaches the surface of the earth. At the latitude of Britain, between 50° and 60°, it would average about 800 W/m² at the top of the atmosphere at midday and rather less at the ground surface, even on a cloudless day, due to atmospheric absorption. Averaged round the 24 hours and round the year the power density is about a hundred watts per square metre or hundred megawatts per square kilometre.

PHOTOELECTRICITY

When light falls on a suitable sandwich of different semiconductors between two conducting layers, a small electric current will flow in a circuit connecting the two conducting layers. Photovoltaic cells based on this principle, using sheets cut from expensive single crystals of highly purified silicon, can already be made to produce electricity at around 1 volt with an efficiency of over 20%. But even if 20% efficiency could be obtained from cheaper materials such as amorphous silicon, allowance must be made for some practical limitations. The direct current output has to be converted to high voltage alternating current for transmission; the photovoltaic cells cannot cover the whole of the surface available without making cleaning and maintenance impossible; and there will be periodical attenuation of sunlight by dirt, algal growth, rainwater, snow and ice. In Britain it would in practice need at least 50–80 km^2 to produce as much electricity as one modern coal or nuclear power station. The peak output on clear days in the summer would be several GW from 80 km^2 and even when the sun was not shining such an array could reach 1 GW over a good part of each day. Unfortunately the middle part of the day in summer is not a time when extra electricity is needed in Britain, although it is in the USA for air conditioning, and, remembering the maintenance cost of the window cleaning of 80 km^2 of glass, and allowing for the area of paths needed for maintenance, even using automatic equipment, I think that this is unlikely to be the best way of using large amounts of solar energy more than about 3000 km from the equator.

In hot countries, where clear skies can be expected for most of the year, and where there are large areas of land not being used for anything else, the situation is different. The photovoltaic cells which are used in satellite applications provide electricity at about a thousand times the cost of that from existing ways of generating electricity on the ground, but the cost will undoubtedly come down as time goes on, and in dry climates there will be little fungal or algal growth which is likely to present the chief maintenance problem in humid areas. The frequent suggestion that we could use the great deserts to produce an important part of the world's electricity is, however, invalid. The major deserts are covered with sand, which has two drawbacks. The first is that is doesn't stay put. Large areas are covered with dunes which drift under the action of the wind and could leave any fixed equipment under 10 metres or more of drifted sand. The second drawback, common to dusty as well as sandy areas, is that any surface in the open receives a regular sand-blasting by windborne particles, imposing a short life on the protective cover over the active cells which otherwise would be expected to last for a long time.

There are certainly places in the world not too far from the equator in which there are large areas of land available with a firm surface, which are of little value for agriculture. A very material advantage of solar power in such areas is that in hot climates the peak demand for electricity occurs in the hottest part of the day, for the running of air-conditioning equipment, and solar cells will give a midday peak output around four times the round the clock average.

I am indebted to Professor Beckmann for telling me of farmers in New Mexico who use solar power plants at 10 000 dollars per installed kilowatt, with diesel back-ups, for pumping water (Beckman, 1988). He also sent me an advertising price list for solar battery chargers, presumably for use in rural areas, from 42 to 376 watts output—which he estimates work at 5–7% efficiency.

I have no doubt therefore that some large-scale solar electricity supplies will develop when the cost of construction of the cells and the cost of policing the arrays comes down far enough to compete with other methods. I have not seen any comment elsewhere on the cost of policing, but the fact that the sunlight is free and the danger to the public zero once the array is established means that a capital cost of more than £1000 (at 1989 prices) per installed kilowatt† would be acceptable. Accordingly a few square kilometres would be worth £1000 million, and even a module of a working area of one square metre could be worth £500 or more and would be a most useful addition to the roof of any private house. Unpoliced areas of tens of square kilometres could be expected to stimulate quite a lot of highly innovative and possibly violent private enterprise, and a correspondingly high replacement cost for missing modules.

It has been suggested that such solar arrays might be set up in the so-called geosynchronous orbit used by communications satellites at 40 000 km out in space. Such an array would be arranged always to face the sun, and has been described in detail by Dr Peter Glaser (1970). It would suffer eclipse by the earth for less than half an hour each night. The electrical energy produced by a major satellite array would be used to produce microwave power at a wavelength of a few centimetres,

---

† The term 'installed kilowatt' means the total of buildings and equipment required per kilowatt of electricity that the system can produce. The price of electricity must cover the interest on this capital as well as fuel and operation.

chosen for its ability to pass freely through the atmosphere, whether cloudy or not. This could be focused by a number of large parabolic mirrors attached to the satellite on to a corresponding set of microwave receivers on the ground, in an area kept clear of people and conveniently placed to supply a large market for electricity.

There are however some quite serious drawbacks, apart from doubts about the cost. The microwave power density at the receiver would be several hundred to a thousand times the legally permissible level for human exposure and if it missed the receiver could form a good imitation of a death ray nearly a kilometre in diameter. Fortunately the piecemeal construction necessary would give plenty of time for experience with systems to prevent such occurrences before the power transmitted reached dangerous levels.

I do not believe that such a station will be built for a long time. No public inquiry could reconcile the inhabitants of areas near the receiving plant, and it would be too vulnerable. I have dealt with this proposal only because it catches the imagination and is one of the more fascinating examples of something that was absurd science fiction a generation ago, but which would be perfectly practicable if we should so desire a generation hence.

While the prospects for major assemblies of photovoltaic cells are poor, the prospects for small assemblies are good in hot countries, though of doubtful use in Britain. Such small assemblies are already widely used to provide power for such things as repeater amplifiers in microwave links on mountain tops that would be extremely expensive to keep running on the basis of heavy batteries needing frequent replacement. Of course only very small numbers of these are needed, and a much more extensive use would be for individual roof units in the many rural areas of tropical countries without cheap mains supplies. Even under light cloud a quite moderate area of photovoltaic cells at the present commercially available efficiency of less than 10% would be useful. Air conditioning in the more humid tropical regions is of enormous importance, both for making possible indoor work in comfortable insect-free conditions, and as a status symbol for those who can afford it in the home or office.

A very useful export market may not be far away, but there would be few applications in Britain unless there is a reduction in cost, unlikely in the immediate future.

An alternative approach to the direct use of solar energy has been developed using a series of mirrors to focus sunlight on to a blackened boiler which produces steam to drive a turbogenerator. With a large enough area of mirrors concentrating light onto a small boiler, very high steam temperatures can be produced, with correspondingly high efficiencies (over 40% being easily achievable). It is of course necessary to adjust continuously the axis of each mirror both horizontally and vertically to compensate for the earth's rotation during the day using a mechanism similar to the heliostat used on astronomical telescopes. There are two solar plants in California, Solar One and Solar Two, with 10 megawatt and 20 megawatt peak power respectively, both in the desert. Their construction was heavily subsidised by the Department of Energy, and I do not know of any plans for further major plants.

For this system the number of hours of direct sunlight per year is most important. In the photovoltaic system the diffused light from an overcast sky, although there is of course less of it, can be used with the same percentage efficiency as direct sunlight; but in the mirror–steam-generator system the light from even a thinly overcast sky is of no value. In no part of the British Isles could it be cost-effective and, though it may well become the system of choice in areas where several favourable conditions coincide, is unlikely to become an important contributor to world power supplies.

In Britain by far the most effective direct use of sunlight will be as a contribution to the heating of living space and hot water systems in individual houses; and perhaps for pre-heating the large quantities of hot water used by some industries. This will obviously contribute nothing directly to the power supplies now needed for industry and transport, but could save substantial amounts of energy which would otherwise be drawn directly from these power supplies. In the most widely used form a few square metres of a blackened sheet-like collector backed by a water-circulation system is exposed low down on a south-facing roof under one or more sheets of glass or plastic. These sheets allow free passage of the chief energy-carrying visible radiation from the sun, but block the invisible infrared radiation from the blackened collector as well as preventing cooling by the wind. The efficiency can be increased by a few per cent by the use of selective surfaces which absorb efficiently in the visible range but which radiate very inefficiently in the infrared, whereas most ordinary black surfaces are efficient for both.

The water circulation system, operating by convection or driven by a small pump will then feed hot water into the top of a thermally insulated hot-water tank at a higher level inside the roof, drawing water from the cooler bottom of the tank in exchange (see figure 5.2). The roof system is thus equivalent to an ordinary flat central-heating radiator mounted upside down so as to return heat to the tank instead of withdrawing heat from it.

In summer such a system would need about one square metre per person, which would supply enough hot water for all domestic needs on most summer days, and even in winter could contribute usefully by prewarming the water fed to a thermostatically-controlled hot-water system, hence saving fuel or electricity.

The versions of such a system now commercially available are unlikely to pay for themselves and their maintenance before they wear out—except perhaps in the southwest of England—unless the cost of fuel increases at least 15% per year throughout the life of the equipment, but they are certainly fully economic and already widely used in, for example, the Middle East. A well designed home-built system using second-hand materials should be fully economic even in Britain.

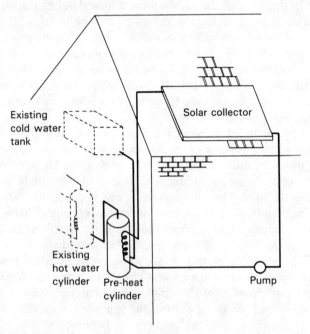

Figure 5.2 Solar heating installation—schematic view (overflow pipes omitted). The pump is not needed if the pre-heat cylinder is placed in the roof.

During the summer a system a good deal smaller than the average roof can on fine days capture a lot more heat than can at once be used, so that some form of heat storage can increase the value very much. For storage over a day or two the answer can simply be a large insulated tank, but the rafters in an average house will not carry the weight required to store enough hot water for long periods. A large well insulated underground tank fed from the primary tank by a small pump of the sort used in most central-heating systems is the usual answer, although it is again doubtful whether this could pay for itself within the working life of the system.

A more reliably promising plan is to use an artificial pond as a heat collector. This is being done on a large scale in Israel and other parts of the Middle East. The method is to establish a two-layer pond over a black bottom to absorb sunlight. The lower layer is a strong salt solution, which even at temperatures of 80 °C or more is denser than the upper layer of cold water. Convection within the salty lower layer brings the whole thickness of this to a nearly uniform high temperature while mixing only very slowly with the cold top layer. Water pipes passing through the hot lower layer can remove the sunshine-generated heat for domestic or other requirements. The cost of renewing the two layers as diffusion and residual convective mixing occur is considerable, but may be much reduced by the use of a low-density transparent gel for the upper layer, covered by a few inches of fresh water which can be circulated to remove dirt and to prevent the gel from drying out. Less hot water being needed in a warm climate than it is in Britain, a quite moderate sized pond might be able to supply all the needs of a compact village community with no access to mains supplies. Large units are also possible. A solar pond at the Dead Sea, opened in 1983, uses a system of the kind proposed for the Caribbean and is generating 5 MW of electricity (Meyers 1984).

With so many possible systems, direct use of solar heat could save really substantial amounts of fossil fuel in the warmer countries; but it is clear that in spite of the vast amount of solar energy pouring down on the earth each day, we are not yet clever enough to make very much use of it directly in Britain.

HYDROELECTRICITY

By far the largest renewable resource so far exploited is the gravitational potential of water evaporated from the oceans by the sun's heat and deposited as rain or snow on land high above sea-level. Water-mills have been used for grinding corn for 2000 years, and formed the main source of power other than human and animal muscle until the nineteenth century.

During the last 100 years the large-scale use of water power has changed almost entirely to its use for producing electricity. The usual pattern is that a fast-flowing river is dammed to form a reservoir with its water level as high as possible above the outflow below the dam. The water is then allowed to flow at a controlled rate through turbines driving electricity generators. The overall efficiency of conversion of the potential energy of the water in falling from the high level to the low can be over 80%, and when the dam can be built in a deep narrow gorge on hard rock the capital cost per kilowatt of output averaged round the year can be very much smaller than for any other method of producing electricity.

A great practical advantage is that the output can be made to match the demand; water being run faster through the turbines during the day and more slowly at night.

In Britain there are no really large examples, but a considerable number of small-to-moderate units in Scotland produce between them about 1% of our electricity requirements at very low cost. In the development organised in the USA by the Tennessee Valley Authority between 1933 and 1943, 32 dams were built, some entirely for electric power production, some also to aid navigation or to provide water for irrigation. Between them they produce some 3 GW of electric power, worth several hundred million dollars per year. By 1978 Brazil had an installed capacity of 21 GW, from a large number of dams.

In November 1982 the largest dam in the world so far was inaugurated at Itaipu on the Parana River, which forms the boundary between Brazil and Paraguay. It cost £9000 million—the cost of six or seven large nuclear power stations—and has backed up the river to form a lake 151 km long (and has displaced 42 000 people, although the population density of the area was low). It will be some years before all of the 18 turbines and generators are installed, but when they are the electrical output will be more than 12 GW. This is close to the output from six new large coal or nine or ten nuclear stations but has no fuel cost and much smaller operational and maintenance costs.

A well-built dam properly maintained on a good foundation should last for many hundreds of years, but the life of the whole system may be much less than this. Turbines and generators could be relatively cheaply replaced every few decades, but the life of the reservoir may be limited by the build-up of silt brought down by the river. In the original fast-flowing river this would have been carried away, but in the nearly static reservoir it will fall to the bottom. It has been said that the big Aswan dam built in Egypt may have its reservoir filled up in less than 30 years by the vast amount of silt carried down by the Nile.

Although, as will be shown in a later chapter, big hydroelectric systems are more dangerous than any other large source of electricity, the power produced is so cheap that they are likely to be used wherever suitable water supplies—and will certainly preserve from burning many hundreds of millions of tons of coal a year.

TIDAL ENERGY

Tidal energy is introduced here because the engineering problems are the same as those of inland hydroelectricity; big dams must be built and the power is obtained by allowing water to run from a higher to a lower level through turbines.

Tides arise as the result of the gravitational attraction of the moon, with a minor contribution from the attraction of the sun, and energy taken from the tides would come from the rotational energy of the earth. The result would be a very slight increase in day length over and above the gradual increase due to tidal friction.

The oceans do not cover the whole of the earth, and the tidal movement of the water is restricted and sometimes funnelled by the shape of the solid land barriers. As a result the height to which the tide rises varies from place to place. There are few places in the world where the tides are sufficiently high and where there are suitable conditions for them to be used for producing electrical energy.

The principle is simple. A dam is built across a bay or estuary, and turbines are built into by-pass channels as in a hydroelectric system. The French have used a tidal plant at La Rance in Brittany for over 20 years, with an average output round the year of 65 000 kilowatts, but the fact that the primary energy is free has not been enough to make it an economic success; and there are doubts as to whether it will even be worthwhile to keep it in operation when the machinery needs to be replaced.

Serious consideration is being given to a very much larger scheme in the Bay of Fundy, between USA and Canada, north of Boston, which has one of the highest tides in the world; and in Britain proposals have been made for a dam across the Wash or across Morecambe Bay in Lancashire. Apart from the cost, the intermittent nature of such large supplies would be a serious matter. A proposal to harness the tides of the Mersey for electricity production has completed a first stage study (costing about £2 million) collecting an appreciable amount of information on existing conditions in the estuary. A Stage 2 study will select the barrage location and study the effects on bird populations and sedimentation of silt, and should be completed by 1990 (Department of Energy 1989).

By far the most practicable area in Britain would be the estuary of the River Severn. Several schemes have been considered in some detail, the largest and most expensive producing up to 7 GW with a dam between Minehead and St Govan's Head, provision being made for the passage of shipping (*Energy Paper* 46 1979). A useful fraction of the cost could be recovered by building on it a toll road bridge connecting Somerset with South Wales. At present it is being taken sufficiently seriously by the Government at least to consider the expenditure of some of tens of millions of pounds on detailed surveys and technical planning, with the advantage—to the Government—that no final decision needs to be made until the results come in.

More recently a smaller and more attractive scheme, which would not preclude the larger one, has been put forward by Wimpey Projects, combining a tidal barrage between Sudbrook Point in Gwent and Severn Beach in Avon, with a new toll bridge to relieve the present heavily overloaded Severn Bridge (*New Scientist* 1984b). This would provide 100 000 man-years of employment and cost under £900 million, with an average output of 280 MW.

Among general advantages is the fact that almost all of the cost will be for labour, much of it unskilled, which would permit a very useful compensating reduction in unemployment relief, saving up to a quarter of the cost by reduction in unemployment pay and the administration costs related to this. Among the disadvantages there is the complete destruction of the long-established and specialised ecology of the estuary. Most bird life will gain from the fact that the less muddy water will allow more light penetration and a larger production of invertebrates; only the largest British flock of dunlin—wading birds feeding on the mud flats—would be badly affected. And we should lose the Severn Bore, which many people regard as an active rather than a passive site of outstanding natural beauty.

CURRENT BIOLOGICAL SOURCES OF ENERGY

Biomass energy includes in principle the traditional burning of wood for fuel, but the term is usually applied to more recent systems. It is already being exploited by some local authorities where the burning of domestic wastes provides heat to warm the buildings around the incinerator. Some sewage farms produce useful amounts of the fuel-gas methane by anaerobic fermentation of sewage.

By far the largest project now operating is in Brazil where sugar cane is being grown on an enormous scale to produce alcohol for fuel. The area eventually to be employed is 170 000 km^2, larger than the area of England. The alcohol is produced by fermentation of the sugar and distilled from the dilute solution thereby produced—the residue of the cane (bagasse) being used as fuel for this. It is hoped by this means to supply 50% or more of the fuel needs for Brazilian transport, thus replacing the corresponding quantities of non-renewable oil. An advantage is that with even 10% of alcohol, four-star petrol can be produced without any lead content. The proposal, already operating on a large scale, is to mix 20% to 50% of alcohol with petrol to make a fuel that would require little modification to most car engines while at the same time building car engines capable of running on 95% alcohol alone (the remaining 5% being water). Production of these had reached 40 000 a year by 1983 (*New Scientist* 1983a). Seven and a half million acres (30 000 km^2) have already been planted with sugar cane for this purpose (private communication from Brazil).

The total efficiency of conversion of solar energy to plant material is small, and many fermentable crops could not aid total energy conservation when used for the production of fuel alcohol, because the

distillation required to separate the alcohol from the watery fermentation medium uses more energy than is contained in the alcohol.
Alcohol production may however be profitable in cash terms because it
gets fuel where you want it, in the engines of cars for example.

As a conservationist I do not like this project, for which huge areas of
irreplaceable rain forest have already had to be destroyed—perhaps
unavoidable for food, but not for motor cars.

A more satisifactory procedure, also being developed in Brazil, is the
production of charcoal from eucalyptus. Charcoal can be used for steel
smelting and for gasifiers in heavy lorries, and eucalyptus can be grown
on land unsuitable for food production. Eucalyptus trees store about 2%
of solar radiation as wood, yielding perhaps 4 tonnes of charcoal per
hectare per year.

The difficulty of keeping so much land away from food production
would not be so important in the USA with its far smaller and still falling
rate of increase of population; but little of the USA is suitable for sugar
cane production, and few other crops will simultaneously produce such
quickly and easily fermentable material as sugar at the same time as the
large quantity of suitable fuel for the distilling plant.

The only country which has a practical long-term plan that would not
use an area likely later to be needed for food production is Sweden, with
only about 50 people per square mile. It plans to be using 4% of its forest
area (one million hectares) for continuous fuel production and, with the
wastes from wood-pulp production, to be producing a quarter of its
needs from wood by 2005.

Unfortunately, in most areas of the world where wood is used as a
fuel, although it is in principle renewable, it is not being renewed.

The growing of special crops is often an unpromising source of large-
scale energy for another reason—implied but not explicit in what has
already been said. The growing, harvesting and processing of the crops
may well use up more energy than the crops produce. The energy value
of the food crops grown in USA is several times *less* than the energy
used in cultivating, fertilising and protecting from pests. This is acceptable so far, food being much more valuable than the energy needed to
produce it, but would be acceptable for energy production only if the
form of energy produced were far more valuable than that of the energy
used up.

WIND ENERGY

Here we have a source of energy that could make an important
contribution to the electricity supply in Britain before the end of the
century.

Before 1939 wind-driven pumps to bring up water from welis for individual farms and houses were a common and conspicuous feature of the Essex countryside within 30 miles of London. Many more and larger ones were used in Holland to pump out the water from drainage ditches below sea level.

With the spread of mains water since 1945 these uses have practically disappeared in Britain; but since the sudden increase in cost of energy supplies following the OPEC price rises interest has expanded rapidly in what are now called aerogenerators or wind turbines, big enough to contribute useful amounts of electricity to the grid.

Sporadic attempts have been made for 20 years or more to build efficient machines, using good engineering principles and the understanding of aerofoil design derived from experience with aeroplane propellers. Early machines produced less energy than expected and rarely lasted for long. It seems likely that increases in size had been made too quickly before enough experience with high winds was obtained. One of the fundamental difficulties in designing a wind machine is that, to be economically worthwhile, it has to give a useful output for a good proportion of the time, which means working with less than average wind velocities for considerable periods. At the same time, if it is to have an 80% chance of lasting for 20 years it must be able to survive the worst storm to be expected in 100 years (which has a 20% chance of occurring within 20 years). And of course, nobody can say with any accuracy just how serious the worst storm in 100 years is likely to be.

I have no doubt at all that such problems will be overcome, but it would be optimistic to assume that we could do so just by spending a lot of thought and a lot of money without a great many years of experience.

It was known to St John that the wind bloweth where it listeth [wills]; it also bloweth when it listeth, which makes it difficult to use for satisfying an inelastic demand which varies for independent reasons. It is therefore essential to have alternative sources of supply for the occasions when it may not blow usefully at all for days on end. Accordingly wind machines do not reduce the need for installed capacity, but they can reduce the extent to which it is used by replacing expensive coal and oil by free wind energy. A very thorough study has been made by Halliday *et al* (1983) of the effects of integrating wind energy with other sources for the national grid. Their plan covers several features. The wind turbines themselves would be distributed in a number of regions round the mainland of Britain to take advantage of the difference of timing of wind maxima and minima. A carefully planned mix of different fossil fuel stations plus hydroelectric storage would optimise the saving of fuel by using diesel-electric plants, cheap and quick to build but expensive to run, to cope with rapid reduction in wind output or sudden increase in demand, thus reducing the need for coal-fired stations to run expensively on stand-by, and giving time for them to start up from cold and in turn to replace the diesel-electric plants if wind power remained low for more than eight hours or so. With the additional help of rolling weather forecasts for 10 hours ahead, and computer control of the stand-by stations, 20% to 30% of UK electricity might be supplied by wind turbines without serious complications.

Lord Marshall has estimated that wind power should come out comparable in cost with the best of the nuclear power units at 2.2 pence/kWh, and looked forward to a possible gigawatt of wind driven capacity by the year 2005 (*Atom* Jan. 1989, p. 23).

The costs are of course highly speculative and probably optimistic for the first machines built, but should come down rapidly as working experience and the number of standard units manufactured increases.

Modern machines have little in common with the traditional windmill in appearance. The sails of early windmills were broad and obtained their power by moving slowly under a strong force. The wind turbines, like aeroplane propellers, move fast under a weak force, which is much better adapted to the needs of a generator which needs to spin fast to give electricity at a sufficient voltage for efficient transmission. Two main types are being studied, one with the conventional horizontal axis of rotation and the other with a vertical axis of rotation. The former is slightly more efficient so far, but has to be turned (automatically) about a vertical axis to face the wind, while the latter obviously works equally well whatever the direction of the wind. The latter has another major practical advantage in that the heavy equipment such as gearing and generator can be at the bottom of the tower rather than at the hub near the top. Both types have cut-out systems, which 'feather' the blades when the wind gets too strong.

A number of industrial firms have been designing wind turbines, and some twenty individual wind generators are in operation in various parts of Great Britain, with outputs from 15 kW to 750 kW. Some of these are already providing useful supplies to islands in the north.

In October 1988 a wind energy demonstration site was opened at Carmarthen Bay. Here wind generators of different designs offered by different manufacturers are being evaluated for this country's wind farms. To begin with, three demonstration wind farms are planned in the UK, each with 20 to 25 machines of 300 to 500 kW apiece; and land tracts measuring 2 km by 2 km have been chosen, one in Dyfed, one in Co Durham and one in Cornwall. The CEGB and the Department of Energy have been studying off-shore wind resources. They have sponsored studies of what might be possible, given turbine generators of 3 to 6 MW on artificial archipelagos; and a test machine is to be erected off the north coast of Norfolk, 5 km out. However, the question arises as to whether the benefit of the more energetic off-shore winds would be frittered away in higher building and operating costs. (*Atom* Jan. 1989).

Since the wind velocity increases with height above the ground, it is advantageous to build machines as large as is practicable. What is practicable at any one time is something not too much bigger than similar machines of which sufficient experience has already been gained. The strengths required of different parts of the structure do not increase proportionally. Thus, if you double every dimension of a wind machine, the force exerted on the sails and on the top of the supporting tower will more than quadruple, since the wind will be stronger at twice the height and the area of the disc swept out by the sails will be quadrupled. The breaking strength of all the girder sections will be only four times as great under a simple stretching force, and they will also be twice as long and will give way if bent with the *same* bending force at the top as would the thinner shorter girders of the original structure. Allowing for all this easy, but when you have to think of the effects of a gust of wind affecting one side more than it does the other it gets very difficult, and it is necessary to increase size by small stages, each stage being studied for long enough to see and understand any unexpected results of the increase in size. It is certainly true that we could build bigger and better machines now if we had started trying seriously 20 years ago, as we certainly should have done; but we didn't, and 20 years' experience of the effects of vagaries of the weather takes just 20 years to obtain.

Holland and Denmark are already producing a useful part of their electricity from wind machines, but Britain has not yet done very much owing to the cheapness and availability of its own sources of coal and oil. A single 55 kW wind turbine has been supplying electricity in Fair Isle, between Orkney and Shetland, for some years, saving very useful quantities of the oil which would otherwise have been burnt by the diesel-electric set which provides the main part of the supply. A 3 megawatt machine on Burgar Hill in Orkney started operation at the end of 1987, but it is thought that numbers of smaller machines may have a better future.

Several considerably larger machines—up to 93 m diameter, with an axis height 62 m above ground—are already in operation in the USA. Owing to the lower average wind speed in the sites used, these give little if any more energy output than the Orkney machine. The California wind farms were doing well while they could be financed as tax shelters, and wind generators around Altona Pass are still running (January 1989) though facing lawsuits about visual pollution and noise; but most other big projects have been shut down as uneconomical.

It is desirable that as much of our electricity as practicable should be produced by non-polluting wind generators, but there are likely to be a lot of problems in siting them. The really windy sites in such places as the Lake District and Snowdonia will not be acceptable, and in many places with no claim to outstanding natural beauty there will be difficulty in getting the necessary planning permission. Local opposition is likely to be based on the (probably exaggerated) fear of interference with television and on the local effects of service roads and transmission lines. With the smaller wind velocities to be expected in East Anglia for example as compared with Orkney, 60 m diameter machines would be needed to give 1000 kW each and would have to be spread over half a kilometre apart. The 1000 machines which would be needed to replace the proposed nuclear reactor at Sizewell would have to be spread over at least 500 km², a large part of east Suffolk. A thousand 40 m metal conductors rotating 30 times a minute will cause a lot of periodic reflection of radio waves used in broadcasting. Each machine will need an 11 kV or 33 kV line to connect it up with the transformer station for feeding the grid at 133 or 400 kV, as well as the service roads for construction and maintenance. This will not only be expensive, but will involve regular payment for 'wayleave' to the people whose land will have to be crossed.

My mind has quite a high threshold of boggle, but it definitely boggles at the problem presented by 1000 applications for planning permission, and for compulsory purchase orders—if the land needed is to be obtained at tolerable cost—each with a different group of TV watchers, farmers and wildlife enthusiasts seriously opposed. To get anything done at all the CEGB might have to be given powers for the overriding of local opposition and for implementation of unpopular plans quite unprecedented for any civilian organisation in peacetime. A mitigating factor has been pointed out by Dr N W Pirie. This is that there are many regions where normal agriculture would benefit by extensive wind-breaks; and that many of the trees planted for this purpose could well be replaced by wind generators, with local support.

I very much hope there will be a lot of wind-generated electricity used in the next decade or so, but the extra technical problems facing wind turbines several miles offshore, even at twice the cost per kWh produced, seem far easier to solve than the non-technical problems inland, and the power then available is limited only by the percentage acceptable to the grid.

Apart from cost, there seems no technical reason why we should not have a prototype 5000 kW wind turbine operating a few miles off the coast of Lincolnshire or East Anglia within five or six years, although we should still need a good deal more experience of North Sea conditions. Then if we could set up an industry expanding exponentially and doubling its output every two years we could have 5 million kW, corresponding to four or five big coal or nuclear stations, in a further 20 years. We should not yet have saved much coal, because the energy cost of construction of each turbine would take time to pay off, and each year more energy might be absorbed in building increasing numbers of new ones than could be obtained from all of those already built. This however is true of any rapid expansion programme for any form of power production; and such a plan, however expensive, would leave the next generation with the tools to make a worthwhile contribution from wind power a real and practical and not just a theoretical option.

During the first two decades of the next century, if all goes well, we should be able to use the wind to save a big fraction of the fossil fuel that we are now using, although we may for a long time need to keep a large capacity in the form of fossil fuel stations to guard against the risk of long periods of good weather. Two hundred 10 MW wind turbines, spread three deep along 100 km of coast, would produce enough electricity to replace one large coal power station. Even if wind power on the scale suggested to save most of the fossil fuel now used in electricity production will not be economically worthwhile for some time, it is clear that when the technology is fully developed wind may be able to make a contribution to our energy needs that will be larger than the contributions that nuclear power is making now.

It has been pointed out by Dr Michael Grubb (1988) that the chill factor will be greater in windy periods, and will therefore increase demand for heating when the wind-generator output is high; so that the effective capacity of a wind plant should be usefully above the 40% assumed by the CEGB.

WAVE ENERGY

Until recently, wave energy seemed to be much the most promising of the sources depending directly on the sun. Detailed study makes it seem less promising and likely to be a good deal more expensive than energy from the wind. Accordingly, less will be said about wave machines, although there is certainly enough wave power within reach to replace a useful number of power stations if this were felt to be more important than the cost. Britain is particularly well endowed with good sites for development, its western coast receiving the full force of waves built up over the Atlantic by the prevailing south westerly winds, themselves arising from the non-uniform heating of the atmosphere by the sun.

Measurements at a number of points round the British coast, reported by Winter (1980), have shown an expected long-term average power of 48 kW per metre of wavefront, at distances of 100 km or more offshore. Most energy available is, perhaps unexpectedly, well out to sea. A lot of it is reflected or dissipated as the water gets shallower, so that the very conspicuous energy of the waves finally breaking on the shore is far less—as well as less easily extracted—than that of the smoother waves of the ocean.

The scale of the collectors would be very impressive. The most promising line round the Hebrides might take 300 collectors each a kilometre long, and the structure and anchoring of each one, as in the case of aerogenerators, would have to be strong enough to survive undamaged the worst storms to be met in one of the stormiest areas around Britain. The cost of replacement would be so high that a longer life than for aerogenerators would be desirable—especially as, even more than in the case of wind machines, a number of units might be expected to fail together in a storm greater than that for which they were constructed. To give a 95% chance of lasting for 30 years each unit would have to be capable of standing the worst storm to be expected in 600 years. Again, we do not have adequate records to estimate this, and large safety factors will be needed.

A second bonus from the scheme, as suggested above, unless the considerable cost forced us to have the units built in Taiwan, would be the rejuvenation of the British shipbuilding industry. Whether it would be obvious to the Government I do not know, but it seems obvious to me that it would be worthwhile to subsidise the construction to an extent considerably higher than the total of unemployment benefit now being paid to the people who would be directly involved in the construction, plus those employed by subcontractors. The greater prosperity of the region concerned would provide extra employment in local shops and other service industries, with a corresponding increase in money coming back to the Treasury from taxation. It is not apparent that this has been taken into account by the Government in respect of the building of the usual types of power stations.

The system for extraction of wave energy best known in Britain was invented and developed by Dr S H Salter in the University of Edinburgh. His plan was based on a series of floats ('ducks') rocking about a spine several wavelengths long lying along the direction of movement of the waves. Energy was taken from each wave to each 'duck' that it lifted, and transmitted down the spine to a generator. Starting with small structures a succession of stages were well financed until a design for a full-scale model for sea trial had been reached. At that point, for reasons which have never been made clear, money for the final—necessarily more expensive—stage was refused.

Norway and Japan have both built prototype plants delivering useful amounts of power for areas far from major sources of supply, each country having built and tested two different systems.

Britain's first demonstration prototype is now (February 1989) being installed near the landward end of a tapering gully on the coast of Islay by Dr T J T Whittaker's wave-energy team at Queen's University in

Belfast, with the help of a grant of £600 000 from the Department of Energy, which is also providing £87 000 for a detailed survey of the potential for small-scale wave power in inlets along the British coastline (*New Scientist* 1989).

The height of the waves entering the gully is increased by a factor of three or more at the site of the equipment. The waves reaching the plant alternately drive air out of the space above the oscillating water column in a cylinder very firmly fixed to the bottom of the gully, and suck air back in again. A two-way turbine invented by Professor Wells, fixed in the neck of the cylinder, spins in the same direction for each direction of air flow, and drives a generator capable of delivering 180 kW. The fact that the system is firmly and permanently fixed to the rock bottom in relatively shallow water reduces the energy to well below that which could be tapped by Salter's 'ducks' in deep water, but makes it easier to construct a stormproof plant, and to carry out routine service and maintenance.

The world's biggest wave-power generator could be providing electricity for the Western Australian town of Esperance within this year or next if an 8 million dollar project announced by the West Australian government comes to fruition. This plant should be capable of generating up to 3 megawatts (*Electronics and Power* 1987).

GEOTHERMAL POWER

Less than 10 miles away from us there is a free heat source large enough to supply the whole planet with electricity for an unlimited time. As has been mentioned already, the composition of the earth includes a number of radioactive elements. Although after its formation the outer crust of the earth cooled down in a million years or so, the interior has been kept hot by the radioactive decay of these, and heat from the interior is slowly but steadily being conducted to the surface. As we dig, there is an average temperature rise of 2 to 4 °C per 100 m, so 10 km down the temperature is likely to be around 300 °C.

In some places, often near volcanoes or over other hot spots in the earth's crust, such as the hot-spring areas of New Zealand, Japan, USA and Iceland (geyser is an Icelandic word for hot spring) temperatures of 250 °C or more can be reached at a depth of a kilometre or less (Curran and Curran 1979).

So far the most useful sources have been in the areas in which large volumes of water lying not too far below the surface in the region of hot spots have been heated to well above 100 °C. At The Geysers in California the hot steam from such a source has been tapped very successfully to drive turbogenerators with a combined electrical output of about 1.1 GW, and it is planned to double this by 1990 (IEE 1984). Similar but smaller hydrothermal systems are operating in Italy, New Zealand, USSR, Japan, Iceland and the Philippines. By the end of 1983 the world's total geothermal output was over 2 GW.

The geothermal areas already containing hot water at a reasonable depth may not always be permanent, genuinely renewable, sources. The hot water present was originally rainwater that soaked down through porous rock, and the hot water removed as steam to drive turbines may not be replaced as fast as it is used. In any case, there are few places where the hot water is already *in situ*, and the long-term prospects are far better in the much larger number of places where hot *dry* rock occurs at reasonable depths. The procedure then is as follows. Two holes 100 metres or so apart are drilled down to reach temperatures of 150 °C or more, and water under high pressure is used to open up natural joints in the rock between the holes. Cold water injected in one hole picks up heat from the rock, and returns to the surface at high temperature. When the pressure is released at the surface this water will flash to steam.

The amount of power that can be got from any particular pair of holes is obviously limited, but as we gain experience we should be able to go to greater and hotter depths, and the total energy available is in human terms entirely unlimited. Costs at present are expected to be high, but will clearly come down, and though the development of any new source of energy from first trials to really large-scale use has in the past taken 50 years or more, this could easily be a significant source of our electricity before the end of the next century. It is not likely to affect very much our options over the next 30 years.

The southwest of England contains the heat equivalent of 6000 million tonnes of coal, stored in deep granite rocks. I am indebted to the Director of the project for permission to report that after a few years of preliminary work the Camborne School of Mines in Cornwall is undertaking studies at intermediate depths (2500 metres) to determine the feasibility of creating a viable hot dry rock reservoir (HDR). The aim of the next phase is the establishment of a prototype HDR system at commercially useful depths. It is hoped that this stage can be reached by 1991.

Apart from power production, geothermal energy may contribute to district-heating systems. Supplies of hot water below 100 °C are of little practical use for electricity generation but can be useful for domestic heating. In Paris and in several other places there are already small district-heating systems in operation using such supplies, and the number of such schemes will doubtless increase.

SUMMARY

Of the genuinely renewable sources of energy, wind power is the only one that seems likely to provide an important contribution to British supplies in the next 30 years, although the direct use of sunlight should become increasingly important in countries less than about $35°$ from the equator. Biomass in the form of crops grown specifically for energy production can be environmentally damaging, and is usefully available only in those few countries which have small populations and large areas of well-managed and renewable forest. By the 22nd century the unlimited supplies of geothermal energy may have become the chief or even the only large-scale source of energy, though large advances in several technologies are needed before it can 'take off', and it is likely to be more than 50 years before it begins to make a useful contribution.

Chapter 6

Energy Supplies—Coal, Oil, Natural Gas and Nuclear

COAL

Although coal is Britain's most important source of energy after oil, and will certainly remain so for at least another generation, its use is familiar and needs much less discussion and explanation than do the much smaller or entirely prospective sources discussed so far.

In 1914 the annual output of coal in Britain was around 270 million tonnes. At that time it was almost the only fuel used for industry, for transport and in the home—for cooking, water heating and house warming. Most of the population used the kitchen warmed by the cooking stove as a living room in the winter. The better off would also have one or two other rooms warmed by open fires during the day; bedrooms would not be warmed at all unless someone was ill. During the next few decades the cooking stove was gradually replaced by gas (itself made from coal) in the towns and by stoves burning paraffin oil (kerosine) in the country, and in the last 35 years home heating has been increasingly converted to central heating by oil or gas.

Until this last 35 years nearly half the coal output was used in private houses, which were the main source of the pall of dirty smoke and the main cause of the dense smogs that affected our big towns before the smoke-control acts were implemented. During the same 35 years the railways have abandoned the use of coal in favour of diesel oil and electrification, to the great advantage of their environment. In 1950, 27 million tonnes of coal were used in gas production, but since the advent of natural gas from the North Sea this use has been postponed until the gas runs out. Now the main use of coal is for the production of electricity, for which nearly 90 million tonnes are now used annually.

PRODUCTION OF ELECTRICITY

The functions of oil, coal, gas or nuclear power used for this purpose are all the same: every one of them boils water into high pressure steam which is then forced through turbines (which are equivalent to a multiple 'windmill' encased in a tube). A simplified layout is shown in figure 6.1. The turbines are kept spinning by the expanding flow of steam in just the same way as a wind turbine is kept spinning by the wind. The steam is under very much higher pressure—170 atmospheres or more—so that a steam turbine is far smaller and spins much faster than does a wind turbine, for similar outputs of power. This makes it possible for the electric generator, which needs to spin fast to produce a high voltage at the right frequency of alternating current, to be mounted on the same shaft as the turbine, without need for mechanical gearing.

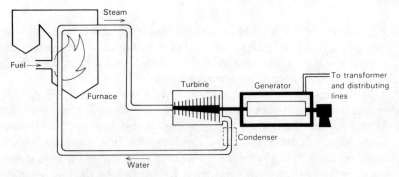

Figure 6.1 Schematic diagram of a coal-fired or oil-fired power station.

A modern boiler no longer consists of a tank with a fire underneath it fed by a lot of sweating stokers stripped to the waist. Instead it is constructed from an assembly of up to 500 miles of thin-walled steel tubes, spread out in a cylindrical furnace and through which water is driven under pressure (IEE 1983a). The furnace itself is far more like a gas burner than a traditional coal furnace. The coal is pulverised to a dust finer than face powder and blown in with the necessary air. This has the advantage that the efficiency of combustion is much increased, and that ash does not accumulate into a slag needing periodical removal but is carried away with the combustion products. These are mainly carbon dioxide but include oxides of nitrogen formed at the very high temperatures involved, and sulphur dioxide from sulphur impurities in the coal. There is little or no smoke as any carbon particles produced are quickly burnt. The mineral ash however remains as a fine dust, and since some 200 000 to 300 000 tonnes of it are produced per year by a large electric power station this must be collected by filters as 'fly ash'. Less than 1% of this—perhaps 200 tonnes a year—passes the filters and is discharged through tall chimney stacks with the hot gases from the furnace. These carry it for a long distance before they cool, and it finally falls back to the ground over a large enough area to be harmless. The oxides of nitrogen and sulphur are not so harmless and will be discussed below. A modern coal-fired power station is shown in figure 6.2.

There are a few places in Britain, for example near Whitehaven in Cumbria, where coal is close to the surface and is obtained by open-cast working, but most British coal lies too deep for this and has to be dug from pits, sometimes thousands of feet deep.

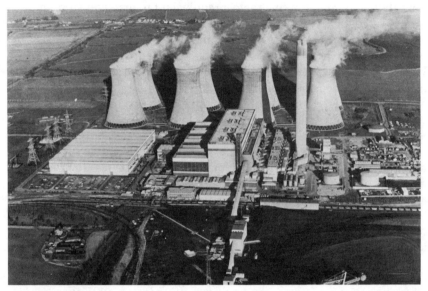

Figure 6.2   2 GW coal-fired power station at Eggborough, Yorks. In the foreground can be seen part of the million tonnes of coal kept in stock. Photograph courtesy CEGB.

PETROLEUM

For some decades, up to the sudden three-fold rise in price imposed by OPEC in 1973, oil-fired power stations produced electricity more cheaply than did coal, fuel oil being far easier to handle as well as producing nearly 2.5 times as much heat energy per tonne when burnt. It carries much less mineral matter than does coal and nearly all of this is removed during refining, so that very little ash is produced when fuel oil is burnt. Its use in electricity and gas production increased up to a peak of 20 million tonnes per year in 1972, with 45 million tonnes used for heating and 35 million tonnes for all forms of transport. Since the big price rise the oil-fired stations have been used as little as possible. The oil stations—which are relatively easily turned on and off—are used only in the periods (such as cold afternoons after dusk on weekdays in winter) when demand is high.

In spite of the present high cost of oil a new, large and very efficient oil-fired power station on the Isle of Grain was completed and brought into use in the summer of 1982. Such stations take nearly 10 years to build and this one has been started before the OPEC price rises. It will of course have made it possible to shut down some of the older, smaller and less efficient oil stations. The use of oil for electricity production is now about 4.5 million tonnes a year (equivalent to just over eight million tonnes of coal).

Before the price rises fuel oil was the main source of energy for the very rapid extension of domestic central heating after the Second World War; this use has however fallen even more rapidly as the cheaper North Sea gas became available.

The most important use of petroleum has been and will probably continue to be its use in transport, which now uses about 65% of the total. For heating and power production it can easily be replaced by a combination of coal, gas, nuclear power and the renewable sources. For transport a fuel reacting with 3.5 times its mass of oxygen that does not need to be carried around cannot be beaten in energy-to-mass ratio by any possible chemical reaction in which all the reacting compounds have to be carried in the vehicle. Oil equivalents made from coal could replace oil from underground, but liquid fuel of some kind we shall want for a long time.

Sooner or later mineral oil will become too expensive to burn. Already the Department of Energy is contributing £3 million to British Coal's pilot plant project on obtaining petrol from coal (*Atom* July 1988, p. 43).

The petroleum that comes from underground or from beneath the sea is a liquid known as crude oil—or, in the trade, simply as crude—and consists of a very complex mixture of volatile (evaporating quickly at room temperature) and non-volatile liquids, and a variety of solids such as waxes and bitumens. Traditionally, the gases were flared off (burnt in a controlled but wasteful manner) as they escaped from the mixture initially extracted, so that they should not form explosive mixtures with the surrounding air as would certainly happen if care were not taken to prevent it. Nowadays the technology has developed for liquefying the gas mixture by cooling, and for transporting it in refrigerator-ships or land vehicles, and it can be sold and used as natural gas. (This is liquefied petroleum gas or LPG as referred to below.)

The partially degassed crude oil is then sent to a refinery, where a process of distillation separates it into a series of fractions with different boiling points. There will still be a considerable gas fraction, largely composed of propane and butane ('Calor gas') but including hydrogen and a large number of light organic (carbon) compounds. The fraction that boils off next will comprise petrol (or gasoline) of successively heavier grades, followed by kerosine (the paraffin oil used in oil lamps and oil stoves), then the fuel oil used in diesel engines, and finally waxes, tars and pitches (bitumens).

In a big modern refinery the crude oil will be passed through a great many complex processes which improve the mechanical or chemical properties of the products, finally yielding the desired properties of a very large number of oils, greases and solids, each having the optimum characteristic for its particular use. The quantities handled by a big refinery are very large. At any one time a single refinery may have a million tonnes or more on the premises, some being delivered or despatched by tankers, some being processed and some stored in large tanks like gasometers, awaiting sale.

NATURAL GAS

Apart from the supplies associated with oil already mentioned, there are very large supplies available from underground gas fields, derived like oil fields from the breakdown of ancient organic remains by heat and pressure. British gas supplies, for a century made from coal, are now essentially all derived from such fields under the North Sea. These are giving us the energy equivalent of 30 or 40 million tonnes of coal each year. Natural gas is the cheapest source now available to Britain, and should last at the present rate of extraction for at least another 20 years. A huge new field—the Troll field—between Scotland and Norway, is in a deeper part of the sea, which will increase the difficulty and the cost of exploitation, but should greatly extend the time for which gas is available. A very big gas field in Siberia is supplying a large part of Western Europe through a 5000 mile high pressure pipeline that is one of the biggest constructional achievements of the 20th century.

In other areas such as Algeria the supplies of natural gas do not justify the construction of a pipeline to distant consumers, but the gas can be liquefied (at a temperature around $-160\,^\circ C$) and transported in refrigerated ships and road or rail tankers. Natural gas is in cheapness and convenience the best source of energy now available to us for everything but transport, for which its bulk (in gaseous form) is inconveniently large. When it runs out—or rather, when it becomes too expensive for large-scale domestic use—we shall presumably go back to gas derived from coal.

A little more than half the gas now used in Britain (figure 5.1) is used domestically; most of the rest in industry.

NUCLEAR POWER

This is now producing 19% of the electricity in Britain. It is less versatile than coal or oil. It produces a great many radioactive materials as by-products. Some of these are valuable in medicine for the treatment of cancer, for insect pest-control using the sterile male technique, for sterilisation of food and equipment, and for such trivial things as luminous dials for watches. The great majority of radioactive by-products however are potentially dangerous nuisances. These add to the cost of power production, and the handling of them will be discussed in detail in Chapters 11 and 12.

Basically, the only important use of nuclear energy on a large scale is the production of heat. Designs were published nearly 30 years ago for the direct use of this heat in desalination plants (Hammond 1962), producing from seawater 40 tonnes of fresh water per second—half the average rate of flow of the Thames. Water on this scale may soon be needed in California and other regions which are rapidly using up the underground stores of water which have been built up over hundreds of thousands of years. In Africa it could be used to make the Sahara flourish, a desalinated river being driven uphill from the relatively flat west coast of North Africa, through a series of large reservoirs, by wind-powered pumps. In Hammond's scheme, the cost per tonne of the water would have been a third of the cost of that desalinated using heat energy from a fossil fuel plant, but the capital cost would have been immense, and no such plant has yet been built. As a large-scale source of fresh water it might not be able to compete with icebergs towed from the Arctic or Antarctic by large nuclear-powered tugs. A combined electricity and desalination plant could use much of the 60–65% of heat energy that is lost from either fossil fuel or nuclear plants.

The only large-scale use of nuclear heat so far is to replace the heat obtained from the burning of coal or oil in electric power stations. Most people feel that they understand well enough how steam can be produced by heat from burning coal or oil. It does not seem to matter that they have not the remotest notion of the complicated physics and chemistry involved in a coal fire; and even the better informed minority, who recognise that the most important overall effect is the combination of carbon with oxygen to form carbon dioxide, might be hard pressed to explain just why the rearrangement of electrons involved releases energy, and why this works in practice only at high temperatures. In living animals the reaction of carbon compounds with oxygen to give energy works at lower temperatures, but the chemistry is even more complicated.

Knowledge of the how and why of coal burning is clearly not needed for the confident feeling of understanding. To understand nuclear power you don't *need* to know any more than that in a reactor it is the nuclei of the atoms that are burning; if you didn't know that it was the rearrangement and sharing of the outer electrons in the atoms which constitute ordinary burning, this should not matter very much. In a sense, even the hazards are quantitatively different rather than completely different in kind. If it were not shielded, the heat radiation from the furnace of a gigawatt coal-fired station could kill you outright at a considerable distance; the combustion products are carcinogenic and mutagenic as well as straightforwardly poisonous, many remain so for considerable periods, and many of them get spread over the countryside.

However, a nuclear reactor which you cannot stay in for more than a few minutes even 1000 years after shut-down *is* importantly different from a coal furnace which you could stay in all day after a week; and the heat radiation from the inside of a coal furnace which could be stopped by a sunshade *is* importantly different from the gamma and neutron radiation from a reactor, which may need 15 feet of concrete as a shield.

The feeling that nuclear power is mysterious as well as just unknown is partly the fault of the physicists, who brought Einstein into early explanations. Hardly anybody expects to understand Einstein. Einstein in fact is entirely irrelevant. He did indeed produce an equation, $E = mc^2$, where E is energy, m is mass and c is the velocity of light, which is correctly described as universally true, suggesting that mass is a form of energy. Einstein's relation is better, and more understandably, described by putting it the other way round as $m = E/c^2$. This states that any kind of energy has mass, as well as just how much mass it has. It has been abundantly confirmed by careful measurements of the masses before and after simple and energetic nuclear reactions, in which the change of mass may be several tenths of one percent. The equation is just as true, however, of the heat produced by burning sticks as it is of the heat produced by uranium in a reactor.

Primitive man did not have to learn Einstein's equation before learning how to use fire, and if he had known it he could not have measured the mass of the heat energy (0.2 μg) given off by the burning of 1 kg of dry sticks.

After Chadwick had discovered the neutron, and *after* Hahn and Strassmann had found that it could cause some of the atoms of uranium to break up, giving off more neutrons, and *after* the heat energy produced had been measured, we could use Einstein's theory to calculate how much lighter the products must be than the original uranium—about 0.1%, which is of no practical importance. And *after* we knew what the products were and what their masses were, we found that these too agreed with Einstein's theory. The main difference from the bonfire is that we can very easily measure a loss of mass of 0.1%, but we cannot measure the couple of parts in 10 000 million difference between the combined mass of the sticks plus the oxygen in which they burnt and the combined masses of the ashes, the smoke and the gases that are produced. The chemists too have a share of the responsibility for the misunderstanding. For generations they have told young students that there is no change of mass in a chemical reaction. What they *should* say is that chemists can't measure it.

The fact is that the larger part of 'understanding' is familiarity rather than theoretical knowledge. All that I can impart in this book is information; familiarity can come only with time.

A simplified explanation of the nuclear physics involved is given in Appendix 3. Here I shall only summarise the essentials. The power-producing reactors now in operation in Britain gain their energy from the fission of the rarer isotope of uranium, uranium-235, when it absorbs a neutron. (The word 'fission' was borrowed from biology, where it means the division of a cell nucleus into two parts in reproduction.) The products of nuclear fission are highly radioactive forms of lighter elements, plus two or three neutrons. In a steadily running reactor, on the average one of these neutrons produces fission in another nucleus of uranium-235, thus maintaining a chain reaction; just as some of the heat produced by burning coal is used in heating the next bit of coal to the temperature at which it too can burn. Of the remaining neutrons some will be captured by the commonest isotope of uranium, uranium-238, to make uranium-239 which after a couple of radioactive transformations becomes plutonium-239, and some will be absorbed by impurities or escape from the reactor core without making any useful contribution. The radioactive fission products from the uranium-235 are hurled apart at enormous speeds—10 000 to 20 000 km/s— and about 90% of the heat produced in the reactor comes from the kinetic energy of these as it is shared out between molecules of the surrounding material. The remaining 10% comes from the successive radioactive decays of the fission products.

Power reactors (the nuclear furnaces used in nuclear power stations) can be divided into two groups. Those in the first group, which includes almost all those now operating, are described as 'thermal' reactors.

THERMAL REACTORS

These use either natural uranium, which has only 0.72% of the fissionable isotope uranium-235 (U-235), or uranium enriched to 2.3% or more in U-235. In all these reactors the uranium fuel must be embedded in a moderator to slow the neutrons down. Neutrons emerge from fission at speeds up to 20 000 km/s and do not easily interact with a nucleus that they do not strike directly. However, a neutron which has been slowed down to thermal velocities† (still around a couple of kilometres a second), can interact with a U-235 nucleus ten or more nuclear diameters away from its initial path, whereas with U-238 it must still make a nearly head-on approach.

Neutrons can be slowed down to thermal velocities by colliding many times with the nuclei of any atoms which do not themselves absorb neutrons too easily, but they need fewer collisions, and therefore lose energy more quickly, with the lighter nuclei. Carbon and oxygen make good moderators, although the oxygen is very corrosive when hot and must be combined with something else to make a solid or liquid capable of resisting the high temperature of a useful reactor. Hydrogen-1, the lightest of all atoms, is an extremely efficient moderator, needing little more than 20 collisions on the average to bring the energy of a neutron down to thermal velocities, against 100 or so with carbon. Unfortunately, hydrogen-1 also absorbs neutrons, so that it can only be used (combined with oxygen in water) with uranium enriched to several percent in U-235 to make up for the neutron loss.

Deuterium (hydrogen-2) on the other hand, which is not very much worse than hydrogen-1 in slowing neutrons down, absorbs 600 times fewer neutrons—fewer even than carbon and oxygen—so that 'heavy water' made from deuterium and oxygen is an excellent (but very expensive) moderator in conditions in which it can be prevented by high pressure from boiling.

Magnox reactors

The first series of British reactors used fuel elements consisting of unenriched metallic uranium in bars about 25 mm in diameter, in tightly fitting sheaths made from a magnesium alloy known as Magnox. The fuel elements are mounted end-to-end in a series of parallel vertical channels cut into blocks of highly purified graphite (pure carbon) which forms the moderator. The assembly of graphite and fuel elements is known as the core of the reactor, and is placed in a reinforced concrete vessel (in the first reactors steel was used) in which carbon dioxide gas is kept at a pressure of 50 atmospheres to improve its capacity for carrying heat. This gas is continuously pumped through the fuel channels, then through a heat-exchanger in which water under pressure is heated to 350–400 °C, the carbon dioxide itself being cooled, and then back through the reactor channels to be reheated (see figure 6.3).

† Thermal velocities are those which are in equilibrium with the range of molecular velocities of the surrounding material, so that in collisions with these energy is as likely to be gained as lost.

In the preliminary explanation of the process it was stated that on average just one neutron from the fission of one U-235 nucleus caused fission in another, i.e. the multiplication factor was exactly 1. If more than one went on to cause fission, the power output would steadily rise; if less than one it would fall, eventually to zero.

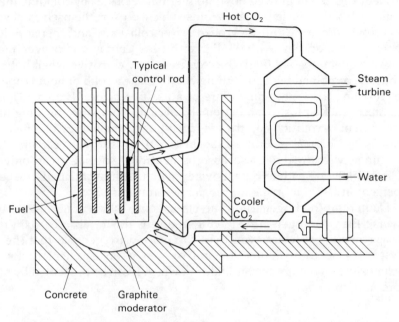

Figure 6.3 Schematic diagram of a Magnox reactor.

It would be very difficult to build a reactor so as to give a multiplication factor of exactly unity. In practice reactors are built capable of a multiplication factor a little greater than 1, and adjustable control rods are inserted to reduce it to well below 1. Control rods consist of good absorbers of neutrons such as boron steel or cadmium alloys. When the whole system is complete and tested, the control rods are slowly withdrawn until the measured neutron concentration begins to rise, and are then adjusted gradually to give the power output desired. With several hundred tonnes of uranium and over 1000 tonnes of graphite the temperature does not change quickly, and because some of the neutrons from fission are emitted after several seconds delay the changes of multiplication factor are not instantaneous. Hence it is easy to establish stable operation, with automatic control, at any level desired, and to switch the reactor on or off as required.

The first two Magnox stations to feed electricity into the grid—Calder Hall and Chapelcross—were designed to produce military plutonium as well as power, which limited the possible annual electrical energy output to a level too low to be economic by itself.

The steadily increasing output of the commercial stations built for maximum electrical power output only is shown in table 6.1. The load factor is the percentage of time over a year in which the station was operating; this compares favourably with coal station load factors in all cases, although the latest Magnox plant at Wylfa took some years to 'run in'. Figure 6.4 shows the Magnox station at Oldbury-on-Severn. The Berkeley power station, one of the two oldest Magnox stations, is being shut down because the renovations required by the Nuclear Installations

Table 6.1 Nuclear power stations in operation.

Power station	Type of reactor	Operator	Net accepted electrical output (MWe)	Commissioned
Berkeley	Magnox	CEGB	276	1962
Bradwell	Magnox	CEGB	245	1962
Calder Hall	Magnox	BNFL	180	1959
Chapelcross	Magnox	BNFL	186	1959
Dounreay PFR	Fast reactor	UKAEA	250	1974
Dungeness A	Magnox	CEGB	424	1965
Dungeness B	AGR	CEGB	690[*]	1984
Hartlepool	AGR	CEGB	840[*]	1984
Heysham 1	AGR	CEGB	840[*]	1984
Hinkley Point A	Magnox	CEGB	470	1965
Hinkley Point B	AGR	CEGB	1120	1978
Hunterston A	Magnox	SSEB	300	1964
Hunterston B	AGR	SSEB	1000	1976
Oldbury	Magnox	CEGB	434	1968
Sizewell A	Magnox	CEGB	420	1966
Trawsfynydd	Magnox	CEGB	390	1965
Wylfa	Magnox	CEGB	840	1971
Winfrith	SGHWR	UKAEA	100	1967
Under construction				
Heysham 2	AGR	CEGB	1230	1988
Torness	AGR	SSEB	1320	1988
Sizewell B	PWR	CEGB	1110	1994
Proposed				
Hinkley Point C	PWR	CEGB	1175 (designed)	1998

[*] Current performance achieved by two reactors at each site (*Nuclear News* 1988).

Inspectorate would cost more than the value of the extra working life that could be expected. It will be dismantled by stages as the radio-activity of each part decays sufficiently.

Figure 6.4 The Magnox nuclear power station at Oldbury-on-Severn. Photograph courtesy CEGB.

Advanced gas-cooled reactors (AGRs)

When enriched uranium with more than the natural 0.72% of U-235 became available, it was possible to afford a larger loss of neutrons from the core. Instead of bars of uranium metal, stainless steel tubes containing pellets of uranium oxide enriched to 2.3% of U-235 are used. These can safely stand much higher temperatures than could the Magnox alloy, giving a higher efficiency, and the high absorption of neutrons by the stainless steel is acceptable because of the enrichment. Carbon dioxide at 43 atmospheres pressure is again used to transfer heat from the core to the boiler. The core, gas circulation system and boiler are all inside a single concrete containing vessel. Each AGR power station has two such reactors, each with its own 660 MW turbogenerator.

The basic plans for the AGRs were excellent, and a prototype giving 33 MW output at Windscale ran successfully for several years. The detailed performance of the full-scale reactors, however, fell below the design target in several respects. The jump from 33 to 660 MW was itself too large; new problems always arise when any structure is scaled up, as was explained in the discussion of wind turbines in Chapter 5. As with the Magnox stations, for the construction of which five separate consortia were set up, too many different consortia (three this time) were given orders, so that the skilled engineers available were spread too thinly; and each team was starting fresh without the direct experience that would have been gained if successive plants had been built by a single combined team. Too much of the precision work had to be done on site instead of in normal factory conditions. As a result, of the five AGRs started at the end of the 1960s, the first (Dungeness B) was many years late. Hinkley Point B in Somerset and Hunterston B1 and B2 in Scotland have been working well, as are Hartlepool and Heysham I. Anxiety to avoid any risk of over-precipitancy after the errors of the early hurried working has probably caused some of the delay, and a mistake which led to corrosive seawater flooding one of the freshwater cooling circuits at Hunterston B took two years to recitify.

After a gap of several years, which caused serious disruption of the engineering organisations which had been set up in the expectation of a steady flow of orders, two more stations were ordered for Heysham II and Torness. Construction of these is complete, and power from them should soon be available. The present situation is shown in table 6.1.

As would be expected, the USA has (as of December 1987) the largest number (106) of nuclear power plants in operation, with 13 more being built. France had at the same date 53 nuclear power plants, including a fast breeder reactor and four gas-cooled reactors, with a total capacity of 48.42 GWe.

The uniform value of the designed output is determined by the use in every plant of a pair of 660 MW turbogenerators of standard design. Most of the reactors could produce more steam than these could handle, but it will not be known for some time how close to the turbogenerator capacity they can run continuously. Hinkley B was running steadily at 90% of the rated output by 1980.

In July 1988 it was reported in *Atom* that by 1987 there were more than 400 commercial nuclear power plants on line world wide. These generated 293.3 GWe, 16% of world electricity production.

Pressurised water reactors (PWRs)

By far the greatest number of nuclear power stations in the world are of this type. Small but efficient units were first designed for nuclear-powered (but at that time conventionally armed) submarines, which were thus given an enormously increased range and operational period. Scaled-up versions were then built for electricity production in the USA, whence they have been exported to many countries, and they have formed the main body of the 106 operating reactors in the USA and the 53 in France. France is currently commissioning half a dozen new stations a year and will in a few years be producing over 70% of its electricity in this way, at the cheapest price in Europe. Time will tell whether this price is justified remembering the large subsidies paid by the French government during construction.

In the PWR the core consists of an array of zirconium alloy tubes, containing enriched uranium oxide fuel pellets, zirconium being used because it absorbs far fewer neutrons than does stainless steel. The whole of this array is placed in a thick-walled steel pressure vessel filled with pure ordinary water, which acts both as moderator and as heat transfer agent from the core to the steam-generator circuit outside. The water in the steam-generator circuit is completely isolated from the primary circuit, because the water in the primary circuit will carry a considerable amount of radioactivity.

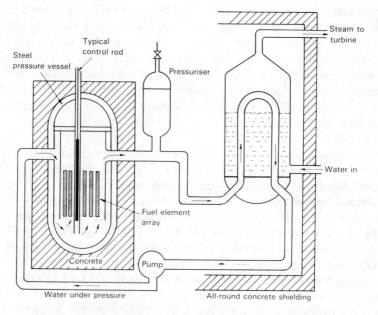

Figure 6.5 Schematic diagram of a pressurised water reactor.

The uranium enrichment is sufficiently large to make up for the loss of neutrons due to absorption by the hydrogen in the water. In the system designed for Sizewell B the paramount need for the safety of the pressure vessel will limit the internal pressure to 155 atmospheres, and the saturated steam will be produced at 69 atmospheres at 285 $^{\circ}$C. This will give a lower heat-to-electricity efficiency than the AGRs, but the reactors will be cheaper and quicker to build. Figure 6.5 shows the layout of a typical PWR.

The PWR being built at Sizewell is of the same basic type as that which suffered a melt-down at Three Mile Island, but it will include a number of safety features designed to avoid this type of accident. A detailed discussion of this is given in Chapter 9.

Several other thermal reactor types have been considered. Detailed studies in Germany have been made of small helium-cooled reactors, in which the enriched uranium fuel units are spherical and coated with refractory coats of carbon and silicon carbide. These are claimed to be inherently safe; old units can be discharged from the bottom of a graphite tube while fresh units are fed in at the top, and spheres cannot kink or get jammed.

Nearer home, the UKAEA and Rolls Royce Associates are proposing to build a 300 megawatt pressurised water reactor near Winfrith in Dorset, the entire reactor pressure vessel of which will be underground. This mini PWR would be called the Safe Integral Reactor (SIR) and uses natural convection, which never fails to work, rather than electrically driven pumps.

A thermal reactor will produce about 1 GWy of electricity from the fission of one tonne of U-235, whereas a coal-fired power station producing the same amount of electricity will need to burn about 2.5 million tonnes of coal.

The established reserves of uranium extractable for less than $26 000 (1975) per tonne add up to about a million tonnes in the non-communist countries. At a price of $39 000 (1975) per tonne, four times this amount would be available (*Nuclear Engineering International* 1975). Since 2.5 million tonnes of coal cost the better part of £100 million this looks like an adequate supply of uranium at a bargain price. The advantage is much reduced by the fact that less than 1% of the uranium is used in a thermal reactor, so that more than 100 tonnes have to be used for each tonne that contributes to the power produced. The cost of fuel must therefore be several million pounds for each GWy of electricity produced. This is still a great deal less than the cost of coal; and a ten-fold increase in the cost of uranium, which would permit the use of very much larger reserves, would be needed to double the cost of the electricity produced. Such an increase would, however, put the cost of nuclear electricity decisively higher than that from coal (at its present price); but even in the unlikely event that no further reserves are found at $39 000 per tonne (which would increase the cost of electricity by about 10%), four million tonnes should last for 30 years. We shall be lucky indeed if the real cost of coal, which is the major cost of electricity from coal-fired power stations, goes up by as little as 10% in the same period.

FAST-NEUTRON REACTORS

The second group of power reactors use uranium so much enriched with fissionable material that they can do without a moderator. This could be achieved by using 20% or more of U-235, but existing prototype reactors use about 20% of Pu-239. In this case there can be no loss of neutrons in the moderator, and if a 'blanket' of depleted uranium†, 10 cm or so thick, is placed all round the reactor, the neutrons escaping will be absorbed in the U-238 to produce more Pu-239.

In a good design the amount of plutonium produced may be more than that burnt up in fission. A reactor so designed is therefore often known as a 'breeder' reactor, because not only does it produce enough plutonium to refuel itself when its initial charge runs out, but there will be a surplus which could be saved up until there is enough for the fuel of a second reactor. The difference from thermal reactors in this respect is not enormous. The plutonium in the spent fuel from a thermal reactor can be enough to replace at least 90% of the U-235 burnt up, while the breeder produces perhaps 103–110%.

It is a mistake to combine the alternative names of 'fast-neutron reactor' and 'breeder reactor' into the composite 'fast-breeder reactor'. This tends to suggest a similarity to rabbits which is not deserved; by the time enough plutonium has been produced, extracted and fabricated into new fuel units, it will take 30 years for a second reactor of the kinds now available to be added to the first. Breeding can be used only for the next generation therefore, not for ourselves.

The world population is currently multiplying at a rate which would double it in about 40 years. Allowing for delays and losses, breeder reactors can barely keep up.

The Dounreay fast-neutron reactor

No commercial fast-neutron reactor has yet been built in Britain, but one (Super Phénix) has been running for some time in France but has had some teething troubles; and a British design for a gigawatt reactor has been awaiting government approval for some years now, based on the very successful operation of two successive experimental reactors. The first of these was a small unit, the Dounreay Fast Reactor (DFR) delivering 60 MW of heat and 15 MW of electric power at peak; but as it was used only for experimental work its output on the average was only a few megawatts. The second experimental unit, the Prototype Fast Reactor (PFR) was delivering 250 MW of electricity by 1977, and provided all the information needed for a full-scale reactor.

† Uranium from which most of the U-235 has been removed in the process of enrichment of a small proportion for use in thermal reactors.

Each load of fuel in a commercial power reactor would be used until about 10% of the plutonium in the core was burnt up, and would then be reprocessed to get rid of the accumulated neutron-absorbing fission products and to recover the remaining plutonium and U-238. The uranium units forming the blanket, in which fresh plutonium would have been produced, would need replacement less often, but for every plutonium atom burnt in the core slightly more than one uranium atom would have been changed into plutonium either in the core itself or in the blanket. At each processing stage some uranium would be lost with the fission products, but eventually more than 60% of the original depleted uranium could be burnt up to produce energy.

Used in this way, the 25 000 tonnes of depleted uranium now stored in Britain would produce more energy than the entire 300 years' supply of coal in all our reserves, and after these 25 000 tonnes were used up the cost of extraction of uranium from the 4000 million tonnes in the ocean would be unimportant. Each tonne of uranium would eventually produce as much energy as two million tonnes of coal.

As a result of the reduction in demand, and the current glut of uranium, the Government decided that natural uranium for thermal reactors is unlikely to run short as soon as had been expected and hence a full-scale fast-neutron reactor should not be built. This is clearly justified in the short run, but in 20 years we may be regretting the inevitable loss of the outstanding team that has been working at Dounreay and the resulting need to import reactors from abroad when we might have been exporting them.

FUSION

While a great deal of energy can be obtained by the fission of a gram of very large nuclei, such as those of uranium and plutonium, even more can be obtained by the fusion of a gram of hydrogen nuclei into heavier ones. This is the process that has kept the sun burning for 4500 million years. Even at the pressure and temperature near the centre of the sun the process is slow for hydrogen-1; on the average each tonne of the sun produces about one fifth of a watt. As the sun contains many tonnes of hydrogen and is not pressed for time, this is quite adequate.

The only practicable method for us is to use the much more vigorous interaction of the nuclei of deuterium and tritium†, to give a helium nucleus and a neutron, with a release of about 10 kWy of heat per gram

† Tritium is hydrogen-3; it is radioactive with a half-life of 12.5 years, and the small amount that is made by cosmic rays is too widely distributed to be collected efficiently. It has therefore to be made, by bombardment of lithium with neutrons in the reaction vessel of the plant itself.

of the mixed hydrogen isotopes (10 GWy per tonne). The conditions required are horrendous. A mixture of the two gases kept at a 100 million °C for a few seconds at a moderate pressure could give a useful output as could a mixture compressed to a very much higher pressure for a shorter time at a rather lower temperature.

I shall not discuss the methods by which we shall eventually be able to get useful energy from fusion, as the development of this approach is still far from producing any practicable plant at acceptable cost. Contrary to popular belief, fusion reactors could be excellent producers of military-grade plutonium by absorbing some of the very large number of neutrons passing through the walls of the reaction vessel; and when running would produce even more radioactivity than would a thermal fission reactor with the same output of electricity. This however is mostly in the form of tritium, with a half-life of 12.3 years, and the amount of long-lived radioactive waste would be much less.

A more attractive and cheaper source of nuclear energy would arise if the nuclear-weapon powers agreed to scrap 20% or so of their bombs, in which case the plutonium released could supply as many of either thermal or fast reactors as we could build for quite a time. There is unfortunately no immediate prospect of this.

'Cold fusion'

Enormous interest was aroused all round the world when, on 23 March 1989, two chemists—Professor Stanley Pons of the University of Utah and his erstwhile tutor, visiting Royal Society Professor Martin Fleischmann from the University of Southampton—announced at a news conference that nuclear fusion had been observed (Fleischmann and Pons 1989). A solution of a lithium salt in heavy water (D_2O) was electrolysed in a cell with a spiral platinum wire anode surrounding a cast palladium rod cathode. Currents up to 0.8 amperes at about 12 volts were used, and after a long 'conditioning' period a continuous production of excess heat was observed for over 120 hours at a time. Besides this, the authors reported that one large sample 'achieved ignition' overnight.

It was claimed by the authors that this could have been due only to nuclear fusion—though a loss of electrolyte in such a system could have led to fairly dramatic effects of a purely chemical nature.

Nuclear fusion of deuterons could have arisen from one of two reactions

(1)   deuterium-2 + deuterium-2 → hydrogen-3 + hydrogen-1 + energy
(2)   deuterium-2 + deuterium-2 → helium-3 + neutron + energy

or from some unsuspected process leading to helium-4.

All of the hydrogen isotopes would deposit their kinetic energy over a range of some tens of microns in the liquid or solid in which they were formed, but only the helium-3 of reaction (2) would do so as the neutrons would have carried away 90% of the energy released. A neutron flux of three times background was observed 50 metres away.

On the following day, 24 March 1989, Dr Steven Jones, a nuclear physicist from Brigham Young University in Utah, reported that his group had also achieved nuclear fusion, the neutrons of which were observed, although the energy released was around ten million million times lower than that observed by Pons and Fleischmann.

Dr Jones had been interested by two unexpected phenomena which had been observed in volcanic eruptions: the presence of traces of tritium (hydrogen-3) and the unusually large proportion of the rarer isotope of helium, helium-3, in the helium in volcanic lava; and had hypothesised that at the high temperatures and enormous pressures preceding an eruption deuterium nuclei might be forced together sufficiently closely to permit a detectable amount of nuclear fusion. It should be noted that reaction (1) as well as reaction (2) will eventually lead to helium-3, since tritium is beta-radioactive, giving helium-3, with the insignificant half-life of 12.3 years (Jones et al 1989).

Since these reports appeared, laboratories all over the world have been trying to repeat and extend the results reported from Utah. Some groups have noted heating effects, but have not been able to rule out the possibility of exothermic chemical reactions. Some groups have detected a few neutrons per hour above background, but the constancy of the background is uncertain. (Indoor radon produces polonium-218 and 214 descendants, both of which produce alpha particles energetic enough to knock a neutron out of light elements such as nitrogen and oxygen. The indoor concentration of radon varies in unpredictable ways with time of day, as people open and shut doors and windows.)

At the time of writing it is not possible to say with certainty that useful energy production by 'cold fusion' is or is not a practical possibility.

I will conclude by pointing out that the term 'cold fusion' usefully distinguishes it from the JET hot-gas fusion described above, but may be misleading. To be useful for large-scale power production it must be able to produce high-pressure and hence high-temperature steam to drive turbines, which may or may not be compatible with the use of electrolytic cells. Even if it cannot produce temperatures much above 100 °C, useful only for district-heating schemes, it could well be economically valuable.

It might not be free of some of the less popular features of conventional nuclear power plants. If the main yield arises from reaction (1) it could make tritium for nuclear weapons far more cheaply and conveniently than in a reactor as at Chapelcross or Hanford. If its main yield arises from reaction (2), the flux of fast neutrons per gigawatt will be comparable with that in a PWR and will need similar biological shields; and the decomissioning of large plants at the end of their lives will have much in common with that for fission plants after the removal of their fuel units. It is not known whether the output of a large operational plant would, as seems likely, increase as the temperature increases as at Chernobyl; or whether it could be built to decrease as in all present Western reactors.

I would have liked to see 'cold fusion' succeeding because it would be far easier to build efficient small plants for small applications. But it seems now that any power from 'cold fusion' is extremely unlikely.

HOW DO WE CHOOSE?

Throughout this and the previous chapter I have concentrated on technical feasibility, availability of resources and cost of the different sources of energy. The choices that people actually make depend also on factors such as cleanliness and convenience, which must have affected considerably the rate at which coal fires in the home have been abandoned.

The cost of electricity from a long-life torch battery is around £400 per kWh, while the cost of a wrist-watch battery is nearly £50 000 per kWh. Both are sold in large numbers. Electricity has an advantage in convenience over all other forms of individually used sources of power. In a well-off community electric cars which have only to be plugged in at home instead of messing about with petrol pumps, and which do not catch fire after accidents, could be popular at significantly higher prices than petrol-driven ones.

However, cleanliness and convenience are secondary effects which are unlikely to affect directly the major decisions on energy policy. Before we can make any firm decisions we need to consider future needs, relative risks and environmental effects.

FUTURE NEEDS

This book is not the appropriate place for a long discussion of future energy needs in Britain, but it is worthwhile to point out some of the difficulties in the way of making any reliable forecast to cover the next 30 or 40 years during which energy systems planned now will be operating.

If throughout the next 40 years we are living in much the same way as we are now, but with increasingly efficient equipment and improved conservation, we could provide a considerably higher standard of living for the poorer part of our population with little if any more power than the total we are now using. If we learn to make garments with the thermal insulation taken for granted by an eider duck—or even a sparrow—domestic house heating will not be needed, and the total energy requirements will be much less than at present. If on the other hand future generations want to wear fewer clothes than they do now, and want whole towns to be roofed over and warmed, we shall need a lot more.

These two bits of science fiction may be of the kind that takes 50 years or more to come true, but there are more serious possibilities. If we are still using liquid fuel for long distance transport, we may need more than our whole present coal output for the synthetic production of liquid and gaseous fuels when, as it eventually must, this becomes cheaper than the dwindling natural supplies.

As will be shown in Chapter 10, air pollution in our big conurbations is causing many thousand cancer deaths every year, and we must surely expect that at some time vehicles burning oil or petrol will be prohibited in urban areas. Electric transport is already being developed on a small scale, and it is likely that the two to three times lower 'fuel' cost will compensate for the greater capital cost. We are good at mass-producing vehicles, and in 10 or 20 years our demand for electricity might be doubled.

Even the maximum practicable rate of construction of wind turbines and nuclear power plants could hardly double our electrical output in less than 30 years, even assuming long-term planning starting now and allowing both for replacement of the stations which have reached the end of their lives and for a probable major diversion of coal to the production of liquid and gaseous fuels.

Of course, we cannot be *sure* that we shall need to build power producers for 30 years at the maximum rate consistent with a proper safety programme. But we have seriously run down our industrial capacity for such construction, and we shall quite certainly need within the next 20 years all that we can start to build in the next 10 simply to replace the aging ones. At the end of 10 years we shall then be able to guess the future 10 years better; and shall have the option in the following 10 of building several times faster, of building at the same rate, or of building more slowly. If we do not start on a major construction plan soon we shall not have the first option and perhaps not the second. The still-existing construction capacity will not survive if it has nothing to do. It would be a pity to have to ration electricity just when a real reduction in city air pollution becomes technically and economically possible.

As the time I wrote the first edition, the demand for electric power in this country had been decreasing for about three years—a period of slump in industry—but since 1983 has been rising rapidly. The best present estimate of future demand is a rise of 3% per year in Britain, i.e. an extra 10 GW in the next ten years. Taking the world as a whole, and considering increase in world population, estimates of a rise in total world demand of around 2% per annum look to me on the low side. However, even 2% per annum means a 60% rise in the next thirty years.

There is much talk of the importance of freedom. Freedom is meaningless if you have only a single option. I would like more of the next generation than of ourselves to be able to opt for a simpler life if they wish to. But it seems to me to be our job to ensure that there are other choices open to them. All choices need some power. It won't matter if we provide our descendants with more power than they need. It will matter if we leave less.

Chapter 7

Conservation of Energy

Conservation of energy means different things to different people. For some it is a statement of one of the most fundamental laws of physics. For rather more it means reducing the appalling rate at which we are using up the earth's resources of oil and gas, accumulated slowly over hundreds of millions of years. For most of us it means conserving money to spend on something else while remaining as warm and comfortable at home, or producing the same quantity of goods in industry, as before. Two of these objectives can often be achieved at the same time, although this is not necessarily so. For example, oil and coal can be conserved by using wind and wave energy, without conserving money, or by using hydroelectric and nuclear power, which may conserve both. All forms of energy can be saved by producing fewer goods and putting up with lower temperatures indoors in winter, and—in most of the richer countries—higher temperatures in summer.

It must always be remembered that we cannot conserve everything at once. Using the winds or the waves involves increased use of steel or light alloys in building the equipment. Both of these require energy for their production, so that we need to check the energy budget as well as the cash before we can be sure of a real energy saving. There will certainly be a net energy profit, but it may be well below the gross profit. If we keep our houses at a lower temperature in the winter we have to wear more clothes. If we use land for special crops for energy production, whether wood or sugar cane, we cannot grow food on it; and even if the land is unsuitable for food production we cannot conserve the wildlife which would otherwise have been there.

There are several ways in which a change from one energy source to another may reduce the total required. Amory Lovins (1977) in his *Soft Energy Paths* has pointed out the importance of some of them.

EFFICIENCY OF USE OF PRIMARY ENERGY

Electricity used to produce hot water and to warm houses does so with almost 100% efficiency since its use involves no extra ventilation to lose heat to the surroundings, and the heat is delivered when and where in the house it is needed. Used in an electric blanket in the bed, or in a pad under your feet as you sit in a chair, it uses less energy for the same gain in comfort than any other source. At the same time, the electrical energy fed to the consumer is at best 35%, and may be no more than 30%, of the heat energy used in the power station to produce it. Accordingly, for central heating of a whole house, the overall efficiency of electricity will be very much less than the efficiency of gas or coal burnt directly. This, in modern appliances, could be 75% or even more when operating at full load; but in practice in Britain the average efficiency of gas burning for central heating is 65% (*Which?* 1982).

In Britain there is not much saving still to be gained by a change from electricity to gas for central heating, since electricity is little used for this now, although a fair number of night-storage heaters are used for individual rooms. Night-storage heaters warm the house best when nobody is awake to benefit, so that the efficient conversion to heat is followed by inefficient use of the heat; while gas heating systems can run at a low level at night and switch themselves on in the morning when they are needed.

The large advantage of gas over electricity in reducing total energy consumption has been calculated on the assumption that the 65% lost energy at the power station is thrown away. This is usually but not necessarily the case. Some heat in a coal power station *must* be thrown away with the furnace exhaust gases, which have to be forced through filters to remove fly ash, and in both coal and oil power stations must be hot enough (perhaps 300 °C) to have a sufficiently lower density than the surrounding atmosphere to drive these gases up the chimney stacks. No heat is lost in this way by nuclear power stations, but instead it is thrown away into cooling water, of which therefore much more is needed than by the fossil fuel stations.

HEAT PUMPS

A heat pump for house heating is equivalent to a refrigerator set to extract heat from the outside air and deliver it indoors where it is wanted. However a heat pump would give no financial advantage in homes using gas for heating; 2.5 to 3 kW of gas-produced heat could be saved only by using 1 kW of electricity costing just as much (Holmes 1981). Simply to go on using the 2.5 to 3 kW of gas heat would be a lot less trouble and would save the cost of installation and maintenance of the pump. There might be a marginal saving of raw energy—but only if the pump is always working at its best and we neglect the energy used in the production and maintenance of the pump and its accessories.

When heating is done by electricity the gain would clearly be great, both financially and in conserving energy, and there would be a real but smaller gain if oil heating were used. While they have great value in special applications therefore—where for example we have a warmed swimming bath close to an ice rink—heat pumps are unlikely to contribute much to conservation in homes other than those in which electricity is the main source of heat.

Heat pumps could be made reversible and could then be used for cooling in the summer. An increasing number of large shops in Britain are using such pumps, but when used for cooling they are increasing comfort rather than conserving energy (Holmes 1981).

DISTRICT HEATING

The reject heat from power stations could be usefully employed in Britain for district heating. It is widely used for this purpose in Germany and Finland and in other European countries; in Czechoslovakia the waste heat from several new nuclear power stations has been used for district heating in neighbouring towns. Britain does have some examples. Battersea power station in London, now shut down for electricity production, was built not only to produce electricity but also to feed a district heating scheme that supplied hot water to a large number of council flats in the area. The losses of heat were small, and the overall energy efficiency of Battersea in supplying heat energy as well as electricity to the flat-dwellers must have been close to the overall energy efficiency of a natural gas supply used for hot water or space heating.

When the Bankside power station came to the end of its life it was not replaced, although the system of hot water pipes that is the expensive part of a combined heat and power (CHP) system was already in place and had operated for some years. This was discouraging but not final. In Leicester a consortium has been established with Government assistance to study the possibility of a CHP scheme in the city of Leicester. Buying an existing gas turbine power station from the CEGB, they are hoping to start construction during 1989 with a completion date in 1991, laying two pipe networks—for hot water and for steam. Electric power will also be sold. Waste heat from the power station should be reduced from 65% to 15%, and buyers for the heating have been found among local industries. There is no intention as yet to spread this to private dwellings.

The cost of the necessary piping is high, especially when it is to be supplied to a large number of individual houses of different types of construction, and old enough to need strengthening or rebuilding of walls and foundations disturbed by the introduction of extensive new plumbing.

It is not easy to see how even half the waste heat from a really large power station can be economically used for district heating. A 1 GWe nuclear power station has to get rid of 1.5 to 2 GW of heat, summer and winter alike. Hot water for washing up and baths is wanted all the year round, but the average household will be unlikely to use much more than 10 kWh per day for this, corresponding to roughly 0.4 kW round the clock, so that the station could supply five million houses—which would need a quite uneconomic distribution network. All winter needs could be supplied to a more practical number of 100 000 houses. These would however use very little of the available waste heat over a large part of the year. All of the 100 000 houses must have alternative heating systems for the periods when the power station is not working—either for maintenance normally done in summer, or for refuelling or fault correction. In spite of these limitations, district heating could help a great deal to boost the efficiency of the smaller and less efficient power stations which now are run only at times of high demand.

Coal power stations, which use a lot of their waste heat to drive the fly-ash laden furnace gases through filters and up large stacks, have proportionally less heat to spare, but the newest stations are larger so that the problems are much the same. This is a strong argument for smaller power stations in areas where the housing density is high and the cost of distribution is not too great.

According to Dr Peter Chapman of the Open University, district heating should be cost effective at a housing density above 50 dwellings per hectare, if we assume that fuel prices will double in real terms by the end of the century. A conference of the British District Heating Association in June 1983 was more optimistic, believing that 16 houses per hectare would make it worthwhile (*New Scientist* 1983c). It should be well worthwhile where there are large areas occupied by terraced houses, and should pay very well in such areas if it were replacing electric heating. Large power stations of any sort are usually built well away from such areas.

It may be that in Britain, with its relatively mild and unpredictable winters, and its families living in separate houses, the benefits are less and the costs per family higher than those in continental northern Europe, where far more people live in big blocks of flats which need heating for a large part of the year, and which are constructionally simpler to supply. There must be many places however in which we ought to be considering district heating. In the new towns especially it seems that a golden—or at least a silver—opportunity to install such systems when large areas of new housing were being constructed all at the same time has been missed.

DESALINATION AND FISH FARMS

Another way of conserving waste heat from power stations would be to use it to produce fresh water by desalting seawater. This would not be of much economic value in Britain, but might have formed the technical base for a valuable export industry for Britain to supply power and water in the Middle East, instead of leaving this supply to other countries.

A quite different way of using the waste heat would be in warming the water for fish farms. In our climate the growth rate of fish and of edible invertebrates such as crabs and shellfish is limited by the temperature, and has been significantly increased by the warmed waste water from our nuclear power stations, even when they discharge their cooling water into estuaries or the sea without restriction to any selected area.

INDUSTRIAL HEAT AND POWER AND COMBINED HEAT AND POWER

Many industrial firms using large quantities of process-heat at high temperatures—for example for melting metal for forging or alloying—could afterwards get rid of the heat usefully by using it to produce steam to drive turbogenerators, and thus produce a useful amount of their own electricity. Sometimes this could give a sufficient surplus to sell to the grid. Such a system has been built at Hereford by the Midlands Electricity Board and Esso in collaboration with H F Bulmer, the cider makers, and Sun Valley Poultry, both of which use a great deal of low-temperature heat. The system uses marine-type diesel engines to drive two 7.5 MW generators to produce electricity for the grid. The total thermal efficiency is close to 76%, saving £1.25 million a year. This has evidently proved to be economically valuable, and it is to be hoped that other large firms will follow suit.

It must be remembered, though this is not directly concerned with conservation, that small units providing a few thousand kilowatts cannot afford the tall stacks and efficient filtering used by a big power station. On the other hand, they could more easily use coal burning in a fluidised bed. This could be charged with a red hot 'sand' of limestone or dolomite through which air is drawn from below, and use finely crushed coal; then 90% of the sulphur and some of the oxides of nitrogen would be removed as non-volatile sulphates and nitrates.

CONSERVATION OF BIOWASTES

Yet another public energy-saving plan is the burning of municipal rubbish to produce hot water or electricity or both. The Greater London Council runs an incinerator at Edmonton to produce electricity. This opened at the end of 1971, but does not make a profit; operation and maintenance costs £6.5 million a year, more than twice the value of the electricity generated. The plan may have other important advantages—to health for example, or land conservation—but not financial ones.

If it were possible to get at least a large majority of the public to separate the burnable from the unburnable refuse for separate collection this might be profitable, but the financial losses in London have discouraged councils elsewhere from starting similar schemes of their own.

Another plan is to ferment anaerobically the organic material from sewage farms. This oxygen-free fermentation leads to the production of methane, an excellent fuel gas. This has worked well on some large individual dairy farms in the USA, which have obtained enough methane from cow dung to run their own milk coolers, with enough methane left over to supply them with a useful amount of gas for domestic purposes. The residue of the dung contains all of the original mineral plant-nutrients and forms an excellent manure. It is said that the Chinese have distributed some hundreds of thousands of such fermentation plants to villages—presumably to operate on human excreta—with much success.

It is not clear that all the energy-saving systems that I have described so far, put together, would make a major impact on Britain's power needs, flexible and useful as they may be in special cases. Amory Lovins himself, who spends most of the first half of his book in pressing the advantages of these and other 'soft energy paths' for the United States, does not offer much in the way of prospects for Britain. In his introduction (p. 15) he quotes Daly as saying that Western Europe uses about half the energy per head that is used in the United States. On p. 36 he states his view that 'North Americans are adaptable enough to use technical fixes' (such as I have been describing) '*alone* to double, in the next few decades, the amount of social benefit wrung from each unit of end-use energy'. That is to say that in the next few decades they should be able to attain the energy efficiency that Britain, which is neither the best nor worst in Western Europe, has now. It is clear from the latter part of his book that he believes that Britain too could improve considerably on its present energy efficiency. I believe this myself, and I am quite willing also to believe that many useful energy-saving schemes have been left unused due to inertia and unwillingness to do something new. Nevertheless I suspect that the gains to be expected from many of the schemes I have described are little more than marginal, meaning that with the intuitive application of the Blackett factor† nobody can be certain that in fact they will be economic. I hope that I am being pessimistic here. The Blackett factor applies only to the really novel, and several of Lovins' suggestions which use only proven techniques look perfectly viable. It is an unsatisfactory feature of our society that very large sums of money from central funds are made available for small improvements in large-scale systems, but that little or none is provided for trying out large improvements in small-scale projects. Any proposal for individually small projects which can lead to conservation on the gigawatt scale nationally must mean that it can be applied in a great many places. People and local energy use vary enormously from region to region. If a conservation scheme is going to pay in a large number of places (which it would have to do to be widely supported) there must surely be some not too rare places where it would pay very well indeed. My pessimism stems in part from the fact that few such places seem to have been found. It might be a useful activity for the Conservation Society, with members everywhere, to search for such places and even to encourage action with a low interest loan. Some members would disapprove of spending money in this way, but I would expect an adequate majority—including myself—would approve it, and indeed would generously support a special appeal. Nobody, since 1985 when the first edition of this book was published, has taken up this challenge.

POWER IN RESERVE IN DINORWIG

It takes a long time to start a power station up from cold, so the CEGB always retains what is known as a spinning reserve; one or a few stations which are not currently required to supply power to the grid have their furnaces running on the minimum input of fuel required to keep their turbogenerators turning at full speed under no load, i.e. delivering no power to the grid. When required, an increase of fuel supply can build up a large power output far more quickly than would be possible if the turbogenerators had to be run up and synchronised to the grid starting from rest. The need for this spinning reserve, which burns quite a lot of fuel unprofitably while all is going well, is not appreciably affected by the existence of a lot of small stations. Many of these, being small, could increase quite quickly their normal output

† Professor P M S Blackett pointed out that after the physicists and engineers proposing any new project have allowed for every factor they can think of that might increase the cost or the time for completion, some outsider who knows nothing about it must multiply both by a factor of two to get the right result.

when needed, but it would be administratively difficult to ensure a sufficiently rapid reaction time.

A big saving of spinning-reserve fuel however can be made by a quick-release system for storage of energy.

The largest pumped storage system in the world has recently been commissioned at Dinorwig in Snowdonia†. This is a hydroelectric plant operating between two lakes, one about 500 m above the other. When the whole British grid system has spare capacity, for example between midnight and 5 AM, the generators are used as powerful electric motors which spin the turbines the other way round so that they pump water from the lower lake up into the higher one. If there is a sudden demand on the system, water from the upper lake is allowed to run back through the turbines, generating electricity. The output to the grid can be raised from zero to 1.32 GW within 10 seconds of demand. The combined losses in pumping and in generation amount to about 20% of the energy stored but the increased value of electricity at the time when it is most wanted more than makes up for this.

While this plant can indeed reduce the need for an equivalent spinning reserve to cope with rare large breakdowns in supply, the real justification for the very large cost of building it is to reduce the spinning reserve needed for the much more frequent sudden increases in demand. This can easily rise by more than a gigawatt in less than a minute if a really popular programme is being shown on television. This does not result from the simultaneous switching on of television sets, the power demands of which are low and many of which will have already been running when the programme began. It occurs when there is a 'natural break' for advertisements, or at the close of a popular film, and a large number of viewers nip out simultaneously to put on the electric kettle (see figure 7.1). A record rise of 2.5 GWe occurred in January 1984 when *The Thorn Birds* came to a close at the same time as a film ended on ITV, and Dinorwig (with its smaller sister plant at Festiniog) came into operation exactly when required.

The CEGB estimates that Dinorwig is saving it millions of pounds a year (by reducing uneconomic operation of older plant) and that its construction cost will have been recovered in eight years. Such a plant for storing energy should have a much longer useful life than a power station, the civil engineering works being designed for an 80 year life and the electrical plant for a 40 year life, with far lower costs for operation and maintenance. So whatever its profitability may be today it will certainly increase as the real cost of coal goes up—as it must

† A pamphlet giving full technical details is available from the Public Relations Section, Generation and Construction Division, CEGB, Barnett Way, Barnwood, Gloucester GL4 7RS.

inevitably do over the life of the plant. Unfortunately there are few places in Britain where adequate volumes of water can be stored at sufficiently different heights above sea level and not too far apart on the map.

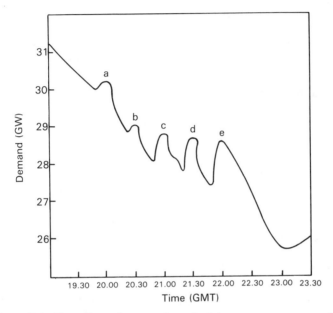

Figure 7.1 The effect of a popular television programme on electricity demand. This shows demand peaks occurring during the evening of 28 October 1975 when the film *Dr No* was shown on all ITV regions. Peaks a and e coincide with the start and finish of the film, peaks b, c and d with commercial breaks.

The Dinorwig plant can, if the upper lake starts full to capacity, keep up 1.68 GW of output for five hours, ample time for 1.68 GW worth of resting power stations to be started up from cold. But it could supply only 25 MW for a continuous fortnight. It would therefore make little difference to the warning by the CEGB that to produce more than 20% of our electricity from winds and waves would give a risk of serious breakdown in supplies in the unusual occurrence of a fortnight's calm seas and light winds. To compensate for a loss of 20% (currently 8–9 GW) of our present electrical output, we should need to bring into action at least eight GW of extra power stations—easy in summer, with about a day's warning, but stretching the system to its limits in winter. To save one of these extra stations for a fortnight would require 40 Dinorwigs, mostly used only very occasionally.

A plan that has been used in West Germany since 1978, and is being planned in America by the Alabama Electric Cooperative, is to store energy by filling underground caverns with compressed air. Excess electricity pumps air at high pressure into caverns made by drilling into a salt deposit and dissolving away a suitable volume with water. This leaves a hard glass-like surface that can retain air at a pressure of 65 atmospheres. When extra electricity is needed the compressed air can be released through a turbogenerator. The Alabama plan will generate electric power at less than a tenth of the Dinorwig output (and about a third of the German one), but might allow fairly small regional electricity companies to build and operate their own storage plants to meet sudden increases in demand.

CONSERVATION IN THE HOME

In Britain there is a large increase in the use of electricity, oil and gas in the winter, owing to the need for warming offices, shops and houses —especially houses, which will usually be kept warm, though perhaps less warm at night, for 24 hours a day. This is not universal; tropical countries do not need indoor warming at any time of the year, and in most of the USA for example the electrical load is greatest on the hotter days of the summer, when air-conditioners form the biggest seasonally variable load. Returning to Britain, however, the extra peak load in winter due to domestic needs is large. No heat is needed to keep the structure of a house at any desired constant temperature; the whole of the heat goes in replacing losses to the environment. A house with perfect thermal isolation from its surroundings would never need any heating at all; indeed if anyone lived in such a house it would need cooling.

Even when resting, the average adult produces and delivers to the surroundings 100 W of heat, and over 24 hours with a normal amount of mild activity must deliver at least 3 kWh, equivalent to over 18 pence worth of electricity, or 5 pence worth of gas. (This human heat is of course what is used in a large 'house-warming' party, to dry out the plaster in a newly built house!) The costs of the more specialised fuel required by people, who cannot digest coal or gas, would be more than £1 a day per person, but the heat is produced exactly where it is needed most.

Large well-insulated office blocks have been built, with controlled ventilation, in which no artificial heating is needed at all in winter apart from a small contribution from the lighting—the heat derived from the bodies of the occupants is all that is required. In an ordinary house of any of the main types built in Britain, and occupied for only part of the day by a small number of people, this cannot be achieved, and though the contribution of human heat is not negligible it is usually a very small part of the heat needed for comfort.

The amount of heat required depends on the acceptable room temperature (different in different rooms), on the loss of heat by conduction through walls and roof, and on the heat required to warm to room temperature the cold outside air as it comes in for ventilation.

Thermal insulation

The methods required for this are now fairly well known—double or even triple glazing of windows; cavity insulation of walls; and lagging of loft floors. Previous generations are frequently blamed for lack of attention to these in the building of houses in the past. In fact this was not due to the stupidity of buyers or the incompetence of architects and builders. It was due to what may be called the open-fire way of life. In the smaller houses of the poorer part of the population no attempt was made to warm more than a single room. The kitchen, which would also be the living room, would be kept warm during the day by the coal-burning cooking stove plus the body heat of an often numerous family, and enough clothing was worn to make this at least tolerable, if not comfortable, in winter. In better-off households one or two other rooms would be warmed by open coal fires, lit each morning and allowed to go out each night.

Open fires were inefficient heaters, and were not used on the continent, but coal was cheap in Britain and fireplaces were cheap to build. Such fires required great quantities of air to carry the smoke up the chimney, which made them superb ventilators, and the heat needed to warm the air coming into the room at several cubic metres a minute was so large that the conduction loss through the walls of the single room concerned was unimportant. Conduction losses to the outside from the other unheated rooms were also unimportant. Insulation of the outer walls and roof would therefore not have been worth the cost.

In the last 30 years there has been an enormous change in our way of life. Cleaner and more efficient closed stoves, often combining space and water heating, replaced the open fires, followed in an increasing number of houses by oil- or gas-fired central heating which warmed the whole house. Important and expensive amounts of heat were then lost through outside walls and through the roof, so that thermal insulation of outside walls and of the ceilings of the upper rooms became well worthwhile. To achieve this a combination of measures is required.

According to *Which?* (1982) the most important is the insulation of loft floors, which often consists of little more than a centimetre or so of plasterboard between roof joists forming the ceilings of the rooms

below. If the space between the joists is filled with expanded ver-
miculite, with slag wool, or with any of a variety of proprietary
insulating materials, the heat loss can be reduced by four to six times.
This is certainly worthwhile; if no insulation is added it would be easy to
lose hundreds of kilowatt hours of heat per 24 hours in cold weather
through the upstairs ceilings of a fair-sized house.

Even with the 10 cm of insulation now recommended, it is likely that
some tens of kilowatt hours of heat will be lost per day in cold weather
in a centrally heated house, even if the upstairs temperatures are kept at
12 °C rather than the 18–20 °C likely to be maintained in the living
room.

Losses from the outer walls of the house will be less; 25 cm of brick
and plaster transmits a great deal less heat than does 12 mm of plaster
alone. The loss however is not negligible. If the whole of a two-storey
brick house is kept at a temperature 10 °C above the outside, the total
loss per 24 hours could easily be 20–50 kWh of heat†. This is several
times less than the loss to an uninsulated loft, and the area to be
insulated is greater, but where the loft has been well insulated and there
is a cavity wall, filling the cavity with a phenol-formaldehyde polymer
foam will probably pay for itself in a few years even if the house is
heated by gas, currently the cheapest source of energy.

The next most effective measure in most houses is draught reduction.
This is cheap and needs little technical ability. Double glazing of
windows is less effective unless the window areas are very large, and is
marginally economic if you have it done by a outside firm. If you can do
it yourself it should pay for itself in less than five years. The effect is to
bring the heat loss through the glass down to two or three times the loss
from the same area of brick wall, while single glazing passes nearly ten
times as much for the same area in the same conditions. There are now
plastic selective reflectors which can improve matters still further,
although the cost is fairly high. Triple glazing is not worthwhile unless
walls and ceilings have been insulated exceedingly well. The area of the
windows is of course much less than the area of the walls except for a
very few architect-designed houses built during the fortunately brief
glass-period of the 1960s (vitromania!)—and in some schools and office
blocks where clearly glass was cheaper than brick for outside walls,
however uncomfortable for the occupants.

Double glazing has two quite significant advantages apart from the
saving of money and energy. The most visible advantage is the reduction of condensation. On a cold day ordinary windows will be heavily
misted in occupied rooms. With double glazing the inner glass is so
much warmer that even in the kitchen misting is usually prevented. The

† One therm = 23.9 kWh.

second is that the convection draught due to cold air running down the window, which can be uncomfortable if you sit with your back to a large high window, will usually be eliminated.

The amount of gas and oil that could be saved nationally by a really effective thermal insulation campaign is very large, and would materially extend the time that our natural gas supply could last. Few of the houses in Britain have been insulated as well as they could easily be, and the possible cash saving has been estimated by the Association for the Conservation of Energy (ACE) as nearly £3000 million a year (*The Times* 14 February 1983). Like all financial estimates in this field, this is highly conjectural. People like to be comfortable, and will often take advantage of improved insulation to keep the house more uniformly warm than before; and therefore save a good deal less heat than would have been possible if the house temperature had been kept at the original average level. Furthermore, when people do save money in house heating, they either spend it on something they would not otherwise have bought, or invest it. In either case there must be some energy content—probably only 10–15%—in the value of the goods or investments. Even half of the saving suggested by ACE however would still be a major achievement.

SUMMARY

There are numerous ways in which total energy can be saved, and some in which oil and gas could be replaced by electricity derived from coal, nuclear power, or the renewable alternatives. Combined heat and power systems and domestic insulation and draught proofing offer perhaps the greatest potential for conservation, although if economic conditions improve the latter will often be used to increase comfort rather than to save energy. Taken all together perhaps a further 10% of oil and gas could be saved, with a considerably smaller reduction in the use of coal and little change in the use of electricity.

The ways in which energy could be conserved involve technical, social and economic considerations. Many technical innovations to make the use of primary sources more efficient have been considered or are being tried out. But some of the possible changes require the cooperation of the public either in using more capital expenditure for more energy-efficient methods, or in direct conservation by willingness to live less comfortably—and in some cases being willing to accept a very small danger to health (see Chapter 10). A lot of the ideas, however, if tested only by the criterion of financial profitability, are non-starters unless subsidised, i.e. financed in part by taxation. A more detailed discussion than I have room for here is given by Currie (1984) with many further references.

Part 3

Quantitative Discussion of Risk

Chapter 8

Accidental Risks to the Public due to Energy Production other than Nuclear

In every industry accidents have happened, and in every industry accidents will go on happening. Most of these involve damage to equipment rather than to people, and the accidents to people mostly lead to injury rather than to death. In this chapter I shall be concerned only with accidents leading to deaths, partly because they are more reliably recorded but mainly because they are more important.

THE RELATIVE IMPORTANCE OF DIFFERENT ACCIDENTS

The problems arise when we try to decide the relative importance of different kinds of accident which do or may cause death. To a naive technologist it might seem that the importance depended only on the actual numerical risk of death to the individual working in the industry concerned, plus the risk to individual members of the public resulting from accidents in the industry. In a large proportion of jobs in a large proportion of industries people are safer at work than they are at home, even allowing for the greater part of their time that they spend at home. In 1985 for example in England and Wales about 2000 people died at home from falls, over 500 died in fires, and 188 choked themselves with food. The risk of death from choking alone is greater than the risks to the public from accidents involving all our sources of energy put together.

Although specific accidents leading to individual deaths are unpredictable—if they could be predicted they wouldn't happen—recorded statistical information from the past will give us quite good numerical estimates of the probable risks. There is a kind of psychological threshold—for familiar forms of risk usually around 1 in 100 000 per year—below which they would be neglected; people will feel themselves safe and will behave as they would if the risks were actually zero. This is perfectly sensible; the risk of being killed on the pavement in Britain by a vehicle leaving the road is a little more than one in a million per year, but it would make life less pleasant if one kept a continuous look-out each way at the traffic every time one went shopping on foot.

THE THRESHOLD OF CONCERN

Accordingly the naive technologist, having intelligently recognised the threshold of neglect, might well decide to forget about any risks below one in a million per year—that is to say, any risks which led to the death of fewer than 50 people in Britain each year. He would be quite wrong. The subjective threshold of unconcern is perfectly real, but it is not based only—indeed it is based hardly at all—on the mathematical odds against being killed. An extremely important factor is familiarity. Where everyone knows the risk, but knows personally nobody who has been killed, nobody bothers. Alternatively, a novel danger is much overrated. When electric light was first introduced, great numbers of people regarded it as far too dangerous to have in the house, regardless of the fact that the traditional candles and oil lamps were much more dangerous. As was mentioned in Chapter 2, extraordinarily small amounts of radiation, which is unfamiliar, can worry people. An erstwhile professor of biology in my university once stated, after the university cyclotron had begun to make radioisotopes, that he was not prepared to accept the introduction by the physicists of a single ion pair into his department. He may not have been aware that the potassium in his own body was producing 100 000 ion pairs per second, and the remark was no doubt meant to be picturesque exaggeration; but it showed a perfectly real concern about quantitatively negligible doses of radiation.

There is still another very important factor which contributes to most people's psychological threshold of concern about risks, which the naive technologist will miss. This is the scale of the individual accident. An accident once a year that kills 1000 people attracts a lot more attention and is much more frightening than a lot of little accidents that kill half a dozen separate people in different places every day. The actual risk to any one individual is more than twice as great in the second instance, in accidents of which type over 2000 people are killed each year.

There are of course special cases in which many people being killed together is genuinely 'worse' than the same number of individuals being killed independently over a considerable range of space and time. The Aberfan disaster, when inadequately stored wastes from the coal industry killed 144 people, mostly children, not only killed a large number together but destroyed the whole life pattern of the surviving villagers in a way that could not have occurred with 144 separate deaths spaced over the whole country.

To me however, the real loss to humanity seems nearly always to depend very much more on the numbers and ages of the people killed than it does on whether they are all killed at the same time and place. For this reason, although I shall give estimates of how many people are likely to be killed together or separately when appropriate, I shall concentrate on the risk to the individual working in each industry and to each individual member of the public living in the environment affected by the industry.

Thirty dying in coal mine accidents each year out of 50 million Britons is trivial compared with the 5000 to 6000 lethal accidents in homes; but 30 a year out of 150 000 coal miners is not trivial at all. At one in 5000 per year it is the most probable single cause of death, and is responsible for nearly a tenth of all deaths among miners over the active part of their working lives.

The biggest difficulty in producing convincing as well as reliable figures of probable risk is that in every field the accidents that people fear most are the major accidents that have never happened. These have every characteristic that can make the public perception of risk greater than the numerical probability of the risk. They are by definition unprecedented; they would—again by definition as major—kill a lot more than the 50 people in a year which looks like the psychological threshold of concern in Britain for familiar risks; and a large number would be killed at once. The not so naive technologist must therefore expect that the most truly realistic figures of risk will be well below those expected by the well meaning but ill-informed public—and uncritically repeated by the equally ill-informed media.

THE RISK OF AN ACCIDENT THAT HAS NEVER HAPPENED

I will begin by saying that I shall not be concerned with the maximum credible accident. This depends on the imagination of the inventor and on the credulity of the receiver—neither of which has any definable limit. There are still people (although, so far as I know, no longer any journalists) who believe that if nuclear power got out of control it could start a chain reaction that would destroy the planet. Doubtless someone could believe that a coal mine in Antarctica might catch fire and go on burning uncontrollably underground until it had melted the whole of the southern ice cap and flooded most of the coastal cities of the world. In the 1960s there were actually a lot of people in Hampstead (and presumably elsewhere, but the ones I knew were in Hampstead) who thought that the accidental dispersion of DDT from crop-spraying aircraft was going to destroy the oceanic plankton and that the resulting loss of oxygen production would lead to the asphyxiation of all animal life on the planet by 1979. (If no oxygen were ever produced again, it would be thousands of years before the present world population could reduce the existing oxygen stocks in the atmosphere to the level now breathed by the people of Tibet.)

On the inventor's side, it is perfectly *possible* for a couple of off-course fully-fuelled Jumbo jets to collide, one of them crashing on Wembley stadium during a cup final, bits of it blocking the exits, and a hundred tons or so of kerosine—a litre each for 100 000 people—being scattered over the crowd and ignited; while the second Jumbo crashes on Canvey Island destroying the LNG storage tanks and a refrigerated supply vessel on the Thames just coming in with fresh supplies, all at a time when there is a suitable breeze to blow the cold gas down into Canvey Town, with its 30 000 inhabitants, before it explodes. There is more than enough LNG stored there to kill the whole population of Canvey and burn the entire town to the ground. Then a fire engine dashing to the spot collides with a chlorine tanker ... and so on and so on.

Unlike the three previous suggestions requiring a little learning and a vast degree of credulity, this is actually possible, merely unlikely.

The fact is that no decisions can usefully be made on the basis of the worst possible case, and in our own lives we never do allow such cases to affect our plans. If we did, we could not leave the house—because in the worst case we should be killed by a car which happened at that instant to mount the pavement. But we could not stay indoors—because in the worst case we would be killed inside the next minute by one of the more rapid accidents that do kill people in their homes.

In considering the large and frightening accidents that have never happened in the power industry we must obviously be thinking of something unlikely. What we have to do is to try to determine *how* unlikely. If an accident killing 1000 people has a probability of one in 1000 years there is a one in 1000 chance that it will happen next year, but there is a far larger probability that it will never happen at all, because the conditions making it possible are likely to be eliminated by competent engineers in a lot less than 1000 years.

The ways in which we can estimate quantitatively the risk of something that has never happened can best be discussed in relation to particular cases, that I will deal with as they arise. I shall now go on to discuss each of the power industries, roughly in order of risk. The numerical estimates resulting, together with the generally larger risks from routine running, will be summarised in table 10.4 on p. 198.

DAM FAILURES IN HYDROELECTRIC SYSTEMS

Here we are on relatively firm ground; a number of large accidents *have* happened, so that people are a bit less worried about them, although the numerical risks are quite large. Since 1930 over 100 dams have failed in the USA. Not all of these were producing electricity; some of the smaller ones were entirely devoted to irrigation or to the supply of water to industry or townships, and most of the failures occurred without loss of life. Nevertheless they killed 355 people in the early 1970s (Ockrent 1981). In Italy the Vaiont dam failed on 9 October 1963 killing 2000 people. Even in Britain there have been deaths due to dam failures. In 1925 the dam at Dolgarrog in Wales broke and killed 16 people. Nearly all of our 2000 or so dams are small. Many have been inadequately inspected. In total they provide less than 1 GWe of electricity. It is unlikely however that the failure of any of them would kill very many.

Fifteen thousand people died when the Gujarati dam in India failed in 1979 (Warner 1981). Obviously the newer and bigger dams now under construction are being built by engineers who have learnt lessons from the dams which failed. This of course is true of all industrial structures, but is of less value for large dams than for most others. The amount of water stored by a large dam may weigh a huge amount. The Kariba dam on the Zambezi River has impounded 16 000 million tonnes of water. Before a dam is built the land is approximately in isostatic equilibrium; that is to say the upper layers are effectively floating on the lower, and the pressures have evened themselves out. If you put several thousand million extra tonnes down on a piece of it, this is no longer true. The deeper rocks are compressed, and are squeezed outwards into the uncompressed surrounding rocks. The result will be a series of minor earthquakes. Nearly all of these will be entirely harmless, and the large area over which the extra weight is spread means that the extra pressure on any small area is small. However, the distribution of quakes is not entirely predictable, and any earth movement below the dam wall itself may obviously be dangerous. Many big rivers follow the lines of geological faults, so that their gorges are preferentially formed in earthquake-prone areas. It has been estimated that several of the Californian hydroschemes could kill up to 100 000 people, and the San Fernando Valley pump storage scheme a staggering 200 000.

In figure 8.1 is shown the relation between the frequency of dam failures and the number of people killed per failure, compared with other manmade causes of disaster (Wash 1400 1975). It is very noticeable that whereas in the case of air accidents, an accident causing a ten times increase in casualties is more than ten times rarer, so that the greatest numbers of people per year are killed in the smallest accidents, the frequency of dam failures went down only about nine times while the number killed per accident went up a hundred times, from 10 per failure to 1000 per failure. Only for the very largest numbers, which have fortunately been extremely rare, does the curve bend down to match the other curves.

Even the building of large dams can cause major accidents. *The Times* reported on 30 July 1983 that 150 people had been buried in landslides at the Guavio hydroelectric reservoir being built 80 miles east of Bogota in Colombia.

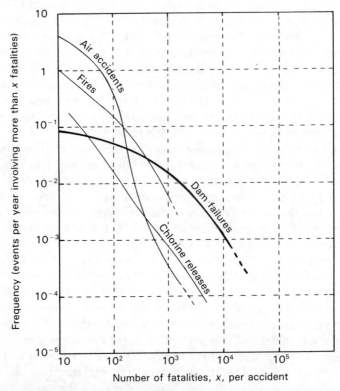

Figure 8.1 Frequency of man-caused events involving fatalities (worldwide) (Wash 1400 1975).

Generally the greater efforts which are made to avoid the more serious accidents are successful in making the total of deaths caused by the larger accidents per year less than that caused by the smaller ones. For dams this has not been successful, and their accident graph is far closer to that shown in figure 2.3 for natural disasters. Here the smaller and commoner causes of trouble have long been understood and guarded against, and it is the larger ones that are out of control.

Unfortunately, data on the electrical output of dams are not so readily available as are the numbers of deaths, but there must have been quite a lot of deaths per GWy of electricity produced.

ACCIDENTS INVOLVING NATURAL AND PETROLEUM GAS

In the first edition I expressed the view that gas from the North Sea is unlikely to be responsible for a major accident affecting the public. It is not very poisonous, and the energy obtainable by burning 1 m^3 of gas is around 1000 times less than the energy obtainable from 1 m^3 of oil. Nevertheless, since 1980 an average number of 50 deaths per year have been caused in Britain from gas poisoning in the home. The total amount in the high-pressure gas grid is large, but there are few concentrations of large quantities at one place. If a gasometer did get holed and caught fire there could be a spectacular fire, but the gas is lighter than air so that the burning gas goes upwards rather than sideways, and since the gas still in the gasometer is *not* mixed with air it cannot explode until nearly all the gas has gone and air leaks in.

I was wrong in assuming that natural gas could not cause a major accident. On 4 June 1989 a serious leak occurred 750 miles east of Moscow in a Russian pipeline carrying natural gas from Siberia. The smell of gas had been detected some hours earlier by local people up to five miles away but nothing was done about it. A drop in pressure in the pipeline was also noticed but misinterpreted, so that instead of shutting the gas off the pressure was increased. As a result, a low-lying area on the Trans-Siberian railway was filled with a mixture of cold gas and air, which exploded violently as two trains moving in opposite directions passed each other—and no doubt ignited the mixture. A fireball of up to a mile and a quarter wide was produced, which removed the branches from trees within $2\frac{1}{2}$ miles.

Ten days later a Soviet newspaper (*Sotsialisticheskaya Industriya*) reported that more than 600 people had died as a result of this accident. Children going to a holiday camp were among those missing or killed.

Although major explosions are few, minor ones are common. Accidents occur where pipe failures start leakages into houses which are unnoticed until considerable quantities have been mixed with air. Any naked light, or a spark from a piece of electrical equipment, can then ignite the mixture and cause an explosion. The leak need not be inside the house. Corroded underground pipes may leak and gas can then run along the pipe ducting, or even through uncompacted earth around the pipe, and emerge into the lower rooms of several houses at the same time. Since 1939 it has been a legal requirement in Britain that where a service pipe enters a house the space between pipe and hole must be cemented and sealed; there must now be few houses in which this has not been done. Nevertheless the British Safety Council has reported that an average of three explosions a week occur in the UK, and it has been extremely critical of the management of British Gas both in respect of maintenance of the pipe network and of its unwillingness to accept responsibility when failures of pipes or equipment cause damage or deaths.

As might be expected, rather more accidents occur in the winter months. In a ten day period over the Christmas and New Year break in 1976–77 there were four major explosions in Britain, at Bradford, Brentford, Bristol and Beckenham (British Safety Council 1980). (Readers who felt that I dismissed too readily (in Chapter 2) the evidence relating lung cancer deaths to position in the alphabet of the hospital region concerned may be struck by the fact that all four explosions occurred in places whose names start with the same letter as does British Gas. B is of course a fairly common letter; about 11% of the place names in the *British Gazetteer* begin with B. Even so, the random chance of all four doing so is only one in 7000, and when one notes further that in three of the four cases the name starts with Br, the random chance drops to one in 540 000†. Brighton appears to have escaped this time!)

During the years 1981 to 1986 there was an average of 32 deaths per year from explosions and fire relating to the supply and use of inflammable gas (Health and Safety Executive 1989). This would be expected to be well below the usual public threshold of concern (around 50 per year for familiar risks), but the noise and material damage involved lead to extensive reporting by the media. If every death from influenza could be heard half a mile away, and got into the television news, people would be more afraid of it than they are now. Furthermore, a gas escape is a risk that people can do something about for themselves, which materially and very desirably heightens attention. There would be a great many more gas explosions if it were not for the fact that most people do notice gas leakages and take sensible precautions such as turning off the gas at the main and opening windows to let it away, and extinguishing cigarettes or any other ignition source.

Risks of really large accidents affecting the public do not come from North Sea gas, or from the few remaining sources of coal gas, but from the large stocks of LNG imported in refrigerated ships from Algeria or elsewhere; or from liquefied petroleum gas (LPG) which is usually one of the products of oil refining. LPG contains a higher proportion of the heavier hydrocarbon gases than LNG, has a higher available energy content per cubic metre, and can be stored under moderate pressure at

---

† Of course, any specific group of four first letters will be highly improbable, just as is any 'ordinary' hand dealt at bridge. A set of the four towns, Jarrow, Queensferry, Ventnor and Yeovil, would be much less probable still, with first letters much less frequent than B. But it would not have caught the eye.

room temperatures. In terms of accident risk however, there is little to choose between them.

Canvey Island

A major British hazard is the large store of liquefied gas on Canvey Island, which, with two oil refineries, covers a couple of square miles about three miles from the town of Canvey, and which was mentioned in the scenario describing a possible but improbable accident earlier in this chapter. The LNG is stored in large insulated and refrigerated cylindrical containers each holding many thousands of tons, and spaced apart to reduce the risk of damage to one container spreading to the others. Containers are surrounded with a 'bund', a wall high enough and far enough away to contain all of the cold gas into which the liquid could evaporate. Apart from intentional sabotage the risk of accident is of course low, but the characteristic behaviour of liquefied gases makes a very serious accident possible.

A very important study of the risks of the Canvey Island complex by the Health and Safety Executive in 1978 was quoted by Niehaus (1983) in a discussion of risk assessment. It illustrates the important fact that the main object of risk assessment is not merely to find out how dangerous a particular activity is, but to locate and remove features of the activity that are responsible for the risks. The study did indeed provide a quantitative estimate of the current risks, from about one in 100 per year of an accident killing 10 people to a chance of one in 5000 of an accident killing 10 000 people (nearly half the population of Canvey Town). Much more importantly it identified a series of possible improvements in safety measures, which had by 1981 reduced the estimated risk by twenty times. It is thought that the methane terminal, which is considered to contribute at least a third of the risk discussed, is also a threat to the 250 000 people in south Essex (Inspector General 1981).

There is a considerable body of practical experience concerning accidents due to liquefied fuel gases. In Spain in 1978 there was a disaster at San Carlos de la Rapito, 120 miles south of Barcelona, when a 38 tonne road tanker carrying liquid propylene crashed in a holiday camp, killing 180 people and injuring 200 more.

If a container of liquefied gas should be wrecked, the contents would pour out over the ground and run downhill like any other liquid, boiling violently as it runs over the far warmer ground surface. Owing to its latent heat of evaporation (the energy needed to turn a liquid into a gas) it does not all boil away at once and, as it freezes the ground over which it runs, it can run for considerable distances before it has all boiled away. It then forms a neutral-buoyancy cloud which is no longer in direct thermal contact with the ground, and can drift a long way with the wind. The gas produced is at first at a temperature little above that of the liquid, and its density is a good deal higher than that of air. It therefore forms a fairly thin layer over the liquid and itself runs downhill in still air, but is sufficiently less dense than the liquid to be carried along on the level, or even slightly uphill, by a moderate breeze. The upper part of the layer will mix with the air above it, so that there will quickly be three regions with importantly different properties. Closest to the ground there is nearly pure gas, mixed with too little air to burn. Above this is a mixture of gas and air with a concentration of gas at which it can burn, the limits again depending on the exact composition (between 2 and 10% for propylene) (Health and Safety Commission 1978). Above this there will be air containing too little gas for ignition.

When this complex system meets a naked light or an electric spark, the intermediate layer will explode but the lower and upper regions will not; and any liquid still present below the system will have its evaporation briefly increased by the heat but cannot burn for shortage of air. The remaining liquid and the concentrated gas therefore continue to run along the ground, for a short time isolated from oxygen-containing air by the carbon dioxide, water vapour and nitrogen mixture which is left from the explosion. Very quickly this, being hot, will rise away, and a fresh three-layer system will be established, continuing to run downhill or with the wind until it meets with a fresh ignition point and has the middle layer again removed by a fresh explosion. It was this unpleasant habit of causing a succession of separate explosions that led to the killing of so many people in the holiday camp. Either a simple fire or a single explosion would have killed far fewer.

It is rarely realised that, tonne for tonne, burning oil or gas liberates a lot more heat than does the explosion of TNT. Hence the burning of 100 000 tonnes of methane, although it will produce very little blast energy, will liberate as much heat as a 1 megaton hydrogen bomb. In clear weather such an H-bomb would set alight dark-coloured clothing at a distance of five miles from an air blast. A hundred thousand tonnes of methane, burning in a few minutes, would radiate a smaller propor-tion of the heat that it releases, but could still kill people and set fire to clothing or curtains at hundreds of yards from the nearest point of the fire.

The risk near a liquefied fuel gas depot is clearly at a maximum when a refrigerator ship comes into dock and transfers its contents to the land storage system. I do not know what are the precautions taken at Canvey, though doubtless these are adequate for any ordinary incident; but at Boston, USA, which has a large LNG terminal, no other vessel above a very small size is allowed to move at all in the harbour from a time well before a refrigerator ship's arrival until it is safely berthed. A crash in the harbour could lead to the destruction of every other vessel in the harbour at the time, and to a ring of damaged and burning buildings round the periphery, which might lead to an uncontrollable fire engulfing a large part of the city.

The hazard is not limited to storage depots. Considerable amounts of liquefied fuel gases are carried by rail, a typical train load consisting of 10 refrigerated wagons each containing 60 tonnes of LPG, totalling as much as would be burnt by over 100 houses in a whole year. Fire and explosion of any one wagon might involve the whole trainload, and 600 tonnes of LPG could form an explosive 5% mixture with air a couple of metres deep over about one and a half square kilometres. Of course it probably wouldn't; a raging fire near the accident with nearly all the energy going straight up into the sky would be far more likely, but a crash in an urban area which broke open even one or two wagons without setting their contents alight could destroy quite an area if it failed to ignite until it had all spread and evaporated.

It is not feasible to make the storage containers proof against aircraft impact, but the most probable large accident must surely be the collision or wrecking of a container ship. With luck there would be few casualties apart from the crew, unless the collision happened in port.

Bearing out my view that decisions should not be made on the basis of worst possible cases, some actual incidents which have occurred since the war are shown in table 8.1. In the worst case any of them could have killed thousands of people. In fact they didn't.

Experience would suggest that the long-term average risk is between 0.1 and 1 death per GWy; considerably less than the risk from hydroelectric power.

Table 8.1 Deaths resulting from liquefied fuel gas accidents.

Year	Place and circumstances	Deaths
1944	Ohio—tanks of LNG split	130
1966	France—tank of LPG leaks	45
1973	New York—LNG in tank explodes	40
1978	Mexico—lorry of LNG exploded in multiple pile-up	100
1978	Spain—road tanker leaks on crash	150
1984	Brazil—LPG pipeline explodes	80
1984	Mexico City—explosion in LPG storage depot	> 500

OIL ACCIDENTS

Fortunately, large accidents have not yet occurred in Britain, but other countries have been less fortunate. The most recent of which I have a record is of one on 19 December 1982 at Caracas, the capital of

Venezuela, where an explosion at the Tacoa oil-fired electric power station killed at least 145 people, injured 1000 more, and caused the evacuation of 40 000 residents. In Brazil in 1984 a petrol explosion following pipeline fracture led to 508 deaths.

The main large-scale risk is from the 15 or so British oil refineries, such as those at Fawley, Canvey or Southampton in England and Leith, near Edinburgh, in Scotland. In each of these there may be nearly a million tonnes of oil at a time brought in by a modern tanker, and another million tonnes or more of refined products in store awaiting shipment. The lighter liquid fractions of these, such as petrol, are a real fire hazard, and the liquefied petroleum gas presents a danger similar to that from the LNG at Canvey Island. The oil refineries need a much larger work force than does the LNG storage depot at Canvey, and considerable populations live close enough to the stores for evacuation to be necessary if a fire capable of breaching large containers should start— and if there were time for evacuation. The potential loss of life in easily imaginable accidents is very large. Table 8.2 shows the estimate given by a report from the Health and Safety Executive of the refinery risk associated with one year's supply of one GWe oil-burning power station (or one million tonnes of heavy oil) (Cohen 1980).

Table 8.2 Estimate of refinery risk linked to an output of 1 GWye from an oil-fired power station with 75% load factor. Figures given are the annual number of chances per million, assuming all recommended precautions are taken, of an incident causing the stated number of deaths. Data from Cohen and Pritchard (1980).

	> 1500 deaths	> 4500 deaths	> 18 000 deaths
	11	4	0.4
With ammonia store	18	7	1.6

Canvey Island has an ammonia store capable of holding 10 000 tonnes of ammonia, but few refineries have such stores. Nevertheless the total risk, without ammonia, can hardly be less than 0.1 per GWye.

Since Britain became an oil producer many people have become more conscious of the dangers to the workers who are actually at the production end. For a couple of centuries the country lived with the continual risk of nasty accidents in the coal mines; now we have even larger risks in the North Sea oil fields. It will be remembered that when the *Alexander Kielland* platform capsized, 123 people were killed. This type of risk does not affect the public directly, and neither this nor the loss of two of the helicopters (one of them with 45 people) carrying oil workers to and from the oil rigs made a lasting impression, because they were felt to be a temporary effect of the lack of experience in a new industry. The *Piper Alpha* disaster in July 1988, when 167 died—many being burnt alive—changed all that. It is clear that very extensive and very visible improvements in the safety provisions, as well as considerable increases in wages, will be needed—with a permanent effect on the cost of North Sea oil.

If the average rate of deaths per year continues, it would add about 0.2 deaths per GWye to the total for which oil-fired power stations must be held responsible.

CHLORINE

An additional hazard in refineries as well as in many individual power stations is the stock of chlorine which is kept to prevent living organisms from growing in the cooling-water systems and impeding their efficiency. A major oil fire could cause the explosion of a chlorine store, which may contain up to 90 tonnes of chlorine, and both these and the road and rail tankers bringing it from the production plants could kill very large numbers of people. Chlorine, which was used extensively as a poison gas by both sides in the 1914–18 war, is lethal if breathed for half an hour in concentrations in air of only 250 micrograms/litre; so that 100 tonnes of chlorine could spread into a lethal cloud 100 m thick and a few hundred metres wide over several kilometres down wind. This would be enough to kill some thousands of the inhabitants of a big city. Even one of the 10–20 tonne tankers of chlorine that we see running along our motorways could liberate enough chlorine to kill a large fraction of the people in a small town if it crashed to windward on a nearly still day and the gas drifted slowly for a long way before it dispersed enough to be harmless.

The numbers that could be killed in worst-case theory are far greater than those that have been killed in practice. In real life chlorine—or any other lethal material—does not spread itself out uniformly at a just-lethal level; it remains in a far smaller—and more quickly lethal—volume, and from the edges of this it is blown away in unpleasant but sub-lethal concentrations. The area within which people are actually killed is therefore far smaller than the possibly lethal area mentioned above. Furthermore, the wind is not usually blowing at the right speed and in the right direction to maximise the effects.

We have quite a lot of practical data to go on, and the numbers killed in real chlorine spills have been small (see table 8.3). I have no record of the numbers left with permanent lung damage. The fact that small numbers are killed is largely the result of an understanding of the risks.

Preparations are made by the police and others for rapid evacuation of people near to chlorine or other toxic chemical stores. ICI, which is responsible for a number of such stores at Thornton Clevely in Lancashire, keeps a five-man team on standby 24 hours a day 365 days a year to cope with emergencies on site.

On the experimental evidence the number of deaths per GWye added by the chlorine at power stations is less than 0.01 and, whatever the feelings of people living near the stores, is trivial when compared with the other hazards of life.

Table 8.3 Deaths resulting from some chlorine accidents.

Date	Place	Chlorine lost (tonnes)	Deaths
13.12.25	St Auban, France	25	19
24.12.39	Zarnesti, Romania	25	60
5.11.47	Rauma, Finland	30	19
5.4.52	Wilsum, Germany	15	7

BHOPAL

For comparison, the world's worst industrial disaster occurred at the Union Carbide chemical plant at Bhopal in Madhya Pradesh in India, when 45 tonnes of methyl isocyanate, probably together with phosgene (one of its precursors), escaped during the night of 3 December 1984. As with phosgene used in the 1914–18 war, deaths were not always quick. Over 2500 people had died in the first year, and at least 200 000 were injured, some very severely. The number is still increasing. It can never be known how many other people will have their lives shortened by less than lethal doses. (Claims for compensation are being made by over 500 000 people.)

COAL ACCIDENTS

Essentially all of the major accidents arising from the use of coal occur in the winning of it by the miners. About thirty coal miners are killed in accidents in British pits annually, and one or two railwaymen in the course of carrying 120 million tonnes of it around the country, but these hazards are voluntarily accepted and hence not the primary concern of the public or of this book. The numbers represent about 0.6 per GWy of electricity, or around 0.2 per GWy of the heat energy obtainable.

There must also be risks to the general public. For example, a train carrying dangerous chemicals might collide with a derailed coal train in a tunnel, and a fast passenger train might then collide with the damaged coal trucks. Up to 600 passengers would be expected to die from injuries or asphyxiation (Cohen and Pritchard 1980). The increase in probability of this resulting from the coal traffic required by an extra 1 GWe coal-fired power station is estimated as 1 in 100 million per year.

The chief accident risk to the public from the coal-fired power stations is probably that from the chlorine stores, as for the oil-fired stations.

ACCIDENTS INVOLVING RENEWABLE ENERGY

Roof-top solar water heaters are likely to cause a lot of injuries and a few deaths of amateur DIY repairers for each GWy of heat that they save, but will not cause multiple deaths. The huge structures envisaged for the exploitation of wave power could cause shipwrecks in stormy weather; if one of the million-tonne floating devices proposed for some systems should break loose in a storm and fail to sink as planned it could drift into the path of a large vessel, causing major loss of life. Again, several things have got to go wrong coincidentally, and though large accidents can be imagined they would be highly improbable. Until full-scale structures have been built and installed, it is not possible to guess *how* improbable. The designers can doubtless be trusted to make the risk a great deal less than one per GWy.

Wind turbines, being a good deal smaller, are likely to cause correspondingly smaller accidents, although an array of such generators in the shallow seas off East Anglia would inevitably be hit by ships from time to time. Being in less stormy seas and closer to lifeboat stations the death rate should be less than for the wave machines. Land-based wind turbines will presumably be placed away from low-flying regular aircraft routes, and although they may cause occasional deaths among the users of light aircraft and even a few deaths on the ground when they get blown over, such accidents should be rare and always small. It is too soon to make any quantitative guesses, but wind turbines are likely to be among the safest sources of electricity.

Chapter 9

Risks of Nuclear Accidents

ACCIDENTS IN MILITARY PLANTS

The probability of a serious number of deaths among the public due to accidental escape of radioactive material from civilian power stations can be estimated only by theoretical studies in Western Europe or the USA, none of which have the design faults that made the Chernobyl disaster possible (see below). Important lessons can however be learned from accidents in military plants in Britain, the USSR and the USA, and I shall begin this chapter by discussing these.

The Windscale fire

The Windscale accident, in October 1957, occurred in a reactor built and operated for the sole purpose of producing plutonium for the British atom bomb programme. The heat produced was swept away by forced air cooling, rather than carbon dioxide, keeping the whole system at a temperature around 150 °C and precluding any possibility of simultaneously producing useful power. As was pointed out earlier, there was no possibility of a nuclear explosion in the natural uranium fuel, but during what should have been a routine operation the reactor core overheated and the graphite moderator caught fire due to a sudden recrystallisation of the graphite which released energy. This melted the aluminium cladding of the fuel rods and the fuel rods themselves caught fire. (In power reactors aluminium is now replaced by a magnesium–aluminium alloy which does not react with uranium.) The uranium, with a lot of extremely radioactive fission products, a little plutonium and a small but important amount of polonium-210, literally went up in smoke.

We were fortunate that Sir John Cockcroft, then the head of the Atomic Energy Research Establishment, had insisted on the building of an expensive filter tower through which the outflowing cooling air had to pass, although the other engineers had thought this unnecessary. This must have absorbed the greater part of the radioactive smoke liberated by the burning reactor, and without it much larger numbers of deaths would have resulted.

Even with the filters, which could do little to stop gases, 10 000–20 000 Ci of gaseous radioactive iodine-131, with a half-life of about eight days, passed into the atmosphere and was detectable as far away as eastern West Germany. Over 300 000 Ci of the radioactive noble gas xenon-133 (half-life 5.3 days) also escaped, being unaffected by the filter, but this dangerous-sounding amount was in fact unimportant. Xenon forms no compounds that could remain in the body, and nearly all of any gas inhaled would have been promptly exhaled. The small amount that at first dissolves in blood and fat will be swept away in a very short time after its absorption, before any significant radiation dose has been received. (I have myself, in medical studies, breathed both radioactive xenon and the similar radioactive gas argon-37, and measurements were made of the rate at which they were expelled—which was satisfactorily fast.)

Numbers of other radioactive substances also escaped from the reactor, the only important one being 230 Ci of the alpha-emitter polonium-210. The iodine-131 was dangerous not only because a lot was emitted but also because it is very efficiently absorbed by the body and is concentrated in the thyroid gland. (This behaviour is in fact used to give radiotherapy in cases of cancer of the thyroid by using radioactive iodine.) Furthermore, much of it was deposited on grass eaten by cows, which passed it quite efficiently into their milk. In the course of a day a cow could ingest most of the iodine deposited on quite a large area of grass, and it was feared that this would present a serious hazard to children drinking milk from cows grazing in areas close to the reactor.

Accordingly, sale to the public of the milk from these areas was forbidden, and the contaminated milk was collected and thrown into the sea. It seemed to me at that time, and still seems to me, that this was stupid and a waste of good milk. Dried, or made into a good local cheese, it could have been kept under refrigeration for nine months, when the iodine activity would have fallen by 8000 million times to an undetectable level which would be much less than the natural radioactivity of the thyroid gland due to potassium. The cheese or dried milk could then have been given away to a large number of young physicists who could have checked the concentrations themselves, and who would have welcomed with open mouths a free issue of a high-quality British cheese. This would have cost the authorities more than it did to buy and dump the lot into the sea; but it would have done more than all the far more expensive published propaganda since then to show people that a radioactive substance which decays can be a lot safer than a chemical poison like lead or arsenic which lasts for ever.

The diversion of the local milk, although essential, was not in time to prevent a lot of people from getting measurable small quantities of iodine-131 into their thyroids. Even Londoners 250 miles away collected tiny but observable amounts. This was only partly due to the ingestion of it by cows in the London area; a nearly similar amount was inhaled directly from the weakly radioactive mass of air which blew east south east over Europe.

One of the valuable properties of radioactivity is its detectability. One can detect a thousand millionth of a dangerous dose. This is not possible with chemical poisons.

The total dose from all sources to any individual added less than his or her annual radiation dose from natural sources, and fortunately 90% of thyroid cancers are curable. However, some 200 million people received detectable amounts, and the total effective collective dose was over two million man-rems, capable on the ICRP assumptions of causing a couple of hundred cancers and perhaps a dozen extra cancer deaths during the few decades from about 1970 onwards (PERG 1981, Crick and Linsley 1983). Polonium-210 may be responsible for 5 or 10 deaths over the same period. (Polonium-210 from natural sources is normally present in our environment, and the measurable quantities in tobacco have been blamed for a significant contribution to the lung cancers due to cigarette smoking.) According to Crick and Linsley, about 33 cancer deaths or serious mutations may have been produced by all of the activities together.

It is of course unreasonable to expect that military plants working on crash programmes can be subject to all of the safety controls expected in any civilian industry, but it is important to explain what were the causes of the accident and to see how far these can be eliminated in the design of power station reactors.

The first point is that the latter run at much higher temperatures, $400\,°C$ or more, at which latent heat energy is quickly dissipated rather than accumulated in the graphite to a dangerous level. The second point is that all of the Magnox reactors used in power stations are cooled with carbon dioxide, not air. Graphite will react with hot carbon dioxide but releases little heat. Magnox alloy may react at very high temperatures, but again releasing very little heat. In AGRs and PWRs there is an additional safeguard in that the fuel is already present as uranium oxide, which can no more be set alight than can the ash from a coal fire.

The Kyshtym disaster

In November 1976 a refugee biologist, Dr Zhores Medvedev, published an article in the *New Scientist* (1976) which mentioned a major disaster in the Urals, between Sverdlovsk and Cheliabinsk, which must have occurred at the end of 1957 or the beginning of 1958. A later article gave details (*New Scientist* 1977). A large area, believed by Medvedev to be around 50 km radius, had been heavily contaminated by fission products, leading to thousands of deaths and the evacuation of many villages. His estimate of the area may be several times too large, but whether 500 or 5000 km^2 were as heavily contaminated as the observations show, the accident remains a major disaster. The size of the area must be known to the Soviet authorities concerned, and will eventually be learned by the rest of the world; it will be measurably contaminated for hundreds of years by strontium-90 (half-life 30 years) and caesium-137.

From the beginning Medvedev maintained that the recorded types of radioactivity proved that the contamination was due to the explosion of inadequately stored waste fission products from a processing plant producing plutonium for bombs. He first suggested that huge amounts of concentrated wastes were buried together in shallow pits and that the radioactive heating over a few years had made them red hot. Water had then percolated into the material and caused a steam explosion. This could quite easily have happened, but it would not have led to anything like the very uniform distribution observed, which required that the wastes had been dispersed into fine particles capable of drifting tens of kilometres in a light wind. To do this the explosive material would need to have been distributed uniformly through essentially the whole mass of fission products. Water does not mix uniformly with red-hot material before exploding into steam. Explosions there could have been, but these would hurl great chunks of material for hundreds of metres, not finely pulverised dust for tens of kilometres. The whole story was therefore disbelieved by many scientists in the west.

Dr Medvedev then suggested that the explosion was in fact a nuclear one, due to the leaching of residual plutonium from the wastes by water running through them, and eventually building up a critical mass that could explode. Such a leaching and concentration of plutonium had in fact been observed in the United States. Again, however, this could not possibly lead to a major explosion. The plutonium would first start a chain reaction in the wet, fast neutrons being moderated by water. Heat would be liberated quite slowly; it would dry out the surrounding ground and the reaction would then stop until everything had cooled down and more water had arrived. In fact this leaching and concentration could produce a particularly safe reactor but not a bomb, the parts of which must be of nearly pure plutonium assembled in a tiny fraction of a second to avoid a mere evaporation and dispersion of the material with at most little more force than a small steam explosion.

A believable theory has recently been developed in America which could explain how the waste explosion postulated by Medvedev could actually have occurred. In the later stages of waste processing to extract plutonium, large quantities of ammonium nitrate are used. Ammonium nitrate even in amounts of kilograms is impossible to detonate by itself, having in effect a critical mass or density analogous to that of a nuclear bomb. In thousands of tonnes, compressed to higher than normal density, it has produced several catastrophic explosions. For example, about 3000 tonnes exploded in Oppau in West Germany on 21 September 1921 and killed 561 people, up to 7 miles away. There could well have been many hundreds or even thousands of tonnes of ammonium nitrate used in the extraction, and still intimately mixed with the fission products when they were buried. A thousand tonnes of ammonium nitrate, once it had dried out, could well have given a pulverising explosion as energetic as a small atom bomb and so have produced the effects observed.

As with the Windscale fire, this disaster was the product of a military operation in a hurry. Wastes from nuclear power stations in Britain are not buried in large quantities with a huge excess of ammonium nitrate. I do not suppose that they are still so buried in the USSR. This particular accident is unlikely ever to occur again in the present bomb countries, in either the civil or military programmes.

In June 1989 the Russians revealed that there *was* an explosion in a container of nuclear waste in 1958.

Idaho Falls

A third accident that did not affect the public but which killed three operators took place in a small military research reactor (SL1) at Idaho Falls, USA, on 3 January 1961. The reactor exploded, one of the men being impaled on a projecting girder on the ceiling, and leaving the bodies and the building interior seriously radioactive.

So far as I know, this accident has never been officially explained, but I am inclined to believe that the culprit was xenon-135. This absorbs neutrons to form xenon-136, with an efficiency about 4000 times the efficiency of absorption by the uranium-235 on which the reactor depends. Xenon-135 is a fission product with a half-life of 9.2 hours, and if it is allowed to accumulate will make it impossible to restart the reactor until practically all has decayed. During normal operation it is changed by absorption of a neutron to xenon-136 (which does not absorb neutrons) as fast as it is produced, so that there is never a serious accumulation. When the chain reaction is shut down for maintenance there should therefore be none left to prevent restarting; but unfortunately it is not produced only by the direct fission of uranium-235. It is also created as a secondary product from the beta decay of the 6.7 hour half-life product iodine-135 (itself a negligible absorber of neutrons).

When a reactor is completely shut down, the xenon-135 will rapidly accumulate as the iodine-135 decays, until both it and the iodine-135 present have decayed to negligible proportions. Until this has happened, which may take a couple of days or more, the reactor cannot be restarted. Accordingly, a complete shut-down is avoided as far as possible, a sufficient chain reaction being allowed to continue to prevent xenon build-up.

It seems likely that the operators of SL1 failed to do this, and that in trying to restart the reactor the control rods were almost fully withdrawn until the reactivity was high enough to start the chain reaction in the reactor in spite of the residual xenon-135. This would go slowly at first, but with rapidly increasing speed as xenon-135 was destroyed—soon far too fast to be stopped by the slow movement of the control rods, and reaching a critical reactivity at which the prompt neutrons emitted in fission were numerous enough to exceed the number needed to maintain the chain reaction without waiting for the small proportion of delayed neutrons. The power doubling time would then have dropped to a millisecond or so and exploded the fuel rods themselves and the reactor vessel, with a huge increase in neutron flux that activated both surroundings and operators.

ACCIDENTS IN POWER REACTORS

Assessment of risk of breakdown

A major study of nuclear risks was made by a team run by Rasmussen (Wash 1400 1975) costing $20 million and taking several years.

The way in which the risk of a reactor accident can be assessed goes roughly as follows. The effect of failure of each unit in turn is considered, together with the probability that failure will occur. For example, the feedwater pump on the feedwater line is a standard item that has been used in many industrial applications and has therefore a known failure rate. Suppose that this averages once a year, and that the pump takes a day to repair or replace. Then the chance that the pump will not be working on any particular day is 1 in 365. If an auxiliary pump is added, the chance of both failing on the same day is 1 in 365^2, or once in 365 years. By using more pumps or more reliable pumps this probability can be reduced as much as desired. The second stage, still concerned with the same pump, is to consider what would happen to the rest of the system if all the pumps did fail. In this case feedwater to the steam generator would stop, the reactor would be automatically shut down and the emergency core-cooling system would be switched on. Then it is necessary to examine what would happen if *both* the pumps *and* the emergency core-cooling system fail simultaneously. Again, the average failure rates of the units comprising the emergency cooling system are known, and by appropriate duplication and upgrading the probability of both going at once is reduced to acceptable levels. This procedure was repeated for each piece of equipment, and included in each case the possibility or possibilities of incorrect response by operators.

Common-mode failures

'Common-mode' failures have to be watched for. Thus if the first feedwater pump stopped because its power supply failed, there would be no protection if the second pump had been installed using the same power supply. We have a great deal of experience of the provision of adequate spare power supplies for hospitals and other establishments in which continuous operation is essential, and the risks of multiple failure can be quantitatively estimated from experience.

It is worthwhile to give an example of a common-mode failure that the designers failed to see (*Nature* 1975). This occurred in the nuclear power station at Brown's Ferry, Alabama, in March 1975, when a workman searching for air leaks with a candle set light to some polyurethane foam insulation. The fire burnt for seven hours and disabled several key safety systems, the supply lines to which ran close together, at the same time. The calculated probability of all these systems failing at the same time by coincidence might have been one in a million per year but the possibility that all might have been caused to fail by a single apparently irrelevant event had been overlooked. Fortunately, none of the systems concerned were called on before the reactor was shut down, and no one was hurt, but the damage was extensive and a valuable lesson was learnt.

A later leak of 16 mCi of radioactivity from the same plant on 16 January 1983 received a lot of publicity because of the earlier accident. The activity, diluted with 1000 tonnes of water, was released to the Tennessee River and there diluted in a further four million tonnes of water. Even the first dilution will have brought the concentration below the lifetime concentration permissible for strontium-90 (the most toxic of the fission products), and the second would leave it 1000 times below the permissible level for the most toxic material known, when ingested, which happens to be the natural radioactive element lead-210, closely followed by botulinus-A toxin and radium-226.

Three Mile Island (1979)

This was the first civilian power plant in the world to have an accident that released radioactivity. The bald facts are that out of nearly 200 pressurised water reactors in the USA and elsewhere, with a combined capital value of several $100 000 million and operated for a decade or more, one has had an accident liberating enough radioactive gas to provide a collective dose of about 3500 man-rems; so that on the ICRP figures there is around a one in three chance that one person will die of cancer as a result. I do not think that any new industry ever has beaten such a good record; and in any other industry this would have been taken as conclusive evidence that, though no safety precautions can be perfect, they at least were adequate. In producing the same number of GWy of electricity the world's hydroelectric systems would have killed many hundreds in much larger accidents.

If this single two-to-one-against death near Three Mile Island goes the less probable way and someone does in fact die of cancer as a result of irradiation, the overall accidental death rate among the public so far would be about 0.003 per GWye.

The suggestion that the Three Mile Island accident was only by most improbable good fortune saved from growing to a really major accident killing tens of thousands of people is tantamount to saying that the probability of an accident killing 10 000 people is almost the same as the probability of an accident which nearly kills one person. This would be unprecedented in human affairs, and is hardly credible, but cannot be reliably refuted by any of the data that I have so far presented. Many aspects of nuclear power *are* unprecedented, and to get any understanding of the likelihood of a major accident developing from the actual minor accident we have to go into details of what actually happened, and of what else would have to happen to cause such an escalation of risk.

How it happened. The actual sequence of events beginning early in the morning of 28 March is believed to have gone as follows. At about 4 AM the main feedwater system malfunctioned, apparently as a result of failure either of the demineraliser or of the air supply to an air-operated valve. With the main feedwater supply system out of operation, the auxiliary feedwater system was to have started automatically. It did not, however, because a number of manual valves in the auxiliary system had inadvertently been left closed after a test of the system in the days prior to the accident. Without feedwater supply, the steam generators dried out, resulting in a rise in the primary coolant temperature and pressure.

The rise in pressure automatically shut down the reactor, stopping the chain reaction, but the intensely radioactive core continued to give out heat. Within seconds the pressure of water rose enough to trigger the opening of a relief valve. This valve stuck in the open position, providing a continuous release of radioactive feeding water from the main steel pressure vessel into a quench tank. This failure and a failure in the main feedwater system (both of a mechanical nature), and the malfunction of the auxiliary feedwater system, were the main initiators of the incident.

The emergency core-cooling system started automatically at two minutes into the accident sequence, and began to raise the coolant level. Within a few minutes the level indicator for the pressuriser misbehaved and went off scale on the high side when actually the pressure was too low. Accordingly an operator manually overrode first one of the two emergency core-cooling water injection pumps and, in a few minutes, the other, believing that already too much water had been fed in. Within two minutes of the emergency cooling shut-off, it appears that at the reduced pressure water in the pressure vessel began to boil into steam, lowering the water level.

Meanwhile, with flow through the pressuriser relief valve unchecked, the quench tank continued to fill, and some 15 minutes into the accident sequence its rupture disc blew open, ultimately releasing upwards of 40 000 litres of primary coolant on to the containment floor. From here water was pumped into a storage tank in the auxiliary building, and the water spilled again when the tank exceeded capacity. Radioactive isotopes of krypton, iodine and xenon began to escape to the environment through the building's ventilation system; vent filters trapped much of the iodine but did not stop the noble gases.

Escape of radioactivity. Six days later the NRC announced some cumulative data from environmental monitoring. No radio-iodine was detected in any of the 130 water samples taken by the NRC, the Department of Energy and the Commonwealth of Pennsylvania since the accident. The thyroid dose to anyone drinking the water was estimated to be less than 0.2 mrem. Eight of the 152 off-site air samples showed radio-iodine present, the largest activity measured being 2.4×10^{-5} microcuries (μCi)/m^3. The thyroid dose to anyone at the site boundary was estimated to be less than 50 mrem over a five-day period. Three plant employees were exposed to radiation beyond the 3 rem per quarter level; dosages of 3.1, 3.4 and 3.8 rem were received when the workers tried to drain and secure the auxiliary building. At least 12 other workers received considerable doses below 3 rem.

Were we lucky? It is widely believed that we were exceedingly lucky that the effects of the accident were not a great deal worse. Except for the fact that manual valves in the auxiliary system were inadvertently left closed and that some of the instrument failures must have been due to inadequate checking and poor maintenance and were presumably somebody's fault, it would be at least as true to say that we were exceedingly *unlucky* that the results were as bad as they were. Which of these views one takes can be just a matter of feeling; in considering the risks for the future we need to know, at least roughly, the actual probabilities, both of what did happen and of what else could have happened. The probability of the exact series of events that occurred was of course very small indeed, but the magnitude of the accident was roughly consistent with the expectation of Rasmussen's report for the number of reactor years that had gone by in the USA at that time.

The automatic responses of the equipment were entirely satisfactory, up to and including the perfectly effective switching on of the emergency core-cooling system, which responded correctly to what was going on, unaffected by the erroneous instrument readings in the control room which led the operators to override and switch it off. Like the Brown's Ferry accident, this was an unforeseen common-mode failure, as no thought had been given to the multiple effects of failure of one unduplicated level-gauge. Again as in the earlier accident, there will not in future be unduplicated critical recording instruments.

The Three Mile Island incident could easily have been several times more serious; a larger proportion of the Zircaloy tubes could have melted and reacted with steam to give a larger hydrogen bubble, exposing more of the core. More of the uranium oxide fuel pellets would then have been freed from their containing Zircaloy tubes, and more of the radioactive gases released. This would have had no direct effect on the pressure in the reactor vessel. There would merely have been a smaller volume of water in a greater volume of the more compressible steam or gas. It is not easy to see how the amount of activity escaping by the available routes to the outside could have increased by more than ten times or so. For a major increase in the activity released there would have to have been another breakdown of some quite independent piece of equipment; i.e. an improbable—but not impossible—coincidence.

Whatever the risk of a big nuclear accident now, it is less than it would have been if Three Mile Island had never happened. In the long run, the biggest financial cost of the accident may well turn out to have been the moratorium on more nuclear power stations in the USA, producing more of a burden on the use of expensive oil and bringing nearer the time when it is too expensive to burn at all.

Chernobyl

The first really serious accident in a civil nuclear power plant occurred at 1.23 AM on 26 April 1986, but was not reported in the west until 28 April when increased radiation levels were reported from environmental monitoring stations in Finland and Sweden, where some dose rates rose by a factor of 10 or more.

The Soviet Union has very properly been criticised for having given no warning to neighbouring countries of the possibility of widespread fallout. When they did give accounts of the accident, however, they were reliable. As the *Observer* two days later put it 'The crass defensive silence in Moscow, however, allowed full resonance to the terrific scream of jubilation that burst from the Western media—the Anglo-Saxon above all.' The Murdoch Press for example gave inimitable tongue, from the *New York Post's* reports of over 20 000 dead through the *Sun's* 'Red Nuke Disaster—2000 Dead Riddle.' I was myself rung up by two separate newspapers—unfortunately I do not remember which—shortly after the first reports of the accident. In each case I was told that the Russians had reported the deaths of only two people, while American experts (unspecified) quoted by the US United Press International estimated that 2000 had been killed, presumably by radiation. I explained that thousands could not possibly have been killed so quickly by radiation, while two could very easily have died from physical injuries caused by flying debris from the explosion—which turned out later to have been correct. My comments, which could have been and presumably were made by any other knowledgeable person, were not reported by any paper that I saw the next day. Instead, the *Daily Mirror* reported 'Many hundreds dead. All is death and fire' and 'Desperate radio message reveals full horror of Russia's nuclear disaster'. The *Daily Mail* headline said '2000 DEAD IN ATOM HORROR' and added that 'Reports in Russian danger zone tell of hospitals packed with radiation accident victims'.

How it happened. The uncontrollable power surge which led to the primary explosion in the RBMK reactor arose, like most major accidents, from a combination of a series of operator faults and a design defect.

A sensible experiment had for some time been planned: to find out for how long the huge amount of rotational kinetic energy left in the turbogenerator could supply a useful amount of electricity for important systems after steam to the turbine was shut off. This could be done only when the reactor was running at low power for maintenance, and this no doubt led to the erroneous assumption that it was safe. To make the study easier, the operators broke six important safety rules, failed to inform the director or chief engineer of their proposed actions, and disconnected all the automatic reactor shut-down devices and the emergency core-cooling system.

The core of a RBMK reactor consists of a moderating assembly of large graphite blocks through which there is a series of vertical channels, each containing a thick-walled tube for high-pressure cooling water and a central fuel rod of enriched uranium oxide pellets in a zirconium–niobium alloy tube. In operation few of the fast neutrons emerging from the fuel would be moderated (slowed) by the thin water layer, but after moderation in the graphite a lot of slow neutrons would normally be absorbed by the hydrogen in the water before they could return to the uranium in the fuel.

In full power operation the cooling channel contains a mixture of water and steam, the latter absorbing many fewer neutrons. In normal running, control rods automatically adjust themselves to compensate changes in the proportion of steam. Believing that manual control would be adequate at low power, the operators neglected to reset the automatic controls appropriately, and left the majority of the control rods so far out of the core as to need several seconds to reach it again. Soon after 1 AM the operators increased the water flow and reduced the power level of the reactor so much that it became badly poisoned by xenon-135. As was explained in describing the accident at Idaho Falls (page 150), xenon-135 absorbs the neutrons required for the chain reaction in a reactor, and after a shut-down can prevent recovery until almost all of it has decayed. There was not enough xenon-135 at Chernobyl to make recovery impossible, but enough to make it necessary to withdraw all control rods to the maximum extent when the operators wished to increase steam production again. At first slowly, but more and more rapidly, the xenon-135 was destroyed; the circulating water warmed up; and as more and more of it boiled and fewer neutrons were absorbed the power output rose faster and faster, soon with a doubling time of less than 1 second, far too fast for the slow-moving control rods to catch up.

As the xenon-135 was burnt away and the proportion of steam in the cooling channels increased, the critical point was reached when the prompt neutrons emitted in fission were numerous enough to exceed the number needed to maintain the chain reaction without waiting for the small proportion of delayed neutrons (see p. 323). The power doubling time then dropped to a millisecond or so; the remaining water exploded into steam; the fuel pellets and zirconium tubes burst apart; and the pressure either of superheated steam or of hydrogen from the reaction of steam with molten zirconium, or both together, blew fuel out of the moderator and the top off the reactor building, leaving the white-hot graphite to burn as fast as the now free access to outside air permitted.

The heroism of the Soviet firemen who fought the fire, knowing as they must have known that the radiation level was lethal, has been well reported. Their action must have very materially reduced the amount of radioactive material that finally escaped into the atmosphere. Twenty-nine of them died from the prompt effects of very large doses of radiation, and 200 others received doses causing serious illness. 11 of these later died, but the rest probably received radiation doses as large as those bomb survivors in Japan who added an extra 2–4% of their risk of dying of cancer within the following 40 years.

For some time there was a fear that the molten fuel, still giving out tens of megawatts of heat from its residual radioactivity, might melt its way through to the large volume of water (stored for emergencies!) below the reactor. This was however pumped away safely and replaced by several metres of concrete, cooled to prevent the still heat-producing fuel from melting its way out; again requiring the highest courage from those who worked against time below the reactor while it melted its way down towards them.

Within six months half of the working RBMK reactors had been shut down for modifications to preclude this pattern of behaviour, and further changes planned should make it impossible for such reactors to operate without waiting for the delayed neutrons produced by fission. No British power reactors can show the rise in activity with rise in output that caused the disaster at Chernobyl.

Lessons to be learnt. In a narrow sense, the Soviet disaster makes no difference at all to our own estimates of risk, since our present risk studies have been made on the basis of the detailed properties of our quite different types of reactor. The accident at Three Mile Island remains the only potentially serious accident in Western types of reactor.

If an explosion in the reciprocating engines of a large (and out-of-date) steamship allowed a piston rod to come adrift and sink the ship by breaking a large hole in the ship's side, it would not change our estimates of the probability of a different type of destructive explosion in a ship driven by a steam turbine. It could even, though temporarily, lead to a reduction of the actual risk for ships using turbines as well as for those using reciprocating engines by causing a general investigation of safety and maintenance procedures, which in all human operations become less thorough and effective after long periods free of accident.

Whatever the risks have been in our reactors, they will be smaller now. The chief reasonable cause for fearing greater risks than those calculated in the West is a very general one: that the Chernobyl accident resulted from something the Soviet nuclear engineers had not thought of, and it reminds any of us who had forgotten that accidents might occur in any of our reactors as a result of something that *our* nuclear engineers have not thought of.

The differences between our gas-cooled power-producing reactors (GCRs) and the Chernobyl reactor are many and vital, and are summarised below:

(1) In our gas-cooled reactors the heat is removed from the fuel by carbon dioxide gas, not by water.

(2) Graphite does not react with carbon dioxide until temperatures several hundred degrees above operating temperatures are reached, and the reaction yields little heat. There is no water in the reactor vessel of an AGR or a Magnox reactor.

(3) A sudden local rise of 100 °C in water at Chernobyl could have

increased the local pressure four times and exploded the water into steam, while an instantaneous local 100 °C rise in a gas-cooled reactor would increase local pressure (very briefly) by less than 20%, without change of state.

(4) In case of complete failure of all cooling systems, the residual radioactivity in the core of any gas-cooled reactor would take much longer to raise the temperature enough to melt the fuel and to boil off the volatile radioactive content; leaving more time for corrective measures, for warnings and for issue of iodine tables in any seriously threatened areas.

It is also impossible, though for different reasons, for the pressurised water reactors to suffer a Chernobyl-type accident. In these, the water itself is the moderator, so that a chain reaction could not continue if the water was lost, and steam bubbles in an overheated core will decrease, not increase, the reactivity. Heat from the existing fission products could cause a melt down if all of the emergency cooling systems failed; but we do not have a water store to which the molten fuel and steel from the pressure vessel could melt their way down. The idea of the 'China Syndrome' that the molten mass might continue down indefinitely is incorrect. Material from the outside of such a molten mass would be freezing out as it went, and calculations suggest that far from reaching the antipodes, or even reaching the centre of the earth, beyond which gravitation would no longer help it on its way, it would melt its way down about nine or ten metres. And there it would stay, at the bottom of a sort of ceramic tube formed by refrozen material from the melted soil or rock. When this had been filled with dry sand, topped by a few metres of concrete, a better primary location for high level waste could hardly be devised.

I am not claiming that serious accidents or serious releases of volatile radio-elements cannot occur in PWRs, though we ought to be able to reduce the risks recognised by Rasmussen's team a decade or more ago, but they could hardly compare with Chernobyl.

Finally, it is worth reiterating that in none of our reactors does the power output increase with temperature, which was the condition for an uncontrollable runaway such as occurred at Chernobyl.

There are some specific lessons that I think could usefully be learnt:

(1) Special attention should be given to periods when reactor power is being reduced, when there is a natural feeling that the dangers will be less than usual.

(2) Existing plans for evacuation in case of accidents at nuclear installations should be systematically overhauled, and nationally co-ordinated with those for chlorine and other chemical stores, liquefied fuel gas depots and oil refineries, perhaps under the aegis of the Health

and Safety Executive. The measures needed must inevitably depend on the weather as well as on the kind and scale of the accident, so that it would be undesirable to report details, for example of evacuation plans; but the existence of plans should be regularly published, together with the statement that the first instruction in all cases will be to stay indoors, shut off all fuel burning appliances since these suck in outside air—*you are safer indoors than in a car in a jam*—close all doors and windows, and listen to the radio or television for further instructions.

(3) It would be sensible to continue to site future nuclear power stations far from large populations, even though this increases transmission losses. In the very unlikely but not impossible event of a major accident, this would both reduce the number affected and increase the time available for warning and for decision on action. It is unfortunate that a similar siting of chlorine stores would be impracticable.

(4) Above all it must be made impossible to withdraw enough control rods far enough for prompt neutrons alone to maintain the chain reaction in any part of the reactor core.

(5) The final lesson from Chernobyl can be thought of as covering public relations after the fallout. Press, TV and radio will be demanding immediate and accurate information, which will never be possible. Accurate information takes time to get. In the worst possible accident however in a reactor without the positive power coefficient peculiar to the Chernobyl type, it will take sufficiently long for the reactor core to get dangerously hot for people at risk downwind of the reactor to be warned to stay indoors and wait for further broadcast instructions. Most important of all, when reliable data do become available, the actual risks involved as well as the raw measurements should be repeatedly broadcast.

I would like to mention here the plan proposed by Sir Frederick Warner, for elderly scientists to be willing to start the first exploration of a nuclear accident site, carrying measuring instruments. Then warning could be given of high levels of radiation and the need for protective clothing, short working shifts, etc, for workers trying to work on the site. He has already had a number of volunteers. Such older people know that the risk of any slow-growing cancer would be less for them than for younger people. A lot of careful planning and coordination with Civil Defence or other existing organisations would be needed, together with well-maintained equipment, but I think the idea is a good one and have added my name to the list. At the time of writing, no official approval has been received.

Death risks due to Chernobyl fallout. The primary explosions had an energy roughly equivalent to, but released rather more slowly than, that of

100–200 kg of TNT. They were not nuclear in origin like a nuclear bomb, but were steam explosions produced by the contact of the dispersed fuel with water. The main problem was to put out the graphite fire and to limit the release of radioactivity. In the following fortnight over 5000 tonnes of material were dropped by helicopter into the reactor: boron compounds to reduce the risk of any further chain reaction, dolomite to give off carbon dioxide and reduce the access of air to the burning graphite, and sand and lead to absorb heat and filter out radioactive particles.

Neglecting the less dangerous noble gases, around 70 million curies were released, about 3.5% of the total in the core, mostly between 26 and 27 April and 2–5 May, of which around 2% (3–4 tonnes) was deposited within 20 km, with the rest going to greater heights and distributing itself over an area containing 400 million people, 75 million being within the USSR.

Large numbers of people were evacuated from heavily contaminated areas. Concern was expressed outside the USSR that the town of Pripyat, close to Chernobyl, was not evacuated for a couple of days after evacuation of more distant areas had begun. This however was because, owing to the wind direction, no activity fell on Pripyat until the wind changed; when fallout began, nearly a thousand buses evacuated the town within a few hours. Many evacuees came back, preferring the radiation to the loss of their homes. On the other hand, a considerable number of women as far away as Kiev, which received some early fallout, demanded abortions for fear of congenital damage to their babies, which had been observed in much more heavily irradiated women following the nuclear bombing of Japan. Fortunately there has been no excess of congenital damage to children who were born either in Sweden or in Russia around Chernobyl (*Swedish Nuclear News*, Sept. 1988).

On the usual ICRP assumption the total radiation released would be expected to lead to a number of hypothetical cancer deaths over the next 40 years: 8000 to 34 000 in the USSR (W), 40 in the UK and 1000 in the rest of Europe. These would not be distinguishable among the numbers of cancers anyway expected due to other causes: 6 million in the USSR (W), 6 million in the UK, and 35 million in Europe. In the UK the number expected is a thousand times less than the number expected on the same assumptions in 40 years from our natural background radiation (Collier and Davies 1987).

A point worth remarking is that the total doses received everywhere except within ten or twenty kilometres of Chernobyl are below the lowest dose (5–25 rems) for which the survivors of the atomic bombing of Japan showed even an indication of an increase of cancer risk (see table 4.1). A great many people would be a great deal happier if they appreciated the fact that the radiation they have received from Chernobyl is within the range in which lies the main evidence for radiation hormesis, which, especially for the young—though not proven—might actually reduce their cancer risk.

A frightening thing about the disaster was the irregular distribution of the fallout. This necessarily depended on the distribution of the winds at the high level to which the hot gases rose before cooling and spreading, which changed during the nine days of their emission. At this point all of the radioactive atoms would have been adsorbed on solid particles or dissolved in small water drops. These would take a very long time to fall to the ground unless caught up in rain drops, so that the distribution at ground level was very small except where rain brought them down. As a result the map of the ground surface activity—by now nearly all iodine-131 and caesium-137—was the map of rainfall during the passage of the cloud, as is shown for the UK by figures 9.1 and 9.2 (Clark and Smith 1988).

A most commendably speedy study of the fallout in Britain was carried out by an NRPB team led by Dr Frances Fry. The surface activity was unimportant; but sheep eating contaminated grass were themselves contaminated by caesium-137 and 134, and by iodine-131. The last was not important as iodine-131 is short-lived and is concentrated in the thyroid gland which is not usually eaten by us. The results were published and broadcast, but the lack of any explanation of the figures given led to fears approaching panic levels, although if properly understood they would have been reassuring. Measurements of the activity of several nuclides were made for cow's milk and for meat from sheep and lambs, and were recorded in becquerels per litre or per kilogram respectively, but unfortunately were published only in this form. Essentially none of the general public, and very few scientists not concerned with nuclear physics, knew what a becquerel was, and naturally took it to be a unit of risk. To the uninitiated therefore the reports of hundreds or thousands of becquerels per kilogram sounded very serious.

Actual figures were of course essential, but there was no excuse for failing to explain their significance at the same time. The body of an average human adult contains 3000–4000 becquerels of radioactive potassium-40, with similar chemical behaviour to that of caesium. The average time for which radiocaesium atoms stay in the body is around six months. (I checked this on myself before the Windscale Inquiry by eating a contaminated fish caught near Sellafield and recording the excretion of Cs-137 over the following year.) The cancer risk resulting from eating 1 kg of lamb containing 1000 Bq is roughly the same as that of smoking a single cigarette.

A more reasonable plan than forbidding sale of lamb would have been

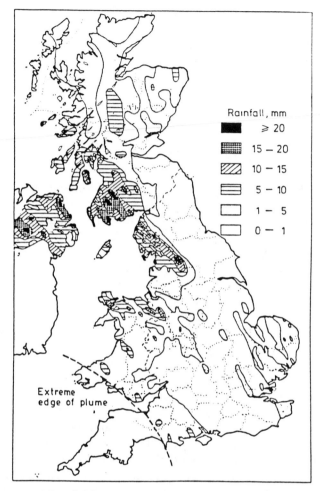

Figure 9.1 Rainfall in the UK during the passage of the Chernobyl plume, 2–4 May 1986, compiled from radar data and measurements at 4000 rain gauge stations. This is not the total rainfall for that period, only that which intercepted the plume, based on air concentration measurements. (Reprinted by permission from *Nature*, Vol. **332**, page 245. Copyright 1988 Macmillan Magazines Ltd.)

for the Government to require the purveyors of mutton and lamb to declare the radioactive content and leave the decision to the purchasers. It seems to me inexcusable that restrictions on movement and slaughter were still being applied in parts of Wales for animals well *below* 1000 Bq per kg; for example to lambs with an average level of 220 Bq/kg in Anglesey (*Y Swyddfa Gymreig News* 3/7/87).

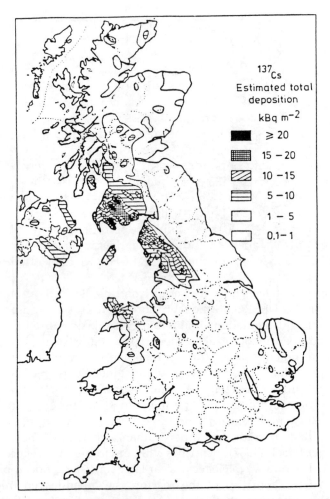

Figure 9.2 Estimated total deposition of ^{137}Cs (kBq m^{-2}) over the United Kingdom due to Chernobyl releases, calculated from a washout factor of 6.5×10^5, the rainfall data and air concentrations. (Reprinted by permission from *Nature*, Vol. **332**, page 246. Copyright 1988 Macmillan Magazines Ltd.)

I am not suggesting that the effects of Chernobyl in Britain will be zero. The extra caesium-137 will be with us for a long time. But the average dose to each of us in Britain is totally insignificant compared with the variations in doses from radon levels around Britain.

The Swedish authorities behaved even worse, putting a limit of 600 Bq/kg on reindeer meat. (Eating a kilogram a day would be equivalent in risk to smoking four cigarettes a week.) This destroyed the livelihood of very large numbers of Lapps, whose whole lifestyle for perhaps 10 000 years has depended on the reindeer; both as food for themselves and more recently as their sole source of manufactured goods bought with the proceeds of the sale of reindeer meat. It may also have led to a major ecological disturbance; the Lapps had for generations limited their killing to leave as many reindeer as could live on their available staple food, reindeer moss, a lichen which if overeaten takes a long time to recover. The deer and reindeer population increased from about 136 000 at the time of Chernobyl to 400 000 by the end of 1986, causing millions of pounds worth of damage to Sweden's forestry plantations (*Times* 18/10/86). Unlike the British however, the Swedes—after the Lapps' lifestyle had been disrupted—raised the permissible limit to 6000 Bq/kg, with a requirement that the amount of contamination of meat for sale should be recorded.

Owing to the vagaries of winds, the reindeer in Finland did not reach the Finnish action level, and the lives of the Finnish Laplanders were not upset.

To conclude this section, it is clear from the figures that outside the borders of the USSR enormously more damage was caused by unreasonably tight regulations than would have been caused if no regulations had been made at all—other than a requirement to report the radioactive content of meat for sale together with the ICRP expectation of risk per kilogram of meat in understandable terms such as the equivalent number of cigarettes or hours of travel in jet planes.

The Soviet authorities are planning to turn the evacuation zone, roughly 60 km in diameter round the Chernobyl reactor, into a national park, in which farming and other human activities will be banned. A report by Izrael (1987) describes the environmental effects of the accident. The only visible effect a year later consisted of dead pine trees within a few kilometres of the reactor, the broad-leaved trees (oak, birch, etc) appearing unaffected. (An experiment in the USA over 20 years ago had shown that the dose to kill pine was about 1000 rads, spread over some months, compared to about 10 000 rads for oaks.) Soil animals living in the forest litter were affected at 7 km from the reactor, but none of the groups showed drastic reduction in numbers. No visible changes were seen in the grassy plants, and no changes were recorded in the animal populations of endangered species (i.e. in the Red Data Books). Doses to the various species of fish in the area were too low to cause noticeable radioecological effects.

Part of this territory is being used for agriculture, with experiments aimed at reducing radionuclide (mainly Cs-137) transfer to man. For example, feeding animals with uncontaminated fodder for 45–60 days before slaughter usefully reduces the Cs-137 content; and large doses of mineral fertilizers, lime and sorbents in ploughed land reduce the take-up of caesium by plants. A few plants, for example cucumbers, take up so little that locally grown cucumbers could safely be offered to visiting scientists.

In another account it was reported that foxes (and presumably small rodents) were doing particularly well; any adverse effects of radiation are clearly smaller than the favourable effects of evacuation of the human population.

A long-term project has been established in Russia to study the 135 000 population evacuated from near Chernobyl. This is one of the largest and most complex epidemiological investigations ever undertaken. It will be a long time before any observable effects can be expected.

Estimation of risks of future PWRs

While I have explained why I believe that a major disaster is highly unlikely, I have not described such disaster as impossible. In the previous chapter I have mentioned only in very bare outline the possibilities of bad accidents in other power production systems. For nuclear power the 'worst possible', though very unlikely, is dealt with in rather greater detail in the rest of this chapter. Most people understand how a devastating forest fire spreads, or how people are killed by a bursting dam, but unfamiliarity with concepts of radioactive nuclei makes them more frightening. This is why I have gone into such detail below, not because it is a more probable scenario than other ways in which sources of power can kill. I may also be biased by emotional factors. I can without difficulty describe the effects of future risks to people of dying in their beds before their time; I would be too squeamish to describe in detail the ways in which people would die in a refinery fire. I shall now go on, therefore, to consider just what damage could be done if a large fraction of the active material did actually escape or become airborne.

The first requirement is to estimate what fraction of the fission products might escape. A very thorough and detailed study of the amounts of fission products that might escape following several improbable but possible breakdowns of effective containment has been made for the proposed PWR at Sizewell (Gittus 1982). A series of possible coincidental failures of systems that could lead to core melting are considered as initiating events; for example a loss of off-site power together with the failure of all of the diesel generators and of the steam turbine driven auxiliary feedwater pumps for a long enough period for residual heat production to boil away all of the water in the secondary and primary cooling circuits.

The possibilities of several ways in which abnormal steam explosions or explosions of hydrogen outside the pressure vessel could damage the containment building, together with the failure of the spraying systems designed to remove airborne fission products, are then considered. Finally the proportions of the more important fission products which might be able to escape from the damaged containment are estimated for four different categories of damage. It is only in the categories for which the containment fails, with a total probability of about 2 in 10^9 per year, that more than 10% of any radioactive material other than the inert gases could be expected to escape. For practical purposes these very small probabilities mean that a serious escape of radioactive material will never happen—and not just because 2 in 10^9 is extraordinarily small. The reason for this is that each of the things which have to fail together for the major accident will fail many times separately in minor accidents. For example, cracks in pipes due to corrosion and vibration were recently found in some American reactors; and as a result all reactors of the same type are being examined and will be repaired, and measures will be taken to prevent such failure next time. In saying this I am relying on the great financial cost of even minor damage to a reactor core. The universal lack of enthusiasm for spending large sums of money unnecessarily is an extraordinarily effective incentive to improvement.

It would be nice to think that all of this shows so conclusively that the accidental death rate from the operation of nuclear power stations is negligible that nothing more needs saying. To a lot of people however, such statements as 'such-and-such a reactor would have to run for 500 million years to make a major accident probable' appear nonsensical. This appearance is right; but such a formulation is nonsense not because one in 500 million for the risk in this coming year was wrong but because the whole of our analysis assumes the people, the equipment and the environment to have the characteristics that we know today. The laws of physics and chemistry are not expected to change in 500 million years; but we shall and so will our machines. Averages over time can usefully be made only if nothing important changes over the time. Thus we can speak of 2000 reactor years whether 2000 reactors have operated for one year or 200 have operated for 10 years—or perhaps even 100 for 20 years—but not of one reactor, or even of a succession of reactors one at a time, operated over 2000 years. The later reactors would be different, and better designed, beasts than the earlier ones.

Apart from the mere form of words, it is dangerous to put uncritical faith in a statement such as 'there is only a chance of 1 in 500 million that a really serious accident will happen'. It neglects the possibility that with such tiny risks the experts who understand and believe the figures will stop bothering to examine the effects of minor changes in operation or design because of the huge safety factors. It is quite unlikely that an unnamed foreign power, or a group of industrial competitors, would use threats or bribes to persuade all the technical people concerned to pass as safe a pressure vessel with serious cracks. But I would not want to bet 500 million to 1 against it.

The figures and the detailed calculations made to obtain them are nevertheless immensely valuable, not for their exact magnitude but for their implications. With far more understanding of the chemical behaviour of the fission products, they back up the conclusions reached in the major study of nuclear risks by Rasmussen referred to on page 151. The conclusions clearly show that the nuclear industry would be wasting time in trying to reduce still further the risk of major accidents, and should instead be concentrating on avoiding the far more probable minor accidents which cost disproportionate amounts of both money and popular confidence—and though trivial in themselves are necessary precursors of larger accidents.

Effects of a hypothetical major accident

While a major accident is exceedingly improbable, no one could claim that it is impossible. Since it is not actually impossible, it will be well to discuss the effects on people if a major accident did occur. Fortunately very extensive published information is available for quantitative estimation of this (Glasstone and Dolan 1977). This information has been used by Fetter and Tsipis (1981) to compare in detail a major release of activity from a big nuclear power station with that from a nuclear bomb.

A reactor producing 1 GWy of electricity will be producing about 3 GWy of total heat energy, which represents the fission of about a tonne of uranium-235 or of plutonium—either of which will leave a tonne of fission products. The vast majority of these are very short lived and will have decayed during the year of operation, but most of the long-lived ones will still be radioactive. Accordingly, the total radioactivity remaining an hour after shut down will fall only seven or eight times further in the next month. The total activity initially in a reactor is very much less than that liberated by fission of a tonne of uranium in a bomb immediately after the bomb explodes, but after a decade or so it will be the same. The fission of a tonne of uranium would represent a very large bomb—about the size of the Bikini H-bomb, which had an explosive output estimated at 15 million tonnes of TNT, half of it due to fission of the uranium outer shell.

Fetter and Tsipis consider a nearly worst case, a situation in which a third of the total activity in a 1 GWe reactor escapes in an hour or so, following a complete melt-down and the destruction of both the reactor vessel and the containment building of a PWR. Isotopes of krypton and xenon and perhaps a little of the iodine would be dispersed as gases; the rest of the iodine and almost everything else would condense into solid particles. The initially high temperature of evaporation would lead at first to all the airborne material rising to a considerable height, but the column of fine smoke-like dust would soon be diluted with cooler air, and the radioactive content would eventually be deposited on the ground over a long distance downwind of the accident.

At this point the expectations of Fetter and Tsipis become unreliable. Using data from the bomb tests in the USA they assume that this would lead to a simple cigar-shaped distribution following near-surface winds. In the bomb tests, near-surface bursts took up with them many tonnes of finely divided soil, to which the greater part of the fission-product atoms were adsorbed. This heavy material fell fairly soon to ground level, before any change of wind occurred. As has been made uncomfortably clear by the Chernobyl catastrophe discussed above, the finer particles and individual atoms released from a burning reactor are carried to a considerable height and may travel a long way before being brought down by rain.

The explosion of a pressurised water reactor would, unlike the RBMK Chernobyl-type graphite-moderated reactors, necessarily stop at once the chain reaction of the core, since the pressurised water is the essential moderator making the chain reaction possible, even if all the control rods had been completely removed. The core would still be hot enough to evaporate volatile fission products such as iodine-131 or caesium-137 and the noble gases krypton and xenon.

The major release of activity under discussion could occur only if a xenon-135 type of runaway reaction produced the needful energy in the fuel before its transfer to the water blew the reactor vessel apart. There are probably simpler and equally effective means of preventing this, but it could be reliably insured against by automatic insertion of control rods, irreversible for several days, whenever the power output of the reactor fell below a critical level—all in triplicate.

However, as I stated above, I cannot prove that a serious accident is impossible—it is difficult to make anything foolproof against a really ingenious fool—and with 800 people per square mile in England the release of a third of the activity in a Sizewell B type of reactor might mean 400 deaths on ICRP figures, spread over a few decades, independently of the distribution within the country.

Gittus (1982) in his final report to the Sizewell Inquiry estimates the number of *early* deaths caused by degraded core accidents at Sizewell 'B' to be about 100 000 times fewer than accidental deaths in the affected area from causes not associated with the reactor. Similarly the individual runs a risk of fatal cancer not due to the reactor which is about 40 million times greater than the risk due to degraded core accidents at Sizewell 'B'. The uncertainties in the results of this study are small compared with these margins.

Since both were concerned with deaths among humans, neither Gittus nor Fetter and Tsipis considered in any detail the effects of contamination on agriculture and fisheries that would follow from a worst case accident. This has now been done by Taylor (1984) who shows that in some conditions the crops on large areas of agricultural land would require drastic restrictions, and that in some of these conditions fishing in the North Sea might have to be greatly reduced. (Industrial crops such as sugar beet and perhaps rape could be grown, as the edible or useful parts would not contain radioactive elements.)

If the wind were blowing out to sea at the time of the accident, the southern North Sea would reach concentrations of caesium-137 ten times greater than that in the Irish Sea off Sellafield, and the total collective dose could be much greater than that so far incurred from Sellafield. In practice of course there would be a ban on fishing for two or three years in affected areas until the natural currents had washed the activity out into the Atlantic and diluted it to an unimportant level.

The report assumes that population doses will be kept below 0.5 rem per year to individuals by the evacuation and the restriction on fishing and crops, which would keep the resulting cancer death rate in the part of the population concerned to about a fifth of the normal excess cancer death rate in British cities over that in the countryside. This would clearly be desirable in our present uncertainty of the risks of radiation at low dose rates, if cost is not considered. It is a pity however that in a carefully quantitative report Taylor exaggerates badly the relative importance of the 1 in 20 000 chance of cancer due to a 0.5 rem dose by stating that the general risk of death from cancer for an average person is 1 in 2000, when in fact it was then 440 in 2000.

Given the measures proposed, it would be unlikely that as many would be killed by these indirect means as would be killed in the more direct ways already described, but the financial costs and the social disruption would be very large indeed.

Figure 9.3 shows the variation of accident probabilities with accident size based on the Gittus study. The very rapid fall of probability as the number of deaths increases results from the very large number of independent faults required for the larger accidents. The largeness of this number in its turn results from the enormous effort put into safety precautions since the work of Rasmussen.

The long-term death rate due to PWRs must be very much less than 0.01 per GWy. The actual risk is thus so small compared with the other risks we run that it hardly matters; but it may be worth noting that even this small risk applies only to the PWR and similar reactors. An AGR, with a gas coolant and a reactor vessel of concrete 14 ft thick reinforced by monitored and replaceable steel cables, has no way of bursting its reactor vessel. Although the pipework carrying the coolant could be damaged or corroded, the total weight of fuel and moderator is so much greater that it would take far longer to warm up and would reach only a much lower equilibrium temperature, well below the melting point of either the stainless steel cans or the uranium oxide fuel. Accordingly, an accident leading to a release of radioactive material on the scale just described is not merely unlikely but impossible.

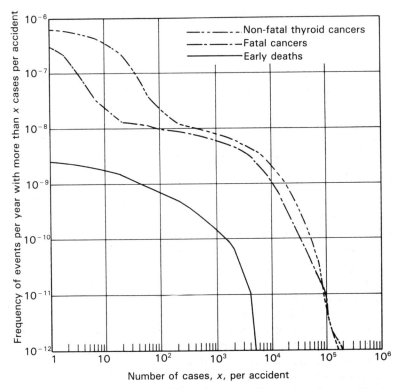

Figure 9.3 The probability of the important effects of nuclear accidents for the proposed PWR at Sizewell.

SUMMARY

The main risks discussed in this and the previous chapter are illustrated in figure 9.4. The probability of killing the number shown on the x axis in the coming year is shown on the y axis, for an output of 1 GWye from oil-fired and from pressurised water reactor power stations. The data for the oil-fired stations are those calculated by Cohen and Pritchard (1980) from the observed records of catastrophic accidents in oil refineries. The curved part of the nuclear risk line shows the relevant part of the

Rasmussen prediction (Wash 1400 1975) while the broken part shows a line through the point represented by Three Mile Island parallel to that of a typical man-made hazard. To compare the nuclear risks with others, I have repeated the chlorine line shown in figure 8.1, but have had to reduce its risks by 1000 times.

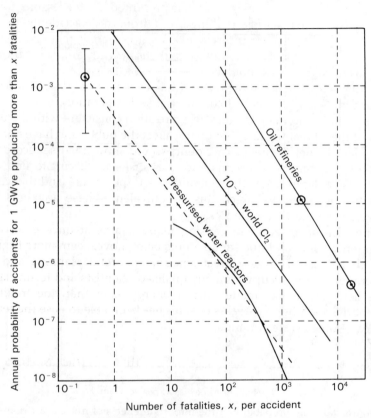

Figure 9.4 The observed accidental risks to the public per GWye from oil-fired stations, and the estimated risk from PWRs. The risk from 1/1000 of world chlorine production is added for comparison. Only one point, derived from Three Mile Island, is available for the nuclear industry; the vertical line through this point shows the 90% confidence limits for the probability of another accident of similar size.

All of the risks are tiny, even oil refineries giving only a one in 1000 chance of killing 100 people in the next year; and as indicated in the discussion above the nuclear data are (fortunately) quite inadequate, owing to the lack of serious accidents. Nevertheless, if PWRs were as prone to lethal accidents as oil refineries we should have had a very considerable number of accidents by now, killing a lot of people.

The most probable estimate of future risks is obtained not by theory but by experimental observation. Over the last 4000 reactor years of Western reactors, one accident (TMI) may lead to one hypothetical death. Several other accidents occurred which gave no significant radiation dose to anyone but which presented useful lessons for the future. If we have learned nothing useful from these accidents, the best estimate of the risk over the next 4000 reactor years is that one accident will occur that may lead to one hypothetical death and several others will occur that will kill nobody.

I know of no major industry that has ever had so good an accident record in its first decades of commercial-scale operation.

Coal mining accidents would give a relation consistent with the upper part of the oil refining line, but do not affect the public and have not been included. Liquefied fuel gas accidents would show a worse record than oil, but the variety of origins and uses make it difficult to find comprehensive figures for the quantity used per year—and hence the number of GWye to which this is equivalent. These accidents have therefore had to be omitted.

The idea of an accident causing many deaths at once is always a frightening one, and nuclear accidents being novel seem more frightening than most. The very great amount of careful planning and consideration that is occupying many qualified scientists and technicians in this field has in fact made nuclear energy safer than the traditional methods; and I hope that this chapter has been able to ease the fears of at least some Western minds.

Chapter 10

The Numbers Killed in Routine Power Production

INTRODUCTION

In the last two chapters it was shown that most of the energy industries are capable of producing very large accidents of types specific to themselves. The probabilities of such accidents are small except for the hydroelectric systems, and electricity from a large hydroelectric scheme is so cheap that the authorities are stoically willing to accept the extra risks (to other people). However the difficulty of deciding reliably on the relative risk is not very important. A GWy of electricity is worth around £300 million so that whether the number of deaths due to accidents per GWy is 0.01 or even 1 (making £300 million per life) the value received per life lost is so large that decisions are unlikely to be made on the basis of the actual differences, however much they may be affected by public perception of the differences.

When we look at the risks involved in routine production, the situation is quite different, in two major ways. The first is that the numbers killed per GWy are much greater and the second is that we can usually tell from past experience how many have been or are being killed. We do not have to rely on theoretical estimates of doubtful reliability and enormous cost.

I shall consider only the risks from those energy systems that are already producing, or which can soon be expected to produce, annual amounts of energy equivalent at least to that produced by a million tonnes of coal.

THE DANGERS ASSOCIATED WITH COAL MINING

Mining for coal is one of the known dangerous occupations. Although, as shown by table 10.1, the death rates are decreasing, about 25 miners are still killed in accidents each year, and 350–400 cases of the incurable lung condition known as pneumoconiosis are diagnosed annually (*H*

Table 10.1 Fatal accidents to employees in the coal industry (*Hansard* 1977 and NCB personal communication 1988).

Year	Number of deaths	Rate per 100 000 employees (total)
1957	396	54
1958	327	45
1959	348	51
1960	317	51
1961	255	43
1962	257	45
1963	254	47
1964	198	38
1965	216	45
1966	160	36
1967	151	36
1968	115	31
1969	100	31
1970	91	30
1971	72	24
1972	64	22
1973	80	30
1974	48	19
1975	64	25
1976	50	20
1977	40	16
1978	63	25
1979	46	19
1980	42	17
1981	35	15
1982	38	17
1983	30	15
1984/5†	22	13
1985/6	27	18

† 15-month period including miners' strike.

and S Statistics 1989). This is a slow killer, and is perhaps as undesirable in reducing the quality of the last two or three decades of life as it is in causing premature death. The number of deaths directly attributed to it on death certificates has been 250 to 300 per year for a long time, but it is only recently that it has been officially recognised as industrially caused and hence qualifying for compensation. In 1974, the relatives of 630 miners (or retired miners) received death benefit under the Industrial Injuries Scheme. The number will have included 300 or so whose deaths were not directly due to pneumoconiosis or any of the less common diseases arising from coal mining, but whose deaths from other causes may have been accelerated thereby. Now most newly diagnosed cases although receiving 10% disability benefit, show no actual detectable disability except in the now rare severe form—complicated pneumoconiosis. Only a tiny fraction of post-mortems show that death was actually due to pneumoconiosis, but if this has earlier been diagnosed it is likely to appear on the death certificate and to be recorded as a pneumoconiosis death in mortality statistics (Dr P D Oldham, private communication 1984). There is no way of telling whether the less serious forms of pneumoconiosis reduce the expectation of life before death occurs from other causes.

The 80 million tonnes of coal, mainly from large collieries, used to produce electricity in Britain each year are estimated to result in about 19 deaths from accidents and perhaps 30 from pneumoconiosis. Besides the miners killed in extracting coal, there will be deaths resulting from the transport of coal from pit to power station (about 2.5 million tonnes per year for each GWye) and a few more from miscellaneous accidents in the generating stations themselves.

Deaths from mining accidents and pneumoconiosis then work out at around 0.7 per GWye, with a further 0.2 per GWye for each of transport and work in the generating stations, giving a total of 1.1 per GWye.

Table 10.2 Industrial accidents to employees in one year (selected industries) (*H and S Statistics* 1989).

Sector	Year	Fatal accidents	Deaths per 10 000
Construction	1986–87	99	1
Coal mining	1987	27	1.8
Agriculture, forestry and fishing†	1986–87	28	0.9
Metal goods, engineering and vehicles industries	1986–87	38	0.17
Oil and gas extraction	1986–87	6	2.1‡
Energy and water supply industries	1986–87	30	0.6
Nuclear fuel production	1986–87	1	0.6
Paper, printing and publishing	1986–87	1	0.02
Transport and communications	1986–87	34	0.25
Construction—self-employed	1986–87	22	4.3

† Excludes sea fishing
‡ For the bad year of 1988 this would be nearer 60 per 10 000

The magnitude of the accidental risk in mining as compared with that in other occupations is shown in table 10.2. It will be surprising to many people that the accidental death rate among miners is not outstandingly high, and does not compete with the risk of establishing North Sea gas and oil or even with cross-channel ferry transport. Typically, public perception overrates accidents affecting a number of people at a time and underrates the quiet individual deaths.

EFFECTS OF COAL USE ON THE PUBLIC

The injuries to the public due to coal burning are less direct, and hence are only very approximately known. Nearly all of them arise from the inhalation of air polluted by the combustion products of coal. After the effective elimination of the major epidemic diseases, and until fairly recent times, this may have been one of the largest single killers in the big cities of Britain. The effects were at their worst in the earlier part of this century, with a steady death rate from respiratory diseases, punctuated with disastrous peaks during and after exceptionally bad 'smogs'. A smog (smoke-laden fog) occurred when the air near the ground was cooler and denser than that higher up, so that the smoke from the chimneys remained near the ground instead of rising and being blown away. If additionally there was little wind the polluted air could accumulate for long periods. In December 1952, 4000 deaths occurred in London in a week of smog and the following week over and above the 'normal' death rate in the same period of previous years—and of course even in these 'normal' periods there had been a steady and hence neglected death rate from respiratory diseases.

The main culprit was the open fire—although industry and the railways produced more than half the total smoke; for each tonne of coal burnt domestic chimneys produced twice as much smoke and discharged it at a lower level. In most of the countries on the continent of Europe, with colder winters and more expensive coal, more costly but far more economical closed stoves were used. These used less coal and burnt it more efficiently, producing much less smoke per kilogram of coal.

A great many injurious chemical compounds, including carcinogens such as benzo-a-pyrene, are produced by coal when inefficiently burnt, but it is thought that the chief killing agent in a smog is sulphur dioxide, or the sulphuric acid and sulphate to which it is oxidised in moist air. Sulphur in various proportions is a component of all coals and fuel oils, the average content in coal for UK power stations being 1.5%, and in fuel oil about 3%. The sulphur is burnt to sulphur dioxide when the fuel is burnt, a big coal- or oil-fired power station giving out about 40 000 tonnes a year. Sulphur dioxide is a gas and goes up the chimney with the rest of the products of combustion. By itself it does not seem to be harmful in low concentrations; it dissolves immediately, molecule by molecule, on contact with the moist surfaces of the bronchi or lungs of people inhaling it, and is almost instantly made harmless by oxidation to sulphate which is a normal and essential component of our blood. It is however rapidly and firmly adsorbed onto the smoke particles, and a smoke particle covered with sulphur dioxide molecules and landing on one of the sensitive cells of the respiratory system may produce so high a local concentration as to overwhelm the local neutralising machinery.

The smoke particles in a moist atmosphere act as nuclei for water droplets and prevent them from evaporating. The result of this was the formulation of the thick, yellow, acid fog known as a 'London particular', although all of the big industrial cities were capable of producing it. A dentist friend of mine who worked in Manchester after the war found, after walking through one such fog, that the acidity of his own saliva was nearly 100 times the acidity at which the enamel of teeth begins to dissolve. Fortunately teeth dissolve very slowly, and our saliva can rapidly neutralise acid—so long as the acid is not constantly renewed (otherwise we would not be able to eat the more acid fruits; lemon juice or vinegar are quite as acid as the Manchester fog).

Although conditions as bad as this were rare, the continuous attack on the lungs and bronchi was far more serious than the occasional and short-lived attack on our teeth. The annual death rate from bronchitis and related conditions in Britain was double that on the Continent— bronchitis was there known as the English disease.

In the late 1940s I tried to make an estimate of the actual death rate by comparing the death rates from bronchial conditions in the cities with those in the countryside. This was made very difficult by two things. The first was the mobility of the British population. The two towns that I found with the highest death rates from bronchitis were Cheltenham and Bournemouth. Neither of these was industrial; they were places to which retired city dwellers suffering from bronchitis were recommended to go to breathe unpolluted air. When they eventually died their death certificates, provided by the doctors who had been treating them, would give chronic bronchitis as the cause of death. The second difficulty was that in a big city where many of the elderly poor had chronic bronchitis as a matter of course, the certificates might well give only heart failure as the immediate cause of death—which of course it usually is—without recording the lung and bronchial conditions which were taken for granted and which added to the strain on the heart causing it to fail. One medical estimate of the time suggested that for every death registered as due to bronchitis it should have been mentioned as a contributory cause in two more.

Accordingly I got the ratios of deaths in large towns to those in the countryside from France and Sweden, where chronic bronchitis was rare and more likely to be recorded; and assumed that the same ratio would have held here if the records had been more nearly complete. From this I calculated a probable excess of deaths of 20 000 a year in British cities above those which would have occurred among the same number of people living in the country. It is unlikely that the result was too large. Indeed, with our greater industrialisation and our open fires it was more likely to have been too small. It could well have been three times too small, even neglecting this bias, if we count in the heart failures for which respiratory responsibility was not recorded.

PRESENT DEATH RATES FROM AIR POLLUTION

The situation now is of course enormously improved. The Clean Air Act triggered by the 1952 London disaster, and the parallel shift away from open fires to central heating and gas or paraffin (kerosine) stoves, has led to a reduction of small-scale burning by at least 100 million tonnes a year. In 1981 chronic bronchitis was still responsible for 15 600 deaths in England and Wales, but there is no useful evidence as to the causes or the distribution between town and country, although the total death rate was a few per cent higher in towns than in the country. It is very difficult to decide how serious are the non-malignant respiratory effects of urban air pollution now, and of these effects how many are due to small-scale coal burning and how many to vehicle exhausts. On balance it would seem that only the more susceptible people are now likely to die of chronic bronchitis as a result of air pollution, and that these may not add up to more than a few hundred per year.

Two thirds of the coal now being used in Britain is burnt in electric power stations. The Battersea and Bankside power stations in London (both now closed) had scrubbers to remove over 97% of the the sulphur dioxide produced; but the bigger and more modern stations have instead used high chimney stacks, to release the sulphur dioxide so high up that it has been diluted and blown for a long distance before it can reach the ground. The modern furnaces, burning more efficiently than the old, produce little smoke—which was largely unburnt carbon—but at their higher combustion temperature do produce oxides of nitrogen which react with water in the atmosphere to make nitric acid. Like the sulphuric acid similarly made from sulphur dioxide, and the sulphur dioxide itself, this is not harmful to human beings in the absence of smoke, and when sufficiently dilute.

Apart from sulphur, there are traces of many other toxic elements in coal, including arsenic, beryllium, cadmium, lead, thallium, thorium and uranium. Some of these are volatile at furnace temperatures, but will be firmly adsorbed on refractory solid particles (mainly aluminosilicates) as the hot gases from the furnace cool down. The smallest particles, below 1 micron (μm) in diameter, which will reach the lung itself when inhaled, pick up a disproportionate share of these toxic elements owing to their large ratio of surface area to volume, and may therefore have a surface concentration of trace elements many tens of thousands of times greater than the concentration in the original coal. Such a particle landing on a single cell of one of the bronchioles or alveoli (minute air sacs) in the lung might do serious damage, although at present levels of pollution a very small proportion of such cells is likely to be affected.

Even the coal-powered stations with high stacks are still killing a few people. Estimates of the risk vary enormously. This is partly because the actual risk due to any one power station depends very greatly on the prevailing local weather and the distribution of population downwind; and partly because there is uncertainty as to the variation of death rate with the concentration of the effluent chemicals. In fact we can put far more reliable limits to the dangers of radiation, new in power station technology, than we can to the dangers of small doses of chemicals that have been emitted for a couple of centuries. As a result, figures varying from 0.1 to 77 deaths per GWye have been given for different stations by different authorities. Even the largest might not be absolutely impossible if the prevailing winds and the terrain were such that all of the effluents from the 3.5 million tonnes of coal used annually by a 1 GWye station returned to the ground in a city built along the floor of a river valley liable to frequent fogs. But 77 per year does not seen even faintly likely for any existing British coal-fired station. In the report by the Health and Safety Commission (1978) the sensibly vague conclusion was as follows: 'However, the small contribution due to the ground-level concentration of pollutants from a large fossil-fuelled power station may cause a few of the thousands of deaths occurring every year from all causes in the surrounding district' (pp. 16–17).

Perhaps the best figure to use in making the decisions that have to be made is five deaths per GWye. This would mean a total of about 200 deaths per year in Britain. Five is a fair quantitative representation of 'a few'. This would avoid the risk of exaggeration, but neglects entirely the possibility that any deaths may result from the tiny doses received by 100 million or so people on the Continent from British power stations, and the synergistic reactions with other sources of air pollution that are to be expected.

The most probable estimate for coal plants in the USA itself has been given as 15 per GWye by Hamilton (1983).

In view of the very considerable and demonstrable effects on wildlife at great distances mentioned in a later chapter, a figure as low as five per GWye suggests that human beings are a great deal tougher in their resistance to poisons rarely encountered during their evolution than are most other living things. This is not a sideways hint that we ought to regard five deaths per year as a dangerous underestimate; I really think it is as likely an estimate for Britain as can yet be made. We probably *are* a great deal more resistant to miscellaneous poisons than are most other living things, just as we can tolerate a far larger variety of environments in which to live.

CANCERS PRODUCED BY FOSSIL FUELS

Here, even more than in the case of the respiratory diseases which may often become important in late middle age not long before they kill, the deaths that we are seeing today are likely to be the result of carcinogens inhaled or swallowed three or more decades ago. It is impossible to distinguish between the effects of pollutants obtained from burning coal and those from burning oil or gas. Nearly two tonnes of benzo-a-pyrene per GWye, sufficient to cause two million deaths from lung cancer, are emitted by coal-fired power stations, but little of this will be inhaled (Atkins *et al* 1979). Nevertheless, the magnitude of the pollution is great enough to produce easily measureable effects.

The results obtained by Frigerio *et al* (1973), which were shown in figure 4.3, are most easily interpreted as meaning that over 40 000 deaths per year in the USA are being caused by some form of pollution in the lowland states, probably pollution by coal burning or by petrol and diesel fumes†. There is now no way of apportioning the 40 000 between different sources of pollution; we simply do not have the necessary information.

What horrifies me is not large numbers; the 40 000 extra deaths are quite a small part of the cancer deaths occurring annually in the USA. It is the apparent lack of *practical* concern among the many organisations which claim to be concerned with the damage we are doing to our environment. It would clearly be expensive, but a large-scale Federal study of the distribution and nature of the known airborne and waterborne carcinogens, together with the associated cancer rates, should be able to

---

†The average death rate from cancer over the whole USA is 150 per 100 000, and that in the seven high-altitude states is 125 per 100 000. If we assume that the difference (25 per 100 000) is due to pollution, then over the nearly 180 million white population this comes to nearly 45 000 deaths.

identify at least the most important compounds and their major sources. The cost would be less, and the time taken might not be much more, than the cost and time of building one more power station—either coal or nuclear.

Considerable efforts were, and some still are, being made to clear up many of the most obviously undesirable pollutants. But the most obvious pollutants may not be the most vital, and the information which the US Environmental Protection Agency uses to choose its objects of attack is sometimes inadequate and indeed hardly justifies the efforts being made.

In Europe too a similar criticism can be made. Several epidemiological studies of the distribution of cancers are being made. These, like the studies of respiratory diseases, show larger proportions in the large towns. An international symposium on air pollution and cancer at the Karolinska Institute in Stockholm in 1977 (Cederlöf *et al* 1978) reported that there were annually five more lung cancers per 100 000 males in the big cities of Europe than in the countryside. This would probably mean an extra 1000 cancer deaths in Britain per year, which, as will be shown later, is likely to be an underestimate. Again, although air pollution is almost certainly responsible, and benzo-a-pyrene is a likely causative factor, there is no way of confirming this or, if confirmed, of determining whether it arises mainly from vehicle exhausts or from the local industrial or domestic burning of fossil fuels. The proportion for which coal- or oil-fired power stations are responsible is likely to be less than one per GWye, over and above the five deaths per year suggested above.

What is important is the very much lower death rate due to the routine operation of power stations of any kind than that due to the energetically more efficient direct use of fossil fuels for small-scale applications. As Amory Lovins point out in his book *Soft Energy Paths*, domestic heating by the direct use of fossil fuel is twice or more than twice as efficient in its use of primary energy as would be heating by electricity derived from a power station of any kind with a thermal efficiency between 30% and 40%. But if electricity alone were permitted for domestic heating and other small-scale heating applications, and for all forms of transport in our big towns, a lot of lives might be saved, together with a good deal of respiratory injury.

We do not have nearly enough power stations to do this at present, and it would cost nearly three times as much as natural gas (which is anyway much less polluting than oil or coal but which it would be good to conserve) which would make it very unpopular. So, if I may put it that way, we are going to have to live with the extra thousands of deaths from air pollution for a long time yet.

DANGERS ASSOCIATED WITH ROUTINE USE OF OIL AND GAS

Some figures for risks incurred by the workforce in oil extraction have already been given in table 10.2. The extraordinarily high rate of 30 per 10 000 in 1974 has fallen considerably in routine working, although even larger numbers were killed in individual accidents in other years; for example 123 were killed in when the *Alexander Kielland* rig capsized in 1980 and 167 when *Piper Alpha* blew up in 1988. The establishment of the North Sea drilling and extraction platforms was known to be dangerous, and the small number of courageous and highly skilled men doing the work were paid accordingly. During the early 1980s the number of deaths fell from 10 per 10 000 to 2 or 3 per 10 000. With more experience it would be hoped and expected that the death rate should steadily decrease. However, the risk of working on even the established oil rigs, including the risks to helicopters in transit between the rigs and the land, will never be reduced to the levels normal in, for example, printing and publishing.

As with coal, the most severe routine risk from the electric power stations using oil arises from oxides of sulphur and nitrogen, and five deaths per GWye is as good an estimate as can be made. Since little electricity is now obtained from oil, and less still from gas, this can hardly account for more than 10 or 20 deaths per year in Britain.

North Sea gas, obtained from shallower and less stormy seas, is a lot less dangerous to the staff, but being in an area with a much greater density of shipping is likely to present a larger risk to sea traffic run by people not in the industry. The risk must however be small, probably less than 0.1 death per GWy of heat.

The chief routine risks to the public other than those just mentioned arise from the distribution to the many small-scale users. Gas explosions were dealt with in Chapter 8. Although equally accidental, the deaths resulting from many minor traffic accidents involving road tankers of petrol, fuel oil or liquefied fuel gases are more appropriately discussed here.

According to the 1978 report by the Health and Safety Commission, already quoted, it is possible to estimate that the road transport of oil for all uses costs the lives of about 12 persons per year, of whom the majority are drivers of other vehicles in collision with oil tankers. Only 1% of the fuel oil carried by road goes to the oil-fired power stations still operative, so that 0.02 or even fewer deaths per GWye can be attributed to this, and about 0.03 deaths per GWy used in transport, heating and miscellaneous smaller applications.

CARCINOGENIC EFFECTS OF AIR POLLUTION IN BRITAIN

Discussion of the effects of larger scale energy production may give a misleading impression of the full effects of the burning of fossil fuels. A third of the coal in Britain is burnt in smaller units, without the high smoke stacks used in power stations; while 30% of the oil consumption is in small heating plants with chimneys 10 m high or less, and over 45% in transport with no chimneys at all, the two together burning over 60 million tonnes of oil per year. The urban smoke level has been very greatly reduced during the last two decades by the regulations requiring smokeless fuel in built-up areas, together with the swing from the burning of raw coal in open fires to the burning of oil and gas in far more efficient systems. Diesel oil and petrol used in transport produce many undesirable compounds; and during the same period road traffic has increased by many times, with a serious output of recognisable injurious materials such as lead and carbon monoxide and dangerous carcinogens such as benzo-a-pyrene.

The recorded cancer rates in Britain show a large effect which is likely to be due to a combination of these. In the *Registrar General's Statistical Review of Cancer* published in 1975, the registration rates for over 100 different types of cancer were separately listed for urban and for rural areas of England and Wales for each of the years 1968, 1969 and 1970. Table 10.3, using the figures in this review, shows clearly that there was a large and increasing excess of cancers in the urban areas. The particularly large excess of male lung cancers could represent a larger proportion of cigarette smokers in urban areas (for which I have been able to find no evidence)—but lung cancer, some of which must be due to benzo-a-pyrene, accounts for little more than half of the male cancer excess, and for hardly any of the female cancer excess.

The populations covered by urban and rural areas were 38.4 million and 10.6 million respectively. About 30% of the men and 40% of the women registered as suffering from cancer survive; the larger proportion of women results from the greater expectation of survival from their high proportion of breast cancers. From the excess number of registrations in urban areas we can estimate the actual excess number of deaths from cancer each year, which rose from 12 400 to 14 300 over the three years. We have therefore an average total of over 13 000 people dying of cancer in England and Wales every year who would not have died if they had all lived in a rural environment. It must be remembered too, that people living in rural areas are also breathing polluted air, albeit less polluted than in the urban areas, so that a figure of 13 000 deaths per year is almost certainly less than the numbers for which the urban environment is responsible. Even if we take the improbable view that

Table 10.3 Comparative cancer rates in urban and rural areas in England and Wales 1968–70. The top table shows lung cancer rates and the bottom table the rates for all cancers.

Urban areas include all country and municipal boroughs, urban districts and London Boroughs as defined by the Local Government Act. Rural areas are the rural districts as similarly defined. The populations of urban areas were 18.6 million males and 19.8 million females. There was little change in these over the three years. The actual numbers of deaths are those expected within 10 years, assuming a 30% survival for men and 40% for women.

	Lung cancer registrations per 100 000			Actual number of urban excess	
	Urban	Rural	Urban excess	Registrations	Deaths
1968 M	103.4	73.2	30.2	5585	3900
F	19.0	13.1	5.9	1170	700
1969 M	104.7	73.1	31.6	5864	4100
F	20.2	15.7	4.5	893	540
1970 M	106.9	72.7	33.2	6160	4300
F	21.9	15.6	6.3	1250	750

	Total cancer registrations per 100 000			Actual number of urban excess	
	Urban	Rural	Urban excess	Registrations	Deaths
1968 M	334.4	280.3	54.1	10039	7000
F	298.4	253.0	45.4	9007	5400
1969 M	340.2	278.2	62.0	11505	8100
F	303.8	257.6	46.2	9074	5400
1970 M	345.9	282.6	63.3	11748	8200
F	310.5	259.2	51.3	10178	6100

the whole excess of lung cancers in urban areas is due to an entirely hypothetical increase in cigarette smoking, we still have an excess of 10 000 other cancer deaths to account for.

Unfortunately no similar records of the urban/rural ratio are available subsequent to 1970, but total cancer death rates in 1986 were over 28% greater than in 1969, and show no sign of effects of the reduction of open fires which was already underway in the 1950s. It must be remembered that cancers take decades to develop; most of those observed must be due to factors acting soon after the war and it will be a long time before we can disentangle the falling effects of coal and the increasing effects of vehicles. We shall assume for safety that the urban excess will not disappear, and each million tonnes of oil burnt in vehicles or other small urban units may be responsible for 100 deaths from cancer. It has already been estimated that the million tonnes of oil needed to produce 1 GWye in modern oil-fired power stations may be responsible for five such deaths; and it could well be that our diesel- and petrol-fuelled transport is causing more deaths from cancer in our cities than it kills in the entire country in accidents.

It may seem a sign of bad judgement that I have taken so much space to describe the tiny risks of large-scale energy production, and then spent so little on risks of energy use a hundred times greater than those of its production. There is a good practical reason for this. The large space has been relevant to decisions that might soon be made and actually carried out. The unproven possibility that our cars and lorries may be adding an extra three or four deaths in 20 years' time to every two that we know perfectly well they are causing today is not going to make anyone consider seriously and immediately the astronomical cost and difficulty of replacing all of our petrol- and diesel-driven vehicles by electrically driven ones. We can however make a start. Light cars suitable for commuters or for shopping in big towns, charging the batteries from the mains at cheap night rates, are certainly practicable. It seems to me that all the 'Greens' should be actively pressing for the bulk production of electrically driven vehicles.

ROUTINE RISKS OF NUCLEAR POWER

As with the other energy sources, I shall begin with a discussion of the risks to the workforce. These have much in common with the risks associated with coal. Before the danger of inhalation of radioactive dusts was appreciated, mining for uranium was appallingly dangerous. The first major source of uranium was Joachimsthal (now Jachymov) in Czechoslovakia, and for a long time after the discovery of radioactivity this was regarded as health-giving rather than the reverse. Even the 1966 edition of *Chambers Encyclopaedia* reported of Joachimsthal: 'The radioactive baths are world-famous for the treatment of cancer, nervous diseases and rheumatism'. This of course was supposed to apply to tourists; it had long been known that many, and possibly most, of the miners in the uranium mines died of lung cancer. In early days this was probably not clearly distinguished from such respiratory conditions as pneumoconiosis, which was doubtless responsible in part for these deaths. This led to no apparent difficulty in recruiting fresh young men into the industry to replace those invalided out to die. Whether we are abnormally neurotic about risks now, or whether the population of Joachimsthal was inhumanly callous then, it is difficult to say, but the uncaring attitude in the early part of this century was not confined to central Europe. The Cornish tin miners in the 1920s also had a virtual 100% incidence of what was then called silicosis (which may have included lung cancers due to radon from the very appreciable amount of uranium in the same area); but when cheap imported tin forced the closure of the Cornish mines there was great distress in the area.

During the war imported supplies of tin were unreliable and the Ministry of Supply considered reopening the Cornish mines. However, somebody looking up the records discovered the appalling health effects. It happened that my sister was the investigator sent down on behalf of the Ministry of Information to find out whether the local people could be induced to face the risks without an unacceptable level of compulsion. For obvious reasons, she found a lot more women than men who had survived the earlier working period, and found that there was general enthusiasm for the proposition, which naturally carried much higher wages than those then current. This was not due to ignorance of the risks. One lady summarised the local feeling by a comment to the effect that 'Of course you've got to be sensible. That Mrs – went and had seven children. Some people never think how they're going to manage when they're widows'. It is difficult to believe that the reaction would be the same today. Cornish mines *are* being worked again, but with effective ventilation.

The risk of lung cancer due to the breathing of radioactive dusts and the radioactive gas radon, one of the descendants of uranium, has long been well understood in the technically advanced countries. Outside these countries the knowledge has not always been applied. Conditions quite as bad as the early conditions at Joachimsthal were reported from the former Belgian Congo ten or fifteen years after the war.

Nowadays the miners of uranium ore in Canada and the USA, for whom detailed information is available, are protected in several ways. Workers in open-cast mines need little more than dust protection, but underground workings will have water sprays to lay the dust, protective clothing, and an individual monitor for each man to measure the radiation doses that he is receiving. As with hospital x-ray workers or workers in any part of the nuclear industry, there are strict limits to the permissible dose per three months and dose per year. In case of accidental overdose the man concerned is taken off work involving radiation until his average dose over the extended period has fallen below the permissible limit.

Many people in Britain, including myself, suspected that in the Namibian mines—the source of much of our uranium—the coloured workers were inadequately protected from radon decay products. This concern has been relieved by a report from James Tye, Director General of the British Safety Council, who carried out an extensive safety audit in the mines, through ore-carrying, engineering, workshops, loss control, 'critical areas', the medical centre and the company hospital, and came away amazed at the management's safety consciousness and wishing that this could be imported into our own country (*The Times* 17/3/89).

The main risk arises from the inhalation of the natural alpha-emitting radioactive gas radon-222 and its descendants. As radon is one of the inert gases, most of that which is inhaled will be immediately exhaled. Accordingly, since alpha particles will not penetrate the skin, only a very small proportion of the gas will decay where it can cause damage. The main danger arises from the fact that its decay products are also radioactive—the whole complicated train of events is shown in Appendix 2. The alpha-emitting decay products which matter are polonium-218 and polonium-214, each with a half-life of a few minutes. Polonium, and the intermediate beta-emitting isotopes of lead and bismuth, are all solids at room temperature, and hence if inhaled will stick to the surface of the bronchi or lungs and will not be exhaled. As the proportion of the various active substances will vary from mine to mine and from time to time in a single mine, the hazard is measured in terms of a 'working level'. This is defined as being that concentration of radioactive substances per litre that will release a total of 2.1×10^{-8} J of alpha-particle energy by the time all of the activity has decayed. Then someone working for 170 hours a month in this is said to have received a dose of 1 working level month (WLM).

The cancer risk involved is difficult to determine with accuracy, although the lung cancer rate is known and, among men who have worked in the mines for a long time, may seriously exceed the 'natural' rate. A detailed study by an international group was reported by Evans *et al* (1981), and it is concluded that 100 deaths for a collective dose of 10^6 man-WLM is probable, after making some allowance for the greater than average number of cigarettes smoked by miners and their greater than average breathing rate. In terms of cancer risk then, 1 WLM is equivalent for miners to a whole-body radiation dose of 1 rem, while for sedentary people, breathing less air per minute, 1 WLM is likely to give a dose equivalent to 0.5 rem.

Data collected in Colorado suggest that the risk of lung cancer to smoking uranium miners is 11 times as great as that to non-smoking miners, and over 70 times the risk to non-smokers in the general population. This factor of 11 is far from certain; a wide range of figures has been given by different groups of investigators (Bodansky *et al* 1987). It seems to me possible that cigarette smoking is acting as a promoter, by inhibiting the clearance of actively contaminated particles from the bronchi when coinciding with or closely following their inhalation, but having little synergistic effect at other times. The difficulty of deriving reliable values for the risk, per unit of dose inhaled, from the reliable and very considerable numbers of excess deaths from cancers observed, was that neither the doses received decades earlier nor the smoking habits of the miners were adequately known. The best estimates that we can make for the risk to present day miners—with protective masks and clothing and working in well-ventilated mines—are probably obtained by assuming that the risk per unit of recorded dose is the same as that estimated by the ICRP for the same partial body dose of radiation.

The maximum permissible dose to uranium miners in Canada and a number of other countries is now 4 WLM per year with not more than 2 WLM in any one quarter. This would give roughly the risk represented by the 5 rems a year maximum legally maintained for registered radiation workers in the nuclear industry and hospitals. In all cases, however, the actual doses received are usually well below the official limit.

The staff maintaining and operating nuclear power stations receive regular doses of beta and gamma radiation, the amount varying a good deal from one type of station to another. The PWRs in the USA may give 1–2 rems a year to their operating staff, while the British AGRs give 200–300 mrem. In PWRs built in Britain, it is intended to modify the design procedures in such a way as to keep the doses down to the same level as is achieved by our AGRs. The annual collective dose at an AGR is around 100 man-rems corresponding to one death every 100 years, or adding 0.01 to the average total of deaths per year per GWye due to the nuclear power industry.

The processing plant at Sellafield (Windscale) gives larger doses to more staff, in total in the region of 3500 man-rems per year. A survey of the 15 000 workers employed at Sellafield has shown a *lower* incidence of radiation-linked cancers than among the general population. This of course is not proof that no extra cancers are being incurred (*Nature* 1983). No published information is available as to how much of the 3500 man-rems per year is due to the processing of plutonium for military purposes, but if this is neglected it represents 3 deaths every decade and, if divided among the atomic power stations, about 0.03 per GWye.

NUCLEAR POWER—RISKS TO THE PUBLIC

Since the amount to be transported is only a few hundred tonnes of natural uranium per year per GWye, no deaths in transport accidents have been recorded. The only casualty (non-lethal) of which I have heard was a member of the staff at Brookhaven, USA. A year or so after the bombs on Japan he saw a lorry labelled URANIUM coming into the establishment and he was so scared that he jumped backwards, fell over a pile of steel girders and knocked himself out.

There are however several routes by which the industry increases the doses of radiation habitually received by the public. Potentially by far the most important in the long run arises right at the beginning. The tailings—the residue left after chemical extraction of uranium from its ores—still contain thorium-230 and radium, and the latter continues to give off the radioactive gas radon which is responsible for the radiation risk to miners. The radium itself is not volatile and constitutes no greater risk in the tailings than it did before the ore was mined; but if the tailings are left unburied, as is unfortunately often the case, the radon can escape into the air far more readily than it could have from the original ore beneath unbroken ground. For each GWye from a nuclear power station, such as an AGR or a PWR using enriched uranium fuel, perhaps 250 tonnes of natural uranium has to be mined. This will be about 80 Ci, and will therefore leave 80 Ci of radium behind in the tailings which will maintain a steady 80 Ci of radon, fresh gas being given off by the radium as fast as the gas already given off decays. In an average 10 mph (16 km/h) wind this will be blown downwind for some 900 km before half of it decays, so that if all of it escapes the tailings it could maintain an average concentration of 10^{-14} Ci/litre over 8000 km^2. The important assumption here is that over most of the distance the radon will be uniformly mixed into the bottom 1000 m of the atmosphere. This assumption may be pessimistic over so long a distance. The windspeed and area covered are unimportant; if the area covered is half that supposed the concentration of radon will be doubled but the population exposed will be halved, giving the same total collective dose.

10^{-14} Ci/litre is a good deal less than is found in the air of most cities, and with an average Canadian population density of two per square kilometre will produce a collective dose of perhaps 35 man-rems per year. This would lead to the death of one person every 290 years. The problem is not urgent therefore, but it could be with us for a long time; the thorium-230 percursor of radium-226 has a half-life of 80 000 years. On the same assumptions very many more over a much longer period would be killed in Britain (with its larger density of population) by the uranium in coal ash. This will be referred to again in the later chapters dealing with wastes and environmental effects.

It must be remembered that none of the activity in the tailings was *created* by the mining. All of the radioactive material in them has been there since the deposits were formed—and has indeed been very slowly decaying ever since. The pulverisation of material will have made it easier for radon to escape but will not have increased its amount. The emission of radon is estimated to be reduced by a factor of 40 for each 6 m of earth covering, or by a factor of 400 by 12 mm of asphalt. Costs of covering the tailings were estimated in 1976 as US $50 000 per hectare for a layer of asphalt 25 mm thick and $2500 per hectare for 1 m of earth (Cohen 1976). Together these would reduce the radon risk to below that before the uranium was mined. About 0.3 hectares of tailings would need to be covered per reactor year supplied with fuel. The additional cost of fuel would be about 6 cents per kilogram of uranium, or $16 000 per GWye. Even if the cost estimates are optimistic, as they probably are, the main problem is likely to be to persuade the mining companies to worsen their competitive position by incurring even minor 'unnecessary' costs, or to persuade the customers to buy the slightly more expensive uranium from the companies who are doing the job properly. Some coordinated local pressure on at least some of the mining companies may be needed.

The US Environmental Protection Agency has made an exhaustive study of methods for reduction of radon emission from the tailings, and concludes that the most cost-effective method is to cover the tailings with a layer of compacted soil stabilised by vegetation, which could reduce the radon emission to normal background levels at a cost in the region of $100 000 (1983) per life saved (EPA 1983).

Nuclear power stations themselves allow very little radioactive material to escape. Traces of argon-41, formed from the interaction of stable atmospheric argon-40 in the air-cooling of the biological shields with the few neutrons that manage to get through these shields, gave doses of about 10 mrem per year (less than one tenth of the natural background) at the site boundary of the earlier Magnox reactors. Later Magnox stations and the AGR stations release considerably less; and the collective dose from all radioactive substances released from all Magnox reactors together is little more than 100 man-rems per year. This would correspond to one theoretical death per century. PWRs release a rather different mixture of activities, but again add fewer than 0.01 deaths per GWye (Cottrell 1981). A much larger amount of radioactive material is emitted by the processing plant at Sellafield. The dose from this rose to a maximum of about 12 000 man-rems per year around the end of the 1970s, but has been falling and by 1987 had gone down to 3000 man-rems. Over a decade this might cause three deaths. In 1987 the maximum dose, recommended by the NRPB, to an average member of the public was reduced to 100 millirem in an average year, or up to 500 millirem in some years as long as the lifetime dose does not exceed 7 rems.

ROUTINE RISKS OF WIND, WAVE AND DIRECT SOLAR POWER

Offshore wind turbines and wave machines will certainly be dangerous to build, and will inevitably lead to significant numbers of deaths among the workforce concerned with maintenance as well as that concerned with construction. Experience with the North Sea oil rigs should be helpful in reducing the deaths to a minimum, but it is not possible to provide even a useful guess at the actual numbers of deaths per GW installed or per GWye of electricity produced. Even in the building industry on land there are 10 deaths per year among 100 000 employees, and the rate among steel erectors and scaffolders is considerably higher. The number killed in building offshore wind turbines will surely be greater than for any of the big land-based power stations, but the work will be interesting and challenging, besides being well paid, and it is unlikely that the larger death rates will either deter the workers or affect the official decisions as to whether or when these technologies should be applied.

Installation of solar heating systems should not involve any greater risks than are usually acceptable in domestic construction.

Not many solar heating systems will fall on passers-by, and DIY enthusiasts who kill themselves by falling off roofs while repairing or cleaning solar panels do so 'voluntarily' and need not therefore be counted. Wave machines and offshore wind turbines will, however, inevitably add to the risks of coastal shipping. Apart from the possible accidents mentioned in Chapter 8, these must add to the unpublicised mass of small accidents that occur at sea just as they do on the roads. There will of course be warning lights and sounds, and each unit will be conspicuous to radar, will be marked on the relevant charts and be easy to see in good weather. Such measures will minimise but not eliminate accidents. As an analogy, if we introduced an additional few thousand sharp bends into an already inperfectly straight main road system, no number of road signs and warnings could entirely prevent an increase in accidents.

THE RADIATION DANGER DUE TO HEAT CONSERVATION

In Chapter 7 it was pointed out that our need for power could be usefully lowered by reducing the waste of heat in ordinary houses. It is not obvious that this has any relation to danger from radiation, but in fact the relation is close and important.

Earlier in this chapter the main specific hazard to uranium miners was shown to arise from the inhalation of radon and its decay products. Most inconveniently, uranium is not confined to well defined bodies of ore, but is widely distributed in every kind of rock or soil. The result is that its decay products are similarly widely spread, and in particular there is always in the air we breathe radon which has seeped out of the ground. The concentrations out of doors are not high, because the uranium forms only 1 or 2 ppm in most soils, because radon from the deeper layers of soil will not reach the surface before it decays, and because it is rapidly mixed with air up to a considerable height in the atmosphere where nobody is likely to breathe it. The radon in the open air therefore adds very little to the radiation doses we receive from other parts of the natural background.

Indoors it is a different matter. Radon seeping through the floor from the soil beneath a building does not get blown away quickly as it does outside. Bricks and breeze blocks also contain radium, but the radon concentrations due to these are usually negligibly small compared with that coming through the floor. The concentration of radon, and its daughter radioactive isotopes of bismuth, lead and polonium, will then increase until the rate at which it is removed by ventilation is equal to the rate at which it is seeping in.

Effect of ventilation rates

Years ago a living room with an open fire would have had so great a ventilation rate that there could be a complete change of air every two or three minutes, so that the room would contain only the radon that had very recently seeped in; and even more important, would contain only the tiny proportion of its actively dangerous solid decay products that had been produced during its short stay. Nowadays radon may be able to accumulate for an hour or more, not only leading to a 15 or 20 times increase in its own rate of alpha emission, but also allowing time for the production and accumulation of more of its radioactive decay products, which would earlier not have been produced until the radon had left the premises. Accordingly, the radiation dose received by the lungs of the occupants will increase more rapidly than the ventilation rate decreases. We have the awkward dilemma that to conserve heat we must reduce the ventilation rate, but that to reduce radiation doses we must increase it.

The problem was first seen to be important in Sweden and Germany, with efficient closed-stove heating and good draught-proofing. In 1980 Sweden issued provisional regulations with annual exposure limits of 0.45 rem for new and 2.5 rem for existing dwellings. There are thought to be 40 000 existing Swedish dwellings in which the dose to the inhabitants is over 2.5 rem a year, leading to a greater than 8% extra chance of death from cancer (i.e. 26% against 24% without the radon).

To see whether the hazard is of importance in Britain, we have of course to look at some more figures. Measurements have now been made in houses in several parts of Britain, the most important having been made by a team from the National Radiological Protection Board

(O'Riordan *et al* 1988), with early but much smaller studies by my own research group in Birmingham, Aberdeen and Orkney—where there are houses built over potentially useful uranium ore. (The houses in Orkney did not show excessive levels owing to exceedingly efficient under-floor ventilation, introduced to reduce the risk of timber rot in a wet climate.) According to the 1988 summary of O'Riordan's team, a systematic national survey of the gamma-ray dose rates and radon concentrations in 2000 UK dwellings has been carried out. This has shown that the mean annual effective dose equivalents from gamma-rays and radon decay products in dwellings are 28 and 100 millirem respectively. When account is taken of exposures elsewhere, the mean annual doses from these sources are about 35 and 120 mrem respectively, the latter being greater than previous estimates due to a reassessment of conversion factors between exposure and dose. The mean annual dose from thoron decay products is about 15 mrem.

More detailed surveys in regions of the UK with elevated concentrations of natural radioactivity in rocks and soil have shown that radon concentrations vary widely, sometimes resulting in annual effective dose equivalents exceeding one rem. Doses from gamma rays and thoron decay products are much less variable, with maxima around 100 mrem in a year. Although the radon concentrations indoors are clearly related to local geology, some weaker influences can also be discerned, such as the storey on which measurements are made and the presence of double glazing. It may seem surprising that double glazing should affect the radon concentration as radon is quite obviously no better able to pass through one sheet of glass than through two. The relevance of double glazing arises because it does what it is intended to do; a house with double glazing will be warmer than one without. The density of the indoor air is then smaller than it would have been if the air were cooler. Accordingly there will be a greater rate of escape of air through the upstairs ceilings and the roof, sucking more of the colder denser air through the ground floor to replace it. And of course, unless there is quite unusually good under-floor ventilation there will be a small but positive suction drawing in the radon-carrying air from the ground.

There are large variations between houses even in the same area, but the radon concentrations in most British living rooms are distributed about a value of 10^{-12} Ci/litre of air (about two alpha particles emitted per minute). Ventilation rates also show considerable variation, measurements by the NRPB suggesting an average of one change of air per hour. On the basis of these figures it is estimated by the NRPB that the average annual dose received in Britain is equivalent to a whole-body dose of about 115 mrem, more than half the total of the natural background (O'Riordan *et al* 1987). If it stays at this level it will be responsible for nearly 600 (hypothetical) cancer deaths each year in

Britain. The average annual dose of 115 mrem covers some very large variations. In parts of Cornwall where there are relatively large concentrations of uranium in the subsoil rocks there are some 30 000 houses with dose rates of 1000 mrem a year or more, and 900 houses with 5 rems or more a year. The NRPB recommends that (very expensive) remedial action should be taken in existing houses if the dose rate exceeds 2 rems a year, and that building procedures should be altered to prevent new dwellings from exceeding 500 mrem a year.

The incidence of lung cancer in Cornwall is less than 75% of the average for the whole country. This however may mean only that there is a reduction of the lung cancer rate due to chemical pollution because, as in the Scottish Western Isles, there is a clean prevailing wind from the Atlantic, and this more than compensates for any excess due to radon.

It is worth noting that no excesses of childhood leukaemia have been reported among the 900 houses receiving 5 rem or more every year, as would have been expected if the excess at Seascale in a similar size of population had been due to the very much smaller radiation doses from Sellafield.

According to Bodansky *et al* (1987), radon may be responsible for 10 000 lung cancer deaths a year in the United States. This is a much higher rate than in Britain, but it may be that the higher indoor temperatures usual in USA go with more suction through the floor and with smaller ventilation rates.

In Chapter 4 evidence was quoted suggesting that small doses of radiation at low dose rates may be less effective in proportion in initiating cancers than are large doses at high dose rates—and 115 mrem a year looks both small and slow. Unfortunately the evidence quoted referred only to beta or gamma irradiation. An alpha particle does too much damage to be easily repaired, and the effectiveness per rem of an alpha dose must be independent of dose or dose rate from zero up to very high dose rates. The 600 cancer deaths to be expected from a year's exposure to 115 mrem of alpha particles will be delayed by 20–40 years, and so the death rate from this cause in Britain will not yet have built up to the full 600 a year to be expected if we continue for a long time to receive 115 mrem every year, as it is not yet very long since the open-fire way of life was abandoned.

There is no way in which we can eliminate indoor radon altogether; but for a given rate of seepage into houses any reduction in the average level of ventilation will increase the number of deaths, and any increase in ventilation will save lives.

Calculations made using the known decay rate of radon show that if ventilation (i.e. the rate at which the air inside the house is removed and replaced by draughts from the air outside, not by sucking in more contaminated air from below the ground floor) is low, then any further

lowering causes considerable increase in the dose to the inhabitants. In a
very draughty house however, a lot of money can be saved by some
draught-proofing without much increase in risk.

It is important to note however that while the numbers of deaths due
to radon are clearly significant, and the deaths (discussed in the earlier
part of this chapter) per GWye of electricity produced by any existing
means are comparatively insignificant, the risks to the individual are not
really large. An extra 1000 deaths per year in 25 million people
represents a risk of only 1 in 25 000 per year (a risk equivalent to the
smoking of a cigarette every five days) starting 30 years after the change
in ventilation rate; and the saving of 3000 kWh of heat per year is going
to save some £70 to £180 per year per household (1988 prices) depending
on the heating system employed. For houses with lots of leaks in which
the initial ventilation rate is much higher to start with, the risk is much
smaller for the same cash saving. If the leaks into a house changing its
air once in 10 minutes are reduced until there is a change of air only
every 20 minutes, six changes per hour would be replaced by three
changes per hour, with a cash gain of £200 per year even in a house
heated by gas, for an increase of risk to the individual of only three in a
million per year, corresponding to about one cigarette a quarter. No
sensible person is going to be deterred from a really useful improvement
to the budget by a risk so small as that.

The object of spending so much time on radon is not to discourage
conservation but to show by contrast the extraordinary triviality of the
risks from any of the large-scale sources of electricity. The only sources
of energy that could give risks comparable with the risks of reducing
ventilation are the small-scale sources, burning coal or oil.

A useful reduction of radon concentration indoors could be obtained
by the use of a heat-exchanger. In this the cold outside air needed for
ventilation is brought into the house through a series of thin-walled
tubes which are in close thermal contact with warm used air leaving the
house. A well-designed system can transfer a large fraction of the excess
heat from the air leaving the house to that coming in, so that this
requires little additional heat from the heating plant. With such a system
even an otherwise completely sealed house could be ventilated at any
rate desired without any serious build-up of radon. Unfortunately an
efficient heat-exchanger is quite expensive to buy, and requires the
continuous use of an electric fan. The combined cost could easily be
more than the cost of the heat saved; and with the risks to the individual
as low as they are it is doubtful whether such a system would ever be
worth the trouble for domestic use. For large office blocks however,
which can readily be sealed well enough to have a complete air change
only once in six hours or more, a specially designed heat-exchanger
feeding the fan-driven internal circulation of air which such blocks

already have, should be quite economic and fully effective in keeping radon concentrations to acceptable levels.

As stated above, I have supported the view that radiation hormesis will occur only after irradiations by beta and gamma rays, and that the effects of the alpha-emitting radon decay products would show no signs of it. Dr Bernard Cohen (1988) has however recently reported a comparison of the risks of lung cancer with radon exposure in 411 counties from all States of the USA. He found a close but *negative* correlation between the average radon concentrations and the average lung cancer rates, i.e. where the indoor radon concentrations were high the lung cancer rates were low. This looks on the face of it to be solid evidence for a large degree of radiation hormesis following alpha-particle damage.

This is pretty convincing, but there could have been an alternative explanation of Cohen's results. It is embarrassingly well known that the big towns of the world have considerably larger cancer rates than has the countryside, presumably due to air pollution. An extreme example in our own country is that the combined death rate from the seven commonest cancers in Glasgow is 40% higher than the rate in the Scottish Western Isles, where the clean air from the Atlantic is enjoyed. Cohen's results could be explained without hormesis if in such large towns the average radon concentrations were lower due to the type of house construction or other factors. However, Cohen himself considered this, and examined the effect of eliminating all counties with populations over half a million; this did not change the overall negative correlation of lung cancer and radon doses. He also considered the possibly confounding effect of cigarette smoking, using data on cigarette sales which were available for states (though not for counties); but allowance for these did not remove the negative correlation. Indeed, the negative correlation with radon levels was as strong as was the positive correlation with cigarette sales.

If we take his results at their face value, we can stop worrying about radon. Against all of my own advice, however, I cannot feel confidence that there are no other factors which could have affected the results in an apparently sound experiment, and shall feel happier if we continue to support measures to guard ourselves against radon in our houses.

To many scientists it seems an irrational anomaly that so many people are still seriously concerned about the tiny radiation doses received from the nuclear industry, while so few show any concern at all about the effects of radon. The average doses to all of us due to this, both in Europe and in the USA, are around a thousand times that due to the effluents from the nuclear industry. This can easily be explained as are the other acceptable risks in Chapter 2. It is clear to everyone that to avoid indoor radon you mustn't be indoors. Living permanently out of doors is unthinkable, and since what can't be cured must be endured we might as well not waste time in worrying. When we begin to exclude radon effectively from the more seriously affected houses, people in the less affected areas will probably begin to worry†.

SUMMARY OF FIGURES

Table 10.4 summarises the risks discussed in this and the previous chapters which arise from the operation of different sources of power. None of them are large even when compared with the other risks which we face in our science-based civilisation. In particular, the risk of incurring cancer from the operation of all power sources put together is less than the risk arising from the reduction of ventilation in our houses since the disappearance of the open fire. And this risk, resulting from the reduction of ventilation in our homes, is less than 1% of the cancer risk from other so-far undiscovered factors in our present environment or diet.

Table 10.4 Summary of the estimated risks to life per gigawatt-year of energy produced.

	Accidents		Routine running	
	Workforce	Public	Workforce	Public
Mediaeval water mill			5000	
Hydroelectric power	0.01–0.1	5	0.01–0.1	—
Solar energy	?	—	?	?
Wave power	1?	1	?	?
Coal-fired electric power	1.0	0.01–0.1	0.6?	5
Oil-fired electric power	0.25	0.1–0.2	0.05	5
Nuclear electric power	0.01	0.01	0.03	0.01
LNG and LPG	0.1–1	0.1–1	no data	
Motor vehicles		200		
Conservation— reduction of draughts			2–20†	

† Per GWy of heat saved by draught-proofing

THE SAVING OF LIFE BY POWER PRODUCTION

Before going on to discuss the further risks that may occur in the future from neglect or mishandling of wastes from the power industries, it will

† The *Householder's Guide to Radon* is available free from the Department of the Environment, Romney House, Marsham Street, London SW1.

help to keep the unavoidable risks in perspective if the positive contributions of energy production to health are considered. So far I have concentrated on the risks and the number of deaths which are associated with the production of power. It is, however, important to think also of the number of lives that are saved by the production of large amounts of energy.

A detailed study has been made by Siddall (1982) of the effects on health of increased national wealth. As a measure of health, Siddall uses the proportion of a population which survives to 65, an arbitrarily chosen point below which the effects of natural aging on death rate are taken to be negligible, and above which they are taken to be preponderant. Between 1870 and 1980 the annual number of premature deaths in Britain (deaths below 65 years of age) dropped from 18 000 to 2500 per million. In both Canada and Britain there was a close correlation with individual income, as shown in figure 10.1; the correlation (corrected for changes in the rapidly growing Canadian population) with gross national product (GNP) is shown in figure 10.2.

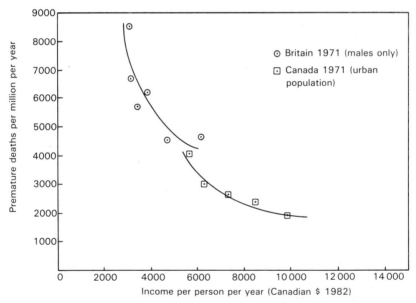

Figure 10.1 Mortality and average income in Britain and Canada.

In 1982 a million Canadians would have used 2 GW of electricity, which would be worth, and hence be contributing to the GNP, Canadian $550 million (1982) per year. Over the last 30 years the annual premature death rate in Canada has fallen by about 1900 per million, while the corresponding annual GNP per million has risen by $8000 million, or about $4 million per life saved. (It must not be thought that Canada has spent $4 million to save a life. The life saved was a by-product, not a primary objective.) Siddall then assumes that only half of the lives saved have resulted from the rise in GNP, so that an $8 million increase in GNP would correspond to the saving of one life. Assuming further that the extra 2 GW of electricity used per million people is responsible for $550 million worth of the increase in GNP , then the 2 GW of power is responsible for saving about 70 lives a year.

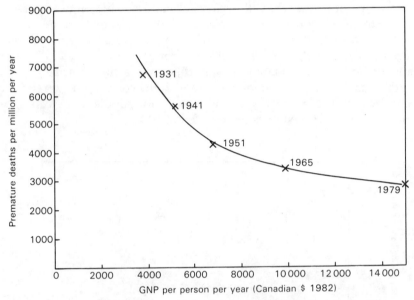

Figure 10.2 Mortality and gross national product per head, Canada.

The exact assumptions made were of course entirely arbitrary; but the nearly exact proportionality of GNP to electricity output has been demonstrated by Starr (1982) for seven different countries, and some connection between GNP and health care cannot be denied. Many people would also claim that the increased use of electricity has had a major influence on the rise of GNP throughout most of the last 50 years. The relation between GNP and health is shakier, but it seems unlikely that the estimate of 35 lives saved per GWye can be many times too large. The figure is likely to be less in the future, however, for many reasons, the chief being the fact that there are now far fewer premature deaths to be prevented, so that we are necessarily limited by the law of diminishing returns.

Over the last 50 years every existing type of power station must have saved more lives than it has destroyed.

While discussing the interaction between wealth and health it is worthwhile to discuss the extent to which we could, by spending money in particular ways, save some of the lives which are being lost in connection with power production. Clearly the right place to begin is with the largest numbers: the people who could be killed by draught-proofing to conserve heat. From data given earlier in this chapter it can be shown that the annual saving in this way of 3 GWy of heat might cost perhaps 500 lives a year. The simplest way of saving these lives is of course not to do the draught-proofing and not to save the 3 GWy of heat. This would be expensive; the cost of producing and distributing 3 GWy of heat from natural gas would be in the region of £300 million—more than £500 000 per life saved. There are in principle much more economical ways of saving lives than this. It would for example be worth spending a great deal to find practical methods of making foundations and walls of existing buildings impervious to radon without causing trouble due to condensation of water.

Comparative figures for the cost of saving lives in a few different ways were given by Kletz (1979). (Inflation since 1979 will probably have added 25% to these, although the uncertainty of the figures must be at least as great as this.) According to Kletz the cost of saving one life by various means in 1979 was as shown in table 10.5.

The figures show no sign of intelligent planning, but the use of £500 000 worth of natural gas or even £2 million of electricity to prevent one death from radon inhalation does not seem outside the accepted range. Whether the average householder would be prepared to pay even a few pounds a year more for 30 years to reduce the risk of death from cancer for himself and his family from 24.03% to 24.00% is of course another matter; people are quite prepared for millions to be spent per life saved if someone else is spending it but even when their own lives are involved they are likely to opt for rather lower figures when they are paying themselves.

Table 10.5 The cost of saving a life.

Cervical cancer screening	£4200
Breast cancer screening	£12 200
Artificial kidney machines	£28 500
Removal of road hazards, up to	£200 000
Precautions in chemical industry	£1000 000
Precautions in pharmaceutical industry	£10 000 000

SUMMARY

To sum up, the risks to human health of large-scale power production of any kind are small compared with the other risks we run, and are far smaller than the favourable effects on health that they make possible.

The largest risks of multiple deaths arise from oil refineries and bulk supplies of liquefied fuel gases, and the largest death rates from normal running are due to air pollution by small-scale fuel burning and by traffic exhaust fumes. Urban areas in Britain have over 10 000 more deaths a year from cancer than would occur if their cancer rate were the same as the rural areas. These could be due in large part to this air pollution. After that come the 600 or so deaths per year due to radon. This radon risk, though it could be much increased by our enthusiasm for heat conservation by draught prevention, should be temporary; effective means of reducing its diffusion through floors and walls should be developed within the next few years. Two or three centimetres of asphalt spread over the foundations, as planned for use on uranium tailings (see p. 190) would be enough if it could be laid without breaks.

The widely discussed cancer risks to the public from nuclear power are in fact smaller than those risks due to any other major power source, with the exception of natural gas, whose cancer risks are believed to be minute. Its main damaging effect on health is indirect and has been due to the diversion of the natural public concern with cancer, by well-meaning and compassionate but often innumerate people, away from the major unidentified environmental causes that are actually killing 80 000 to 100 000 people every year in Britain alone.

Chapter 11

Wastes from the Nuclear Power Industry

A thousand stories that the ignorant tell, and believe, die away when the computist takes them in his grip.

Samuel Johnson

By far the largest *potential* risks due to the wastes from energy production arise from the nuclear industry. The high-level radioactive wastes stored in water-cooled tanks by British Nuclear Fuels Ltd (BNFL) at Sellafield represent the largest and most concentrated assembly of lethal materials, other than nuclear weapons, that has ever been collected together in this country, and is surpassed only by the similar stores in the USA, the USSR and France. I shall therefore begin with the nuclear industry although, as will be seen, the potential risks are very much larger than the actual risks. In all four countries it is likely that, at least until recently, the greater part of the wastes has been derived from the plants producing plutonium for bombs, but the problems of handling the wastes are the same whether from power stations or bomb plants.

WASTES COMING DIRECTLY FROM POWER REACTORS

Waste water, and water from the cooling ponds in which spent fuel is kept before transferring it to Sellafield, contains some tritium and caesium-137 and traces of other substances. These are usually discharged in dilute form directly into the sea, where they have a negligible effect compared with the discharges from Sellafield to be described below; but the Magnox station at Trawsfynydd in North Wales discharges them into a lake where some of the caesium-137 is taken up by fish. Enough is absorbed for a regular fisherman eating large catches of fish from the lake all the year round to receive about 15 mrem a year. This is equivalent in effect to about 5 cigarettes a year; or to being exposed to the non-radioactive carcinogenic risks faced all the time by the British population for one extra day every three or four years. The total collective dose from the annual discharge of water from all Magnox reactors is about 50 man-rems, causing fewer than 0.002 theoretical deaths per GWye.

TRANSPORT AND TREATMENT OF SPENT FUEL

When the reactor is working, fission of the U-235 gives the initial energy, but at the same time some of the U-238, present in much greater quantity, is converted to plutonium-239 which is also fissionable and contributes more and more to the energy as time goes on. By the time 1 tonne of U-235 has been used up there will be 2 or 3 tonnes of fission products from the fission of the U-235 and plutonium-239, mixed with perhaps 0.8 tonnes of plutonium itself. The fission products themselves absorb neutrons, and when too high a concentration of them has built up there will not be enough neutrons remaining to keep the chain reaction going, even when the control rods are at the allowed limit, and the fuel units which contain the U-235 have to be replaced. Different types of reactor produce different proportions of energy and of fission products before the fuel needs replacement, but you get the same amount of heat energy per kilogram of radioactive fission products out of all kinds of reactor, the difference in electrical output depending only on the thermal efficiency, i.e. on the amount of electricity produced per GW of heat. The efficiency varies with the type of reactor, from about 33% for a PWR to 40% for an AGR, the latter producing just 1 GWye per tonne of fission products.

The intense radioactivity of the fission products is accompanied by great production of heat as the fast electrons and gamma rays emitted are brought to rest and absorbed; in round figures a million curies of mixed fission products will produce a kilowatt of heat. A tonne of spent fuel, even from a Magnox reactor, will at first have an activity of hundreds of millions of curies and will produce hundreds of kilowatts of heat. Accordingly, when it is first removed from the reactor, the spent fuel is placed on racks in a tank of water five metres or so deep. This provides both cooling and protection of operators from radiation while the short-lived radioactive nuclides decay. The leakage of activity from corroded fuel rods into the water and hence to the environment has already been mentioned in Chapter 10. Spent uranium oxide fuel from the AGRs, and later from PWRs, could be kept in this way for at least five years, and after that could be stored in air in a suitable shielded store. The stainless steel or Zircaloy, used in the AGRs and PWRs respectively, are not appreciably corroded in either condition, but the magnesium–aluminium alloy in the Magnox fuel rods will corrode excessively if kept too long under water. Such rods are transferred to Sellafield after six months. Transfer is usually by rail and has aroused a great amount of well meant but ill-informed concern.

The transfer is in heavy containers, called flasks, typically with 14 inch (36 cm) steel walls. These are mainly taken by rail but in a few cases by road using low-loader lorries with a maximum speed of 15–20 mph (24–32 km/h). A typical rail flask will carry two tonnes of fuel rods, and is filled with water to carry heat from the fuel rods to the flask wall, which has cooling fins to improve the transfer of heat to the air. According to IAEA regulations, the loaded flasks are required to survive four conditions.

(1) A drop onto steel-faced concrete from a height of 9 m—the height of a three-storey house. Flasks so dropped would strike the ground at 30 mph (48 km/h).

(2) A drop onto a 15 cm diameter steel rod from 1 m in the most damaging attitude, to simulate the possible effects of a collision.

(3) Exposure to a heat-radiation and convection environment at 800 °C (red heat) for 30 minutes, to simulate the effects of a fire following an accident.

(4) Immersion in water for at least 8 hours at a depth of 15 m.

The drop test is not in fact done on each flask, but on one-tenth scale models. The effects of changing the scale of simple steel structures are very accurately known, and such a scale model made of the same material and weighing 50 kg is not only easier to handle but can be more effectively brought to a very sudden stop. It would be very difficult if not quite impracticable to make a stopping platform that would not be heavily and irregularly dented by a full-scale flask, and such a flask might be less damaged than it would be if the platform had not yielded at all. Nevertheless, the same test on a full-scale flask was carried out by the CEGB in 1984; not because the engineers doubted the reliability of the model tests but to encourage the non-engineers.

If two similar objects each moving at 30 mph collide head on (i.e. with a relative velocity of 60 mph (96 km/h)) the effect on each is exactly equivalent to the effect of one of the objects striking an immovable barrier at 30 mph. In each case the object is brought to rest from 30 mph in a distance depending only on the compressibility of the object itself. A railway locomotive is a very flimsy affair compared with a flask, and the crushing of the locomotive's front will cushion the impact for a flask at a relative velocity well above 60 mph.

Again, to satisfy the American public rather than the engineers, who were satisfied already, full-scale 27–70 tonne flasks have been pushed at various speeds into heavy railway engines or large concrete obstacles at the Sandia Transportation Technology Centre in Albuquerque, USA. The collisions were filmed and the films studied in slow motion. The impact of a locomotive hitting the side of a flask on a road vehicle astride a railway line was also studied. At the higher speeds the flasks were spectacularly thrown around, and most conclusively smashed up the vehicles that hit them, but themselves survived the impacts as well as or better than had been expected from the model tests. Flasks on remote-controlled road trailers were run into a massive concrete barrier at 84 mph (135 km/h). The cab of the tractor and a shock-absorber were crushed, reducing the speed of the flask to about half its initial value, and the flask suffered only superficial damage to the external cooling fins. When a 120 tonne locomotive travelling at 81.5 mph (131 km/h) was made to strike a 25 tonne flask placed sideways-on across the lines, the frame of the locomotive buckled and formed a ramp which lifted the flask onto the superstructure of the locomotive. The flask tumbled end over end three times and suffered only a small leak in the lid seal—which would not have affected its ability to contain and shield its contents. A similar test, said to have cost a million pounds, was made in Britain. A 140-tonne locomotive (unmanned) moving at 100 mph crashed into a full-scale flask laid on the track in the most vulnerable position. The crash destroyed the locomotive but the flask remained intact.

The effects of heating are less easily scaled up from a model. For some years a full-scale container could be tested only to temperatures lower than 800 °C, because much technological effort had to be expended before conditions of this severity could in fact be applied to the whole of the container for as long as 30 minutes. This has now been achieved, but was sufficiently difficult to make it barely possible that such conditions could be reached in an accident.

A 99 tonne flask being considered for Sizewell is 6.2 m long and 2.4 m in diameter, and has a steel wall thickness of 89 mm with a lead liner 216 mm thick. Typical rectangular flasks in current use may weigh from 40 to 70 tonnes. When conveying its load of radioactive spent fuel a flask must give a dose of less than 10 mrem per hour at a metre from the flask surface.

Several flasks have been harmlessly derailed in the last ten years, but to be involved in a fire a flask would have to be involved in an accident with a second freight train carrying inflammable liquids. In recent years in Britain there have been only about three derailments per year of trains carrying inflammable liquids. The probability of another train colliding with any derailed train was 1 in 192, and the probability of this being a train carrying filled flasks is about 1 in 7000. The final probability of any flask well back in a train striking a fuel tanker well back in another train and being engulfed in a fire is thus around 1 in 1000 million per year.

Of course, as is often pointed out, this *might* happen tomorrow, so it is worth thinking a little further. Not only does the flask have to be engulfed in a fire, but it has to stay engulfed for a long time. The flame temperature of a petroleum fire can reach 1100 to 1200 °C, but the efficiency of transfer of heat from an unforced flame to an initially cold steel surface is very low; the surface of the flask is covered by a layer of air and petroleum vapour too cold to burn and forming an insulating layer between the flame and the steel. Radiation will gradually heat the flask, but it will be the flames from the oil below the flask that will be most important. Then the relatively thin steel girders forming the base of the carrying truck will warm up much faster than the multi-tonne flask, and long before the flask gets hot enough to matter the girders will have softened and collapsed. Whereupon the flask will fall through and put out the fire below it, leaving itself in the position of a kettle with the fire round the sides and on top—a poor arrangement. In the fire tests the tested flask was mounted on concrete pedestals, but it is stretching the probabilities a lot further if we have to postulate a third train carrying concrete blocks colliding at exactly the right moment with the first two trains. Finally, it is difficult to believe that the fire would be left alone. Even if the railway line could not be reached by fire engines, one good water jet from a distance, though quite ineffective in putting out an oil fire, would be quite effective in keeping the flask temperature well below the danger level; and it is unlikely that tank-loads of liquid mud dropped from helicopters would have to be called for.

Naturally, few people realise how long it takes a large mass of steel to warm up or cool down. In 1982, on the very appropriate date of 1 April, it was reported by the *Birmingham Post* that a 32 tonne steel ingot became white hot after its wooden supports on a lorry caught fire near Lichfield, and had therefore to be escorted by firemen. This was passed on to me in good faith as proof that the heavy waste-containers could be made white hot by a timber fire. Knowing this to be impossible, I telephoned local fire services until I found what had happened. The ingot was one of a regular series of similar loads, and was travelling from Sheffield to Kidderminster where it was to be hot-forged. It had been heated to white heat in a Sheffield furnace designed for such jobs, which took many hours, and was loaded into a vermiculite-insulated steel carrier case in which it did not cool enough to matter during the four or five hour journey. Throughout the journey it was escorted by a succession of fire-service vehicles. On this particular occasion, owing presumably to a local failure of part of the insulation, the ingot set light to some baulks of timber between its steel supports on the lorry. The timber fire was brought under control by the Lichfield firemen who were then responsible for it until it was handed over to the Worcester brigade; they saw it safely to its destination—still hot enough for the operations of its recipients. So far as I know, no one has yet demanded a non-white-hot-ingot zone.

To get a genuinely dangerous radioactive cloud, the flask has to be broken open *and* the temperature of the fuel taken up to 1100 °C or so. I don't want to keep harping on chlorine, but it would make more sense to worry about even such minor hazards as chemical tankers than about

spent nuclear fuel. I don't want to suggest that even these are excessively dangerous, only that they are less safe. The real dangers on the
railways arise from the interaction of large heavy vehicles carrying a lot
of people and travelling fast.

LIQUID WASTES FROM SELLAFIELD

After reaching Sellafield the spent nuclear fuel will be kept for a further
period in a water-filled cooling tank before processing. This is of some
practical value because it leaves time for iodine-131, the most dangerous
of the volatile fission products, to decay to negligible levels. The activity
of this will already have fallen nearly 10 million times in the first six
months before transporting to Sellafield, and a further six months will
leave less than one permissible dose for one person in the whole
consignment.

On 4 October 1981 some spent Magnox fuel which had been 'cooled'
for only two months was accidentally included in a consignment to
Sellafield. This was not detected before processing, with the result that 7
or 8 Ci of iodine-131 was discharged to the atmosphere (Parliamentary
written answer, 19 October 1981). This was less than a thousandth of the
amount released in the Windscale fire, and neither in the air nor in the
local milk did it exceed the permissible limits laid down for BNFL. It has
however led to the establishment of monitoring equipment at Sellafield
that will give warning in the future of overactive spent fuel before
processing.

In both the existing Magnox processing plant and in THORP, the new
plant under construction, a release of radioactive material (inside
heavily shielded tanks) occurs when the spent fuel elements are
dissolved in acid. Almost all of the fission products and all of the
actinides (uranium, neptunium, plutonium, americium and curium) are
non-volatile. However the fission-product inert gas krypton-85, together
with tritium (hydrogen-3), carbon-14 (as carbon monoxide or dioxide
arising from impurities in the fuel) and part of the iodine-131 and
iodine-129† is volatile. Some of the iodine and tritium, most of the
carbon-14 and all of the krypton-85 will leave the solution in gaseous
form, pass through the filters, and emerge into the atmosphere from the
existing high chimney stacks. Since onshore winds are commoner than
offshore winds, most of this active gaseous effluent will be blown inland
and will become sufficiently mixed with the surrounding air to reach the
lungs of a large proportion of the Cumbrian population.

The amount of each gas leaving the works is monitored, and is quite

---

† Iodine-129 has a half-life of 17 million years, and hence is only very weakly radioactive.

accurately known. The annual radiation dose to people from the Magnox processing now going on is extremely low. Krypton-85 is the chief hazard, and even for people within 10 km of the plant exposure is less than one ten-thousandth of the permitted annual dose. The amount of this dose will rise as the total material processed rises and may perhaps increase by five or ten times or so when the oxide processing line is in full operation, depending on the amount of foreign fuel handled. But the dose will still be negligible; even at ten times the present level the annual dose to the whole population of Cumbria would be only 15 man-rems, theoretically giving rise to about one death in 700 years. Over the whole world there could be one extra hypothetical death every hundred years.

When all the fuel is dissolved, the solution is passed on to a second processing vessel, where as much as possible of the uranium and plutonium is separated from all the other radioactive elements; they then pass through a further treatment to separate them, still in solution, from one another.

The uranium and plutonium extracted are purified, and the plutonium stored as oxide in sealed cans within sealed cans. It is stored (for future use in breeder reactors) mixed with uranium oxide, the plutonium oxide forming perhaps 20–30% of the whole. At this concentration no amount of the material could be made to explode, although a sufficient quantity mixed with water would certainly start a chain reaction. A steam explosion could then be produced which would disperse the material and stop the reaction after quite a small production of fission products. Very little radiation penetrates the sealed cans, and the stored material presents no danger to the public.

The rate of production and storage will increase by several times in a few years time, when THORP reaches its full capacity, and there has as yet been no decision as to when stored material will be removed for use in further reactors. I have no information as to how or where the plutonium made in reactors designed for bomb production is stored.

STORAGE OF WASTES

After removal of the uranium and plutonium the solution containing the main bulk of fission products, from both the civil and the military programmes, is concentrated and stored in large stainless steel tanks. Each of these is surrounded by an outer stainless steel tank with a sump at one end to collect any solution leaking from the inner tank. The space between the tanks is monitored to detect any leakage should it ever occur, and pumps are available to transfer such leakage from the sump to a faultless tank. Each tank may have 2000–3000 million Ci of activity, the decay of which produces 2000 kW or so of heat. This heat must be removed by a water cooling system. Seven independent circuits are used in each tank, any one of which could provide adequate cooling in an emergency, and the necessary pumps can be driven by power produced by independent diesel engines in case of mains failure (see figure 11.1). In the unlikely event of all pumping systems failing, provision is made for cooling with water derived from fire-engines; 100 gallons per minute per tank should suffice. It would not matter if it took a few hours for the fire-engines to arrive—it would take more than 12 hours to bring any of the tanks to the boil.

There will be smaller amounts of activity to be stored from AGRs or PWRs than from the Magnox reactors, as the oxide fuel elements in their stainless steel or Zircaloy tubes could be allowed to 'cool' radioactively for a decade in ponds—either near the parent reactors or after they are sent to Sellafield.

Figure 11.1 Water-cooling coils for high-level waste storage tank (BNFL photograph).

LOW-LEVEL LIQUID EFFLUENT

The most important hazard to the public is that presented by the very large volumes of comparatively weakly, but in absolute terms strongly, radioactive solutions, washing water etc, which are discharged through 2.5 km pipelines into the deeper part of the Irish Sea. The 1960 annual activities permitted to be discharged were 6000 Ci of alpha-emitters (mainly plutonium) and 300 000 Ci of mixed beta-emitters (consisting almost entirely of fission products). Very considerable improvements have been made in the last few years, reflecting the successful perform-ance of the new Salt Evaporator and the new Site Ion Exchange Plant (SIXEP). The releases for 1986 and 1987 were as follows: 1986, alphas 119 curies, betas 3190 curies; 1987, alphas 59 curies, betas 2400 curies. This represents a reduction of around a hundred times since the 1970s. The improvements made at Sellafield have however been expensive. The Chairman of BNF stated at one point that up to two cancers would be saved over the next 10 000 years, at a cost of £120 million per cancer (compare table 10.5). For comparison, there are more than 1000 million Ci of natural alpha-emitters in the oceans of the world; including 150 tonnes (150 million Ci) of radium which is over 100 times *more* dangerous per gram than plutonium†. The total quantity of natural beta-emitters in the oceans is about 500 000 million curies, mostly potassium-40.

At first sight then, the Sellafield discharge appears negligible, even when released into the Irish Sea, which takes a year or so to disperse its burden into the Atlantic. This is not so, however. Owing to the absence or extreme rarity in nature of the elements providing the main radio-active isotopes in the discharge, these may be very much concentrated by various physical, chemical or biological processes.

For example, in oxidising conditions in shallow water, plutonium will be rapidly adsorbed by fine particles of silt so that the latter may attain many times the activity per gram of the original sea-water. Ruthenium-106 is similarly adsorbed. The particles will eventually sink to the bottom, but the conditions off the Cumbrian coast are such that silt is continuously moving into the Ravenglass Estuary, along the edges of which deep deposits are building up. Alpha and beta particles are not highly penetrating, so that the external level of these does not increase with thickness, but most of the beta-emitters emit gamma rays as well, which are highly penetrating and which can emerge from some inches below the surface. The result is a gamma radiation level well above background—in a few places high enough to give close to the maximum permitted annual dose (see p. 191) to anyone who remains in close contact with the silt day and night throughout the year. This is unlikely to happen, as wet silt is unattractive to sleep on and is usually under water; the greatest number of contact hours actually maintained by the

---

† Plutonium is more dangerous per curie than radium once it has been absorbed into the body; but is 15 times less radioactive and only two to five parts in 10 000 are absorbed from the digestive system, whereas 20% of radium is absorbed.

most persistent of the salmon fishers is about 400 per year, giving perhaps 10–15% of the maximum permissible dose (0.1 rem). This gives a chance of about 1 in 600 000 per year of developing a cancer as a result, or about the effect of smoking five cigarettes every year or living a few extra hours in normal British conditions. The sandy parts of the coast have picked up very much less activity, so that children and tourists are unlikely to be affected by even this small risk.

In November 1983 a significant amount of contaminated Purex (the organic liquid used to extract uranium and plutonium from spent fuel) was accidentally transferred to the pipeline that normally carries dilute solutions of soluble wastes out into the deeper parts of the Irish Sea. Since Purex is insoluble in sea-water and is also less dense, it rose to the surface forming an oily slick, first detected by members of Greenpeace. Apparently the management had expected that this would be broken up by wave motion, and no attempt was made to disperse it with a detergent. A second batch of the same material seems to have been run out a day or two later, again with no attempt to disperse it. The floating slick was brought in by an onshore wind, and contaminated both the silt in the Ravenglass Estuary and the sandy beaches further north with ruthenium-106 and rhodium-106 (with a half-life of about a year) which emit both beta and gamma rays. The beta emission would not be of great importance even to children running around with bare feet, and the gamma emission would add little to that due to radioactivities already accumulated and giving up to 40 times the natural gamma emission in a few areas. Natural gamma rays provide about a third of our natural background (neglecting radon indoors) so the total effect of the contamination was to raise the radiation dose rate to someone walking on the beach by 10 to 15 times, perhaps to 100 mrem a month. This does not represent a serious hazard. One would have to spend two or three hours walking on the beach every day of the year to increase one's annual dose to the level normal in Colorado or Wyoming, and this would be equivalent to the extra cancer risk of living for a month in an average British city.

Much more important than the general contamination was the discovery of some small patches of dead seaweed and pieces of rubber tube which were far more contaminated with ruthenium-106 and rhodium-106 than was the beach. These emit energetic beta particles capable of penetrating a few millimetres into the body, and the most contaminated materials found could give a surface dose of up to 30 rems in an hour. It is unlikely that anyone would remain in contact with dead seaweed for long periods, but bits of rubber tube could easily be picked up and played with or pocketed by a small child. To produce surface damage leading to a slow-healing sore might take only a few hours, but children do keep very odd objects for at least as long as this, and no one would want to risk their children playing alone on a beach on which such things could be found. Since only perhaps a ten-thousandth part of the body would have been irradiated, the risk of a subsequent cancer would have been very small, and it may well turn out that BNFL did not exceed the legal annual limit for causing irradiation of the public; but there is no question whatever that they grossly breached the ALARA principle (as low as reasonably achievable). As a result BNFL was prosecuted on six charges, pleading guilty on one, being acquitted on two and being found guilty on the remaining three. It was fined £10 000 and ordered to pay prosecution costs of up to £60 000.

WIND-BLOWN PLUTONIUM FROM THE SILT

Plutonium in the shore silt is now of no direct consequence for the reasons given above, if it stays on the shore; but its half-life is long and it could therefore accumulate indefinitely as the thickness of silt builds up. The surface concentration is unlikely to change by any large factor, although the depth of the contaminated layer will increase since the coast in this area is being built up rather than eroded. The plutonium, of which 95% will be in the silt, is an alpha-emiiter, and will therefore affect no one who does not eat or inhale the silt.

Detailed measurements of the airborne activities from blown spray droplets or resuspended silt particles in Cumbria have been made by a team of investigators from Harwell (Eakins *et al* 1982). Collectors in the form of muslin screens, 5 m long and 1 m deep, with their lower edges 1 m above the ground, were set up at nine different points along the coast, from Walney Island in the south to Powfoot on the other side of the Solway Firth in the north, together with a screen at Hayling Island in Hampshire to record the background activity. Each muslin screen was set up at right angles to the wind direction for a measured time of two hours or more for each observation. Measurements were made both with onshore winds to record spray-carried material from the sea, and with offshore winds with a screen close to the sea at low tide to look for plutonium in resuspended particles from dry silt or saltmarsh.

The amount of activity found in the offshore wind on the Cumbrian coast was no higher than that observed at Hayling Island, so that air resuspension of plutonium directly from silt must be very small. The air concentrations of plutonium blowing inshore were never large; the largest observed on any one day was about 500 times lower than the maximum permissible concentration in laboratory air breathed for 40 hours per week for a working lifetime. The smallest, from 35 km north of Sellafield, was nearly 1000 times lower than the largest. The permissible air concentration for members of the general public is of course lower than for laboratory workers, but it is unlikely that anyone could be exposed to more than 1% of the ICRP limit for members of the public. An interesting feature was that the proportion of plutonium in the salt caught on the screen was much larger than the proportion in sea-water itself. This showed that the activity observed was due to plutonium adsorbed on particles of the more radioactive silt which had been entrained in foam from the breaking waves and picked up by the wind from this (Eakins *et al* 1982).

CONCENTRATION IN LIVING ORGANISMS

Besides the physical concentration in silt, there are two important forms of biological concentration—one of ruthenium-106 and other activities in the seaweed *Porphyra umbilicalis,* and the other of caesium-137 in fish. The concentrations are far too low to affect appreciably the growth and well being of the organisms themselves, but are important to us because each of them is eaten by people. Until around 1974 the *Porphyra* activity was the more important. This seaweed was collected and sent to South Wales to be mixed with local weed and made into laver bread, which was eaten by some 50 000 people in the Swansea area. With laver bread containing about 20% of Cumbrian weed, these people will have received a regular annual collective dose of radiation to the gut which totalled around 120 man-rems a year, and which if continued indefinitely could have led to the development of malignant disease by a little more than one person per century. Collection of Cumbrian *Porphyra* has now ceased, so that this risk does not at present exist.

In the years following 1972 there was a rapid increase by nearly ten times of the amount of caesium-137 in the discharge, due to longer storage and more corrosion of Magnox fuel rods in the storage ponds. The concentration of caesium-137 in edible fish has become the most important source of radiation dose to the public for which the nuclear industry can be blamed. Whiting, cod, and bottom-dwelling fish such as plaice or brill near the outfall build up the greatest body burdens, and a person who ate one fish from the most contaminated area every day of the year could build up a body burden of 20–30% of the permitted maximum. (Actual measurements of the body burden of several local fishermen have shown none with more than 10% of this maximum.) While this is by definition within the acceptable range, the rapid build-up after 1972 gave cause for concern. Discussions between BNFL, the Ministry of Agriculture, Fisheries and Food (MAFF) and the Nuclear Installations Inspectorate (NII) have led to measures producing an improvement. BNFL is now operating an extra caesium-removal process by means of a zeolite ion exchanger which has led to a levelling-off of the caesium discharge, and has led to a useful reduction. When THORP is in operation, an improved system using an ion-exchange column should be available, but the zeolite system already in use should keep the concentration in fish well below the permitted limit even with the increased caesium-137 throughput expected. Other radioactive elements taken up by fish, being concentrated in their bones or gut, are not important to human consumers.

COLLECTIVE DOSE FROM SELLAFIELD EFFLUENT

The collective dose to the whole population from the radioactive caesium in fish can be estimated from observations made by research staff at the MAFF's Lowestoft laboratories. Landings of fish round the coast are examined and the radioactivity of samples measured, so that the total amount of radioactive caesium landed can be estimated. Assuming that all of the edible parts of the fish are eaten by people, and not given to the cat, the annual collective dose can then be calculated.

Collective doses recorded by the Fisheries Research Laboratory for total beta radioactivity, mostly caesium-137 and caesium-134, are shown in table 11.1.

Table 11.1 Collective dose to the public due to effluents from Sellafield (MAFF reports, 1975–88).

Year	UK annual collective dose (man-rems)	Rest of Europe (man-rems)
1972–3	1 500	1 900
1975	8 300	5 700
1977	8 900	8 000
1979	13 000	17 000
1981	13 000	15 000
1983	7 000	11 000
1985	5 000	8 000
1987	3 000	6 000

If the 1987 figure is maintained, it would mean that for each decade of working an extra three hypothetical cancer deaths in the UK would be expected, plus six in the rest of Europe. The collective dose each year is in part due to radioactive material discharged in previous years, but the contribution from earlier years will not be great because the soluble elements such as caesium will be swept away from the Irish Sea in about a year and out into the vastly greater volume of the North Atlantic in a further year, where it will be diluted some ten thousand times, and the amount taken up by edible fish will be negligible.

The total number of deaths from the material processed up to 1987 would work out at about 0.15 deaths per GWye, spread over the present and the next few decades if all the processed fuel came from power production. In fact the total fuel processed has been more than that produced by the electric power stations, and must have included a lot derived from bomb production. The rate for which power production is responsible can hardly be more than 0.1 deaths per GWye, and is being improved as the AGRs produce a greater proportion of the output.

It is noteworthy that the chief alpha emitter in the edible parts of fish is the naturally radioactive element polonium, which has about 50 times more activity per gram in fish everywhere than the small plutonium-plus-americium activity in plaice taken near Sellafield (Pentreath *et al* 1979). As the maximum permissible regular intake of polonium is less than that of the plutonium and americium combined, the presence of these is of little importance. There is a larger uptake of plutonium by edible shellfish such as mussels and winkles on the coast near Sellafield, but this is not very important for two reasons. Firstly, it is forbidden to collect mussels or winkles for sale from the shore around Sellafield because untreated sewage is discharged there; and secondly at most five parts in 10 000 of plutonium are absorbed by the mammalian gut. All of the caesium swallowed is absorbed, but this does not stay indefinitely in the body; half of any quantity in the body is excreted in every two or three months.

Much natural concern has been aroused by five cases of leukaemia in 30 years among children or young people in Seascale, a small seaside town of 2000 inhabitants a couple of miles south of the Sellafield works. Leukaemia is not common, and if the risk were the same everywhere one such case in a hundred years in such a small place would have been expected. As was pointed out in Chapter 3, the risks of cancer are not the same everywhere, and a few such concentrations of cases might be expected somewhere in Britain. Nevertheless the close proximity of Sellafield naturally aroused fears that its radioactive effluent was responsible.

A thorough study of this possibility was made by an independent advisory group chaired by Sir Douglas Black (Black 1984), with the help of the NRPB which made a thorough study of the radiation doses received by children in the area, by a study of the local radioactive contamination of sea-food and of the environment together with measurements of the body burden of children at Seascale at the present time. This showed an addition of 20 mrem a year to the usual background of 180 mrem a year to be expected from natural causes, and this is no more than the variation of natural background found in other parts of Britain. After the publication of the Report it was found that an extra source of exposure arising from Sellafield had been missed, but this was not large enough to invalidate the conclusion.

The *Black Report* accepted that it could not be rigorously proved that there was no undetected route by which much larger amounts of radiation could have reached the population of Seascale throughout the 30 years concerned, but regarded this as exceedingly improbable.

The amounts of radiation needed would have been large. At Hiroshima a group of 2000 survivors would have needed more than 200 rems, which the younger children in Seascale would have had to accumulate in two or three years, to produce this number of leukaemias among under-fifteens, and these would have been accompanied by 16 further cases among the over-fifteens, which were not observed.

If the cause was not pure chance, which is highly unlikely, the most likely cause is an infective agent. Virus-induced leukaemias are currently an active field of research, and it may be relevant that Seascale was one of the few townships discharging raw sewage on to the beach—at a point 150 yards *above* spring tide low water mark. A research campaign was clearly needed to look for other causes of 'clumps' of leukaemias and other cancers without the current prejudice which looks only for radiation as the cause.

Since my first edition several concentrations of child leukaemias have been found in small wards in areas remote from any nuclear establishments, and on Tyneside a large excess was found—165 cases between 1968 and 1985 (*The Listener* 26 November 1987). What seems to me a convincing explanation for the cases at Seascale and Thurso (often blamed on Dounreay, 13 km away) has been found by Dr Leo Kinlen (1988). He suggested that large influxes of people to a small and isolated community brought with them virus infections to which leukaemia was a rare response. To test this idea he looked for a community where the population had rapidly increased; in the New Town at Glenrothes, far from any nuclear establishment, the population had increased 20 times between 1950 and 1971, and here he found there was a significant excess of child leukaemias (7 deaths observed, 1.49 expected). Such a successful prediction must carry a lot more weight than the quantitatively unsupported attribution to ionising radiation. It is worth noting that not only does Kinlen given an explanation for leukaemias but he has also provided a reason why many cases appear near large nuclear establishments whether or not they release appreciable amounts of radiation;

these are regularly built in rural areas, and all of them bring in large numbers of people from elsewhere.

This convincing explanation should be a lesson to those who have continued to insist that Sellafield was responsible for the Seascale leukaemias, without any evidence of significant radiation doses. Their unreasonable persistence must have delayed by five years or more the development of a vaccine to protect children where small, hitherto isolated, communities are rapidly increased by apparently unaffected people from elsewhere.

As the Magnox stations come to the end of their lives, the discharge of radioactive material into the Irish Sea should be very much reduced, owing to the longer storage of spent fuel from AGRs and PWRs before processing. For each five years of storage before processing any ruthenium-106 that is released will be reduced by decay 32 times, the caesium-134 by 5.5 times, and even the caesium-137 by about 6%.

Iodine-129 has caused exaggerated concern. The amount released per GWye is only 1 Ci; but this isotope of iodine has a half-life of 17 million years, and in one million years or less it will have spread uniformly round the world. The collective dose over a world population as big as it is now could cause five hypothetical deaths in the first 17 million years, and in the 1000 million years before it has all decayed there could be 10 such deaths. This is the same increase in risk as could be incurred if everyone put one sheet of newspaper on or under the carpet, thus increasing the cosmic ray dose by getting closer to the top of the atmosphere.

This seems to me to be taking arithmetic too far. I am interested in the future or I would not be writing this book; but there is a limit to the amount of the future that we can sensibly try to influence.

ACCIDENTS

It is clearly insufficient to consider only the hazards of normal operation. Careful design and well planned routine monitoring can reduce the risk of accident a great deal, but it is not humanly possible to guard against or even to think of every possible accident. It is, however, possible to make some reliable statements about the *kinds* of accident which could occur.

To take an analogy: if one kept a lion in the garden there would be a number of ways in which one's children would be at risk. If one kept a shark in a pool in the garden there might be equally serious risks, but one would not have to guard against the shark jumping onto the outhouse roof and climbing through the children's window. Sellafield handles vast quantities of highly lethal material. If someone fell into one of the high-level storage tanks he would receive a lethal dose of radiation in a few thousandths of a second (and would be as dead as if he had fallen into a vat of sulphuric acid). But if the material in the storage tanks were poured over the ground it would not evaporate into the air. As I stated earlier the bulk of the small amount of volatile material in the wastes is already allowed to escape at an earlier stage, with a small and calculable risk. Any kind of container breakage can therefore cause damage only over a limited area.

Externally caused accidents are not very probable. The processing lines and storage tanks are protected by buildings and by 60 cm or more of reinforced concrete. The only accidental means of breaching them would be to fly a heavy aeroplane directly into one of them. Crashes of large aeroplanes in Britain average less than one a year apart from crashes on take-off or landing. RAF crashes may be one or two per year. The effective target area of lines plus tanks is at most 10 000 m^2 or 0.01 km^2. The area of Britain is about 230 000 km^2. The chance of an aeroplane hitting the process line or a storage tank next year is therefore less than one in 10 million. If such an accident did occur, it is probable that most if not all of the people in the building concerned would be killed, but neighbouring buildings could be evacuated in plenty of time and there would be no serious spread of active material outside the site.

A break in the pipeline discharging low-level effluent into the sea is less important but much less improbable. A mechanical digger operating in the wrong place could break it open, but the resulting effect would be small. The discharge of beta emitters authorised represents an average rate of only a few hundred curies an hour in working periods; and a discharge of even a few thousand curies onto the beach or the site could be handled without difficulty and without any escape inland.

Major fires, even if assisted by crashing a petrol tanker into a building, would not collapse the relevant buildings, breach the concrete protection of the processing lines, nor transfer enough heat through the concrete shields to damage the stainless steel pipelines or boil their contents.

Internally produced accidents, especially those arising from human error, will of course occur. It is not always appreciated that in establishing reliable safeguards against major accidents one may actually *increase* the probability of minor ones, since reduplication of safety measures means that there are more items which could (harmlessly) go wrong.

One much publicised accident occurred when 9000 litres of liquid containing 100 000 Ci of fission products were found in March 1979 to have leaked from an unmonitored storage tank, and it was some time (later found to be at least eight years) before the staff at BNFL even noticed it. The reason was that the leak contaminated only a layer of soil about 1 m thick at a minimum depth of 3 m. This was undetectable at the surface by the various contamination detectors deployed around the site. The leak was found only when some workmen excavating near the building concerned had their hands contaminated (to a level well below the quarterly permissible dose), and this was detected in the routine check at the end of their shift. A series of sampling holes drilled in and around the area affected showed that the activity had spread along the length of the building through the disturbed earth round the foundations, but had not—and had not in the following four years—spread appreciably beyond this. A deeper hole was drilled near the centre of the active area, and water seeping into this has been continuously pumped into a safe storage tank. The effect of this is that the water table is lowered at the central point to a depth well below that in any other part of the contaminated area, all of which therefore drains towards the pumped hole rather than spreading further.

The NII made a detailed study of the circumstances of this incident (NII 1979), and did not consider the BNFL was in breach of its general duties under the relevant acts, but did consider that it breached several conditions attached to its nuclear site licence. The NII however considered that prosecution of BNFL was not appropriate; but that to ensure that BNFL took the necessary remedial action, of which details were specified, as well as to publish the details of the incident, would be more appropriate. The report accepted that BNFL's radiological protection arrangements were such that any potentially hazardous spread of radioactivity would be detected and prevented.

At no time could there have been a risk to the public, either directly or from cattle drinking from the river along the southern boundary of the establishment site. The water table slopes downward away from the river—there it is of course close to the mean river level—and away from the rest of the land-side boundary towards the sea, into which, even if all the leaked activity had gone there during the eight years from 1971, it would not have exceeded the total of radioactive material that BNFL is allowed to release. Accordingly, not only was there no danger to the public, but if the leak had never been found there would have been no danger to the public.

The newspapers which attacked BNFL for not having a fail-safe system were unfair; the system did fail, and it was safe. This however does not absolve BNFL of all blame. It is clearly important not only to monitor all areas in which leaks of radioactive material would be dangerous—or even undesirable—but to make regular checks and records of the whereabouts of all significant amounts of radioactivity. If 100 000 Ci turned up in the wrong place, unnoticed for several years, there must for several years have been 100 000 fewer in the right place. It can be a useful precaution to wear both belt and braces, either of which can fail safely, but if you cannot tell when the braces have broken you may unknowingly continue for an indefinite period with nothing but a belt, which can no longer fail safely.

The only special type of internal accident that appears to need detailed comment is a so-called criticality accident, in which enough plutonium accumulates somewhere in the system to start a potentially explosive chain reaction. As was explained in the discussion of the Kyshtym disaster in Chapter 9, this can never lead to a bomb-type explosion and is unlikely to lead to a steam explosion. Due to an unexpected chemical precipitation, an abnormal quantity of plutonium did, a few years ago, accumulate in a vessel which should not have retained it. The effect noticed was an unexpected rise in temperature of the liquid contained, and an increased emission of neutrons. This was observed and cleaned up before enough plutonium had accumulated to cause any damage. If a really serious excess were directed into a particular vessel the solution would boil violently, steam bubbles would reduce the volume to below the critical value, and steam pressure would prevent access of any further solution before a major explosion could be developed. If the steam pressure did rise fast enough to rupture the vessel, the reaction would automatically stop as the boiling solution dispersed itself. I do not see therefore how an explosion large enough to disrupt the concrete shields could occur. I don't however like saying that things are impossible, and apparently fool-proof systems may not be proof against a really innovative fool. What *is* clear is that, even if such a disruptive steam explosion could take place, plutonium is not volatile and little would spread outside the site or even the building, so that the public would not suffer extra contamination.

I shall now return from the accidents that have vanishingly small probabilities of ever happening at all to the other waste disposal procedures that take place regularly. Radioactive wastes are rather arbitrarily divided into three groups: low level, intermediate level, and high level, according to how concentratedly radioactive they are.

WASTE DISPOSAL PROCEDURES

Very weakly contaminated material with less than $0.1\,\text{mCi}$ per m^3, which is much shorter-lived and has about a quarter of the natural radioactivity of the potassium sulphate used as a fertiliser, can often be buried on local authority tips. If the wastes are sealed in containers and buried two metres deep, ten times this activity is permitted at such tips. (Even human ashes from crematoria will be naturally more radioactive than this and will be very long-lived.)

Low-level wastes

Low-level waste disposed of by BNFL in Cumbria consists of a great variety of weakly radioactive wastes, including such things as protective clothing, which may not be radioactive at all but which have been used in conditions in which contamination might have occurred. The material will usually be compacted, or the organic part of very-low-level material may have been incinerated to leave a radioactive ash still within the limits which control the classification as low level. These limits are listed below.

(1) The alpha activity on average does not exceed 20 mCi per m^3.
(2) The beta activity on average does not exceed 60 mCi per m^3.
(3) The dose rate at the surface of the unshielded wastes does not exceed 0.75 rads per hour (which for the gamma rays concerned is 0.75 rems per hour).

A practical target level to ensure that the limit is not accidentally exceeded would be 20% or so of the maximum permissible surface dose. At 20% of the permissible dose, if you slept for nine hours on a 1 m thick layer of the waste, you would increase your risk of eventually dying of cancer from the currently normal 24% to 24.01% and would probably be breaking some law.

Any other industry with similarly hazardous material would simply contract to have it removed by a hazardous-waste disposal firm, together with the far more dangerous cyanides and similar chemical wastes, and the materials contaminated with infinitely long-lived elements such as arsenic, lead or cadmium, which such firms already handle. It is not clear to me why low-level waste is not dropped down a worked-out coastal coalpit, in which the water bound to be present is known to be moving towards the sea. The amount of radioactivity to be got rid of each year is about the same as the amount brought out from coal mines. In fact these wastes are being placed in a shallow land-burial site at Drigg in Cumbria. Here trenches are dug down to the underlying boulder clay, some 8 m below the surface, filled with waste to within 2 m from the top, and then back-filled with granite chippings and top-soil. The land at Drigg slopes down to the sea, and the impervious boulder clay ensures that all ground-water movement runs that way. The present disposal rate is about 50 000 m^3 per year from the whole country, and the site should be capable of continuing to take waste at this rate for a further 8 years. It would clearly be sensible however to establish a second site soon, leaving the Drigg site to be used only by BNFL, only a few miles away. The site could then handle all of BNFL's low-level waste for two or three times as long.

Intermediate-level wastes—land disposal

German material has for some years been stored underground in salt deposits, but British intermediate-level wastes have been disposed of in the ocean. Preparations for land burial have however been made, and will have to be implemented since the National Union of Seamen has (July 1983) refused to handle such waste ostensibly because of its possible effects on the sea-bottom environment.

Such waste may contain up to 1 Ci of alpha activity and up to 100 Ci of beta–gamma activity per cubic metre. Well-prepared land burial is suitable for this, in which nearly all of the activity consists of relatively short-lived beta–gamma emitters. The activities of even the two important longer-lived materials, caesium-137 and strontium-90, will have fallen by 1000 times in 300 years. Intermediate wastes for land burial would be solidified or contained in cement or plastic. Those with relatively low gamma activity not requiring thick shielding during transport would be packed in drums lined with 8 cm of concrete and those with higher gamma activity would be transported in returnable shielding containers. It has been proposed that this level of waste should be disposed of in a concrete-lined trench 19 m deep with walls of reinforced concrete, filled to a depth of 8 m with waste drums and covered with a metre of reinforced concrete. This would be water-proofed with a layer of bitumen, itself covered with a 3 m layer of compacted clay followed by yet another metre of reinforced concrete to prevent accidental intrusion, and a further 5 m of rolled clay which is impervious to rain water.

This all seems to be quite a bit safer than a 90 tonne chlorine storage tank above ground. In Spring 1984 two deep burial sites in worked-out mines were found to be suitable, but nothing has been done owing to much doubtless well-intentioned but ill-informed opposition.

Intermediate-level wastes—sea disposal

Until recently most intermediate-level UK wastes have been deposited in the deepest part of the north east Atlantic, selected as being free from upwelling currents and from minerals of economic importance. Packaged solid wastes have been disposed of in this way since 1946. At the London Convention of 1972, proposals for regulations governing disposal at sea were drawn up, and have been ratified by over 50 countries. The Convention not only controls radioactive material but also forbids the dumping of some serious and persistent chemical poisons such as mercury and cadmium, and requires approval for proposed dumping of a second class of materials such as arsenic, lead and copper. The International Atomic Energy Authority (IAEA) was designated as the competent body to define the types of radioactive wastes for which ocean disposal is forbidden (which includes all high-level wastes), and which types require special permits, such as intermediate-level wastes.

Since 1978 the IAEA has set limits of 1 Ci per tonne for alpha emitters (but limited to 0.1 Ci per tonne for radium-226); a million curies per tonne of tritium and 100 curies per tonne of all other beta–gamma emitters. These limits are fixed on the assumption that not more than 100 000 tonnes per year will be deposited over a period of 40 years at a single site. Over ten years the UK in fact dumped a little over 2000 tonnes a year with around half the permissible radioactive content.

For the purpose of this limit the North Atlantic is regarded as a single site for alpha emitters, but may be treated as up to ten sites for beta–gamma emitters. These limits are based on the pessimistic assumptions that all activities will escape in a time short compared with their half-lives and that none are adsorbed by sediments, into which the drums containing wastes may penetrate significantly. The intention is to ensure that the bottom waters of the ocean should not be contaminated to levels higher than would be acceptable in surface water.

The area currently recommended is about 800 km southwest of Land's End, and has an average depth in excess of 4 km. The exchange of water at that depth with the surface is very slow, with an average time of perhaps as much as 1000 years, so that as far as the effects on human beings are concerned the beta–gamma limitation seems nearly pointless; the most important caesium-137 and strontium-90 activities will have gone down over 1000 million times in 1000 years. It is unlikely that any radium-226 at all will appear in nuclear power station wastes—though it will do so in coal-fired station wastes—but if it should be required to be removed from the tailings from uranium production there would be some thousands of curies a year to be got rid of. Ten thousand curies of pure radium would occupy about two litres, which should be possible to seal up in an adequately permanent manner without adding too much to the cost of extracting it.

Natural radioactivity in the ocean

There are already 500 000 million curies of potassium-40 in the water of the ocean, and 40 000 million curies of rubidium-87, both elements being concentrated in all living cells. A further 150 million curies each of radium-226 and polonium-210, all of natural origin, are present. Concentrations of the natural alpha emitter polonium-210 in the digestive tracts of sea creatures are enormous by any standard. A tuna fish, of the type often canned for us to eat, has been found to be receiving 80 rems per year in its pyloric caecum (a kind of sac like our appendix) (IAEA

1982) but we do not eat this. In 1986 I tested tuna from a supermarket tin for alpha emitters, and the test showed no measurable alpha emission.

The concentration of radium and other activities is already quite high at the bottom of the ocean. There are over 1000 million curies of uranium in the ocean, and if its descendants were in equilibrium there should be over 1000 million curies of radium. Thorium-232, with a 14 000 million year half-life, is two or three times more abundant than uranium in most rocks and soils but is much *less* abundant in the ocean. These anomalies are accounted for by the fact that the thorium (but not the uranium) reaching the sea is precipitated and sinks to the bottom. The direct parent of radium-226 (see Appendix 2) is the 80 000 year half-life thorium-230, which has plenty of time to sink to the bottom after it is produced from its parent uranium and before very much of it decays. As a result, a large quantity of the long-lived thorium-232 and nearly all of the thorium-230 is to be found in a layer, which may be only a few centimetres thick, at the bottom of the ocean† (Woodhead and Pentreath 1980). The 'missing' radium-226 is also there; about five-sixths adsorbed within this layer and one-sixth in the water near the bottom. Together with its alpha-emitting precursor and four alpha-emitting descendants some 6000 million Ci have started life in the bottom layer. This represents an average of 60 Ci per km^2, much larger than the amount added in the wastes, which were limited to a maximum of four drums per km^2.

All sea-bottom life must necessarily have evolved to cope with this. Even at the surface some invertebrates seem able to cope with very large alpha-particle doses. One of the small planktonic crustaceans, a penaeid shrimp, normally concentrates natural polonium in the hepatopancreas (part of its digestive system) to a level which delivers a local dose of up to 500 rems a year. Presumably the predators which live on these creatures can also stomach this, so that it gets passed up the food chain to the fish—such as the tuna mentioned above.

The permissible addition of 10 000 Ci per year for 40 years in the north east Atlantic could (if it escapes from its drums) double the alpha activity over 7000 km^2 (less than one ten-thousandth of the ocean area), with presumably the same depth distribution as the natural radium, also coming from the bottom. The actual alpha activity deposited in the five years 1978–82 was 7000 Ci (*Atom* 1983), which, if it is quickly released from the drums, will increase by 1% the alpha activity over about 12 000 km^2 of the 80 million km^2 of the Atlantic.

The natural alpha activity at the silt surface at the bottom of the ocean has been measured at several points (Woodhead and Pentreath 1980), and is actually *greater* than the combined alpha activity due to

---

† The build-up of silt on the ocean bottom is slow, usually less than 1 cm every 1000 years.

plutonium-239, plutonium-238 and americium-241 in the silt near the discharge pipe from Sellafield which has caused so much local concern (Pentreath *et al* 1979).

Woodhead and Pentreath calculated the effect of depositing *high*-level wastes on the ocean bottom. They assumed about a million times more of both alpha activity and beta–gamma emitting fission products than are permitted by the intermediate-level dumping convention, distributed in canisters over 10 000 km^2.

They found that the total exposure of deep-sea fish due to the combined absorption of all of the radioactive elements taken together, assuming that all activity would escape from its containers, would rise to a maximum of 87 mrem a year after a few decades—close to our own present natural background—and would take over 1000 years to drop ten times. One millionth of this (from intermediate-level wastes) would not be important to either fish, crustacea or ourselves if we should ever develop a very expensive taste for deep-ocean sea-food.

It is clear that the recent rate of intermediate-level dumping incorporates a very good safety factor, both for us and for the dwellers at the bottom of the ocean (as indeed it should). I am glad to note that many members of Greenpeace and other organisations concerned with the world environment are beginning to modify their concern with the wastes of nuclear power, and to put more effort into the problems of acid rain and carbon dioxide production and of the destruction of virgin forests.

As a result of their earlier enthusiastic efforts however, British intermediate-level wastes are likely to be buried more expensively, though equally safely, on land.

In figure 11.2 is shown the repository already constructed by Sweden for low and intermediate wastes. Near the entrance to the tunnel there is a building housing ventilation machinery for the repository, a dock for a ship specially designed to carry nuclear waste, and a causeway connecting to the Forsmark power station. The repository is 60 m below the sea bed and consists of four horizontal rock caverns 160 m long and 14–18 m wide, for low-level waste packages, and a huge concrete silo, built in a rock cavern, 50 m deep and 25 m in diameter, for intermediate-level waste. It is proposed that after careful monitoring of the behaviour of the packaging and of the concrete, the repository will be extended to take all the wastes from Sweden's nuclear power stations.

It seems to me that suitable holes in the floors of the horizontal caverns in such a repository would be entirely suitable for our high-level wastes in their steel canisters, after their fifty years 'cooling off' period described below. Careful monitoring of the behaviour of the intermediate-level wastes during this period will provide us with the data needed to decide whether a similar repository with a separate hole in the floors of the horizontal caverns for each canister of high-level waste will be entirely adequate, or whether something larger and deeper would be desirable. Well before the end of the fifty year period the low-level wastes will have an activity equal to or less than that of a human body, and could be removed and decently buried any-where—perhaps in the area designated for the spoil heap of a coal mine.

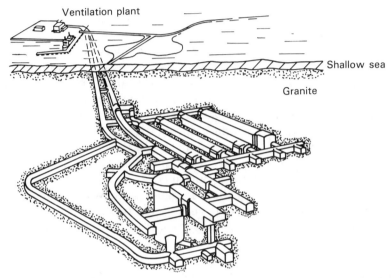

Figure 11.2 Undersea storage caverns constructed off the coast of Sweden (*Atom* September 1988).

High-level wastes

The problem of disposal of highly radioactive wastes is to the public the most disturbing aspect of the development of nuclear power. It is also a problem which has received intensive study by the scientists and engineers involved in the industry. It is clear that the measures which are satisfactory for low- and intermediate-level wastes totalling hun-dreds of thousands of curies are quite inadequate for the high-level wastes totalling thousands of millions of curies. As was stated at the beginning of this chapter, these wastes are potentially the most lethal assembly of material that this country has known. Just after processing, the waste is a great deal more dangerous than any chemical poison, apart perhaps from the nerve gases and some bacterial toxins such as that causing botulism. But it is not in a different class of toxicity from some quite well-known poisons, and unlike the metallic poisons the toxicity gets less with time. Ten years after it has left a thermal reactor, the weight of high-level waste that would have to be swallowed to give an even chance of dying is still around 25 times less than the weight of potassium cyanide that would give the same chance; but after 500 years about 250 times *more* of the wastes than of potassium cyanide would have to be consumed (Cohen 1977). This does not mean that the wastes can be taken lightly. They can cause a significant chance of death from cancer in very much smaller amounts than would give even chances of killing outright. But the scale of the problem is not very different from the scale of other problems that we have been able to treat with success, even if we have not always done so.

The deep-trench system described for the intermediate-level wastes would be quite satisfactory for high-level wastes if we could be sure it would be left alone. The radiation from 1000 million Ci of gamma emitters coming up through 10 m of clay and concrete, and being reduced by a factor of two for every 4–5 cm, would be unlikely to deliver a single gamma ray to the surface during the millions of years elapsing before the whole activity had died away. On the other hand, although the beta–gamma activity in 1000 years' time would have fallen by nearly 100 000 times to a comfortable intermediate level, the alpha activity would have fallen by less than 10 times and would still be many thousands of times above the level for which the trench system was designed. Furthermore, if records were lost and someone did dig down through the multiple barriers in a couple of generations' time, under the erroneous impression that all this concrete must conceal some treasure, drums of intermediate-level waste would not cause him much damage, but high-level waste could deliver a prompt lethal dose in a matter of seconds. Something better than a trench is therefore needed.

Glassification of wastes

It is proposed to transform the thousands of tonnes of lethally radio-active solution in the Sellafield cooling tanks into something that cannot be spilt and cannot boil dangerously if abandoned or mishandled. The radioactive materials are combined with appropriate amounts of silica and additional metallic salts and melted to form an insoluble glass—a dark glass, but like ordinary window glass a permanent solid, impenetrable by water. The melting will be done in a canister with walls 12 mm thick and made of a special stainless steel. When cooled it will become a solid cylindrical block of glass still firmly held in the stainless steel case. This canister will have a lid of similar material welded on, after which there is no way the radioactive contents can escape.

This was done on a pilot scale at Harwell over 20 years ago (the Harvest process), and has been done for some years on a production scale in France. The Windscale vitrification plant is expected to come on stream in 1990. The canisters will be 30 cm in diameter and 1.5 metres long, and each will at first contain just under a million curies of activity and will be giving out 2–3 kW of heat. The canisters will therefore be—as they already are in France—spaced well apart in an air-cooled underground chamber. The air cooling will be run by convection, driven by the heat of the wastes themselves and needing no pumps which might go wrong.

It is at present proposed that the canisters should eventually be used for permanent disposal underground, but it would be more difficult to get rid of the heat produced in a deep underground repository, and it is proposed to keep them for perhaps 50 years in the air-cooled store already described, by which time the heat output per canister would have dropped to less than 0.5 kW.

Final disposal of high-level wastes

A number of means of disposal have been proposed. The most drastic and final is to put the wastes in a rocket and shoot them into the sun. The snag is that rockets are not perfect, and an unacceptable fraction—one in 10 000 would be an unacceptable fraction—are likely to crash or to burn up in the upper atmosphere, spreading their radioactive load in some entirely unpredictable fashion over some unfortunate part of the globe. In the USSR it is proposed to pump the high-level wastes in liquid form into a water-bearing level between two impervious layers deep underground, where the activities will be dispersed slowly until they decay. I have no data on this, and shall discuss in detail only the British proposals.

Deep burial of wastes. The most probable method of disposal proposed would be burial at an adequate depth in suitable rock on dry land. Apart from other advantages, this would make it possible for later generations to recover the wastes in the unlikely case that they should wish to do so. This would be very difficult after deep ocean burial. One possible site would be old hard rock which has been compacted for 100 million years or so under a pressure of 1000 tonnes per m^2 from the overlying rocks, and helped to settle by the minor earthquakes which may occur every few hundred years even in the most stable surroundings. Salt or clay at similar depths are also being considered. Either of these has a degree of plasticity so that the cavities made for the emplacement of wastes would fill themselves up in a few decades.

Sealed, and 500 to 1000 m underground, the wastes—however radioactive—would seem at first sight to present no risk, except that if the shaft were left open somebody might fall in.

Concern has been expressed about possible earthquakes, but apart from the fact that all the old rock formations considered must have survived hundreds of thousands of earthquakes, it is difficult to see what an earthquake could do other than to make the roof of the repository fall in. Earthquakes do not bring up material from 500 metres down; the only natural way of doing this is for a new volcano capable of melting and ejecting both the canned wastes and the surrounding rock to develop. A volcano capable of this would be likely to disperse the wastes through tens of millions of tons of lava, and would certainly not leave anyone alive in the regions inundated by this to worry about whether or not the lava was radioactive. In any case, new volcanoes appear along known lines, points on which would not be chosen for a waste repository. The probability of a volcano appearing in any particular square kilometre in Britain is thought to be about one in a million million per year and new volcanoes give warning signs.

The reason why a deep repository more or less anywhere cannot be regarded as permanently safe is that sooner or later water might penetrate the repository and dissolve away the containers and their contents. If the water then found its way to the surface, or the sea, and if the radioactive elements were not adsorbed and bound to the strata through which the water passed, and if enough radioactive contents to matter then found their way into a source of drinking water, or were absorbed by some sea organisms that we eat, and if it were not noticed and avoided, this would certainly present a cancer risk to the consumers. With such a succession of ifs it may seem stupid to worry, and maybe it is stupid to worry, but it would be quite irresponsible to plan permanent storage in any deep repository without a serious examination of the probability of every if, and a quantitative calculation of the risks that might be run.

Starting at the beginning, it may seem sufficiently unlikely that rock which has been compact and water-free for 100 million years should suddenly allow water to enter the repository. Rocks of the kinds which have been sufficiently impervious to keep vast quantities of oil and natural gas underground for hundreds of millions of years should be waterproof enough. In fact, however, even the most solid-seeming rock is rarely quite impervious to water; and it can be impervious to gas simply because all of the fine channels between the mineral crystals composing the rock, and along which water can move molecule by molecule, are normally filled with water under pressure from the water-bearing strata above. Water moves very slowly through such channels, but there is another important factor. The rate of such water flow varies greatly from place to place in the same rock formation, so that trial borings will be needed to select a site for an underground repository into which the water flow will be negligibly small. Unfortunately, a site which seems suitable may not necessarily remain suitable. A repository is likely to contain the waste output from many gigawatt years of electricity production, and at least for the first 50 years this will be producing many thousands of kilowatts of heat. Rocks are poor conductors of heat, so over the first few decades the rocks surrounding the repository will get warm, will expand, and may form cracks over a region a good deal larger than the repository. This *might* open a route to a much more rapid flow of water than was reaching the repository while it was being dug out.

Experiments are in progress in Cornwall in which electrically heated dummy canisters have been placed in holes in hard rock and the effects on the surrounding rock measured. This study will be continued for many years before any canisters containing wastes need to be buried.

The heat output would be a lot less serious in clay or salt. Clay is self-sealing and for practical purposes impervious; the clay immediately round the canisters may be dried out and cracked, but the surrounding clay will follow any changes of size and remain impervious. Several cubic kilometres of salt in an underground salt dome, such as that at Gorleben in Germany, cannot contain water. If the repository were forgotten by future generations, someone might establish a new salt mine and bring up the salt by pumping down water and pumping up brine, but the amount of salt that can be removed this way is not great. If the entire River Elbe with eight times the average flow of the Thames, which passes close to Gorleben, were redirected through the salt deposit, and came out twice as salty as sea-water, it would take 100 years to dissolve the deposit away.

The wastes can of course be protected from water for a long time, but not for ever. Even the special stainless steel used for the wastes will not last indefinitely in hot salt water, and underground water which has percolated through rock will never be pure and may be salty. Accordingly, the Swedish engineers who have made a major study of the problem propose that, before permanent disposal, the canisters should be cast into a 10 cm thick copper sheath or a similar lead sheath, surrounded by a 6 mm shell of the chemically very inert metal titanium. The lead is not only resistant to corrosion—Roman lead articles sunk into salty silt in the Mediterranean have survived for 2000 years with little loss—but effectively absorbs the gamma radiation from the canned wastes which otherwise would ionise surrounding water, making it more corrosive. Under the environmental conditions expected to prevail around the canisters in the final repository, local corrosion of titanium has not been observed at all. In sea-water at a relatively high temperature, however, titanium is slowly corroded at a rate between 0.0001 and 0.0005 mm per year. Underground water may contain salts, but at concentrations much lower than in sea-water, and the canisters would be kept at a temperature more than 10 $^\circ$C above their surroundings by

their contained radioactivity for only a few hundred years. A period of
10 000 to 60 000 years is therefore required to penetrate 6 mm of
titanium. The corrosion of the lead will depend on the availability
of oxygen. At the maximum probable rate of corrosion of lead it would
be expected to take over a million years to oxidise all of the lead on the
canister. Corrosion is never uniform, and very slight changes in
microstructure or purity may lead to corrosion pits which corrode faster
than average. The stainless steel canister however will not be attacked
until nearly all the lead has gone, since the lead will act as a cathodic
protection, just as does a sprayed zinc or magnesium coating on a car
from which the paint has been scraped. Once unprotected, the stainless
steel would last for only a very short time compared with the titanium
and lead. (All this is on the far from probable assumption that there is an
unexpected water flow in the repository.) Eventually the radioactive
glass would be exposed, and this too would be leached slowly away in
the same waterflow. Again, the rate would be slow. Accelerated experiments in a fast flow of fresh hot water show leach rates of several
milligrams per square centimetre per year, but in the real case, by the
time the metal coverings have gone the glass will be only very slightly
warmer than its surroundings, and the water will not be fresh. In hard
rocks it may already be nearly saturated with silica, which is the main
constituent of glass. The rate of solution is therefore likely to be very
much slower than the leach rate in the experimental conditions.
Nevertheless, if the water flow continues, the glass will eventually be
dissolved.

An alternative to making the wastes into a glass has been developed in
Australia; this is to produce a synthetic material (known as Synroc)
similar to extremely insoluble natural minerals (Lee 1983). This could
certainly be better than the glass but the extra cost will not be
worthwhile unless the leach-rate is reduced by at least 100 times. There
is an uncertainty of at least 100 000 years over the time required to
remove the titanium and lead protection, and to add an extra 5000 or
10 000 years for leaching is hardly worthwhile. Only if the leach rate is
reduced so far that Synroc in canisters unprotected by lead or titanium
would last as long as glass protected by these would there be an
important gain. This is not impossible, and active research on Synroc is
proceeding in several laboratories.

Access of water to the titanium- and lead-sheathed canisters can be
very much slowed by surrounding them in their holes with a mixture of
powdered quartz and bentonite. Bentonite swells when it absorbs water
and thus creates a plastic sealing substance which is almost impervious
and will admit water only exceedingly slowly. It would already be
impacted dry to improve thermal conductivity (Pusch and Jacobsson
1980).

We are not of course considering a repository with only a single canister. The type of repository currently being considered for hard rock in Britain supposes that several thousand canisters, of which up to 10 000 should have been made by the year 2000, would each be placed in holes of a suitable size in a rectangular array. The canisters would be far enough apart to prevent overheating of the intervening rock in the first 50 years or so while a lot of heat is still being delivered by caesium-137 and strontium-90.

To achieve a significant water flow past a large number of canisters, water would have to pass through such a repository at a rate of millions of litres a year, which requires quite sizable or very numerous cracks to let the water in and another lot to let it out again. Since the rock will have been selected for its *lack* of water-bearing cracks, it is unlikely that large cracks induced by the construction or heating of the repository itself will extend very far. However, large quantities of water *can* percolate through surprisingly solid-looking material if there is a big enough area available. Although all of this is very improbable anyway, a continuous flow of water into the repository *cannot* occur unless there is somewhere for it to go; and this means that the water, carrying with it any dissolved fission products and transuranic elements, must eventually reach either the surface or the sea. From a repository 500–1000 m down it must travel at least some kilometres before it can reach the surface.

The radioactive elements will not travel as fast as the water. Individual atoms striking the rock through which the water is passing will be adsorbed for a time, redissolved and carried a little way, then readsorbed, and this will be repeated over and over again. The mechanism is well understood, and the effect is that each element will on the average move at a characteristic fraction of the water speed. After being dissolved, therefore, different elements will move at different average speeds and will reach the surface at different times. From laboratory experiments we can calculate the approximate rate of movement of each of the important elements for any particular flow (Hill and Grimwood 1978).

The basic assumptions are that a single repository will contain all the high-level wastes from reactors which have produced 330 GWy of electricity, a reasonable estimate of the amount that could be ready for glassification by the end of the century. The quantities of all significant isotopes in this amount of waste 10 years and a million years after processing are recorded. The important ones are shown in table 11.2.

Using these data, and assuming that all of the dissolved material will eventually reach the surface and get into drinking water, Hill and Grimwood show that over a period of a million years an extra 150 000 deaths from cancer could be produced.

Table 11.2 Important radioactive nuclides in high-level nuclear wastes (Hill and Grimwood 1978).

Nuclide	Half-life (yr)	Activity from postulated power programme (Ci)	
		10 years after reprocessing	10^6 years after reprocessing
Uranium-235	7.1×10^8	0.95	0.95
Uranium-238	4.5×10^9	18	18
Neptunium-237	2.1×10^6	3.8×10^3	2.5×10^3
Plutonium-239	2.4×10^4	1.8×10^4	zero
Plutonium-240	6540	5.0×10^4	zero
Americium-241	433	1.8×10^6	zero
Strontium-90	28.6	6.6×10^8	zero
Zirconium-93	1.5×10^6	2.1×10^4	7×10^3
Technetium-99	2.1×10^5	1.5×10^5	6.6×10^3
Palladium-107	6.5×10^6	1.2×10^3	1.1×10^3
Iodine-129	1.6×10^7	4.2×10^2	4.1×10^2
Caesium-135	2.3×10^6	3.2×10^3	2.5×10^3
Caesium-137	30.2	9.4×10^8	zero

These figures really look extremely serious, but could be reduced a great deal by a quite practicable method, to perform a second separation of the uranium and the heavy alpha-emitting transuranic elements before incorporating them in glass (Hill *et al* 1980). The result of this would be to reduce 100 times the doses from neptunium and radium.

The remaining few thousands of deaths would not seem serious to the people of the future. In the unlikely event that people in 100 000 years time have the same death rate from cancer that we do, an extra 1 per 200 years in Britain would be no more detectable than it would be today. If, as is much more likely, cancer has been conquered, none of the extra deaths calculated will occur anyway. Nevertheless, if extra deaths do occur, many people feel that it is morally wrong to be responsible for even the few extra deaths that the calculations predict. And we have to consider future repositories, perhaps in some numbers, if the nuclear power industry expands.

Even if I believed that events would in fact follow the course portrayed by the NRPB for the far future however, I would feel that we have a far larger moral responsibility for the car-less people dying now from air pollution in order to let the rest of us live in comfort with our cars.

Around 2000 BC, the early neolithic inhabitants of Britain maintained an extensive mining industry for flints in the chalk areas near Thetford in Norfolk, leaving behind them the series of pits known as Grimes' Graves. If the concerned and thinking people of the time had known how fast the population of Britain was growing, and had calculated the rate at which the resources of flints were being used up, they would have seen that society must collapse in 1500 years or so, when the flints ran out and there would be no defence against the wolves and bears. I do not know what discount rate should be applied, but it does not seem sensible to regard deaths thousands of years in the future as mainly our responsibility, any more than we can regard such fathers of the industrial revolution as Boulton, Watt and Murdoch as mainly responsible for the 6000 people we now kill each year on the roads. If we are leaving an extra source of cancers for the far future, we are also leaving the means to avoid both it and the much larger natural radioactive sources. The rare elements whose radioactive isotopes would do the predicted damage could easily be removed by water purification processes that we have developed to cope with chemical pollution today. The water supply of Los Angeles is already passed through ion-exchangers; while the inhabitants of Los Angeles have trouble with breathing, they are not at risk from traces of radioactive material in the water supply.

Fortunately there are reasons for believing that the NRPB study is unrealistic.

THE RISK FROM NATURAL URANIUM AND ITS DECAY PRODUCTS

According to the theoretical study discussed above, the largest number of deaths would arise from neptunium-237 and radium-226. In the glassified wastes there are 3800 Ci of neptunium-237 and only 18 Ci of the uranium-238 from which the radium is derived. If a higher proportion of the uranium had not been removed, the radium would have been about ten times as dangerous.

We have a direct and excellent check on the actual behaviour underground of uranium and all its descendants including radium. Uranium is not a common element, but it is very widely distributed in the earth's crust, at an average concentration of about 4 ppm (4 grams to the tonne) (*Handbook of Chemistry and Physics* 1981–2).

The uranium and the other radioactive elements in the wastes are of course much more concentrated than they are in soil or rock, but the concentration in the wastes does not increase the probability that any given radioactive atom will find its way up from the repository and into a person on the surface.

The total amount of uranium in the top 1000 m of the 230 000 km^2 of Britain is about 800 million Ci—40 million times the quantity in the high-level wastes that according to the theoretical study led to a peak annual collective dose of 1000 man-rems a year. This, as was stated above, would lead to about a death every 10 years for a million years or so. If the uranium and radium distributed below our feet did in fact behave as the theory suggested, we in Britain should now be having a collective dose of 40 000 million man-rems a year, giving four million cancer deaths a year from water-borne radium alone, neglecting the radon and polonium decay products. Without an extra dozen children per family the British would be dying out.

Even the top metre of British soil, through which all of our drinking water must percolate before it reaches us, contains 800 000 Ci of each of uranium and radium. Both uranium and radium do occur in some water supplies. The deep-well water supply in Wolverhampton used to—and probably still does—contain 30 mg/m^3 of uranium, nearly 1% of the maximum permissible concentration. London drinking water from the rivers Thames and Lea, which may be close to the national average, contains enough radium to give about 13 alpha particles per litre per hour (Hesketh 1980). If everyone in Britain is drinking water as heavily contaminated with radium as this, it could be causing two or three deaths from cancer per year in the whole population. The long-term effect of uranium in the wastes 1000 m down must be at least two million times less than that presented in the theory cited above.

Drinking water in Britain is mostly derived from surface water collected in reservoirs or directly from rivers; i.e. derived from recent rainfall rather than from deep wells. It is probably a coincidence, but if just half of the 60 Ci of radium annually distributed into the environment from the burning of coal, and perhaps not yet permanently locked into the soil, were dissolved in the 200 000 m^3 of rain falling on this country each year, the concentration would be exactly that found in the London water supply derived from the Thames and Lea.

It could be that the effects arising from uranium naturally present in the soil are even less than I have calculated. Our geological experience shows that nearly all of the atoms of nearly all substances underground stay harmlessly where they are. Even without coastal placement and extra heavy-element separation, neither of which has been applied to the 800 million Ci of uranium under our feet, it is likely that the original simple plan described by Hill and Grimwood would be more than good enough. The whole series of NRPB reports however did a difficult and very valuable job in exposing the physical and chemical factors that matter, and in highlighting the important features of their behaviour.

This has been of importance in enabling us to optimise our waste disposal plans, even if a procedure far short of the optimum would have been adequate.

Apart from the assumption that all of the canisters (derived from all the wastes expected from 330 GWye up to the year 2000) lose their titanium-plus-lead protective cases together in a thousand years, for which I can see *no* possible mechanism, the first set of assumptions made by Hill and Grimwood were all perfectly possible, though not always probable. These assumptions were:

(1) That there should be a flow rate of 10 million tonnes of water per year (0.3 m³/s) through the repository in a rock formation chosen as being as free from water as possible.

(2) That the leach rate for glass would be equal to that due to fast unimpeded flow in the laboratory of water not already saturated with silica.

(3) That none of the important substances would become permanently bound, as sulphides or otherwise, in the course of their journey.

(4) That radioactivity would not be removed from the drinking water.

(5) That we could still not cope with cancer in the far future.

These are not even the worst *possible* assumptions. But none of them is likely, and it is a pity that the authors have not made a serious attempt to estimate just what the probability of each of these assumptions might be. The tiny effects of the relatively huge quantities of unshielded natural radioactive materials, which have been part of the subsoil of Britain since Britain began, suggest that if the most likely value were taken for each of the five assumptions the most probable number of deaths in the far future, even for a repository far from the sea, could well have turned out to be zero.

I have sometimes been asked how we should have liked it if *our* ancestors had left *us* a lethal heritage. Well, they left us a heritage far more dangerous than are radioactive wastes hundreds of metres underground. The appallingly unhygienic cities in which civilization developed permitted the evolution of epidemic diseases—cholera, smallpox, typhus and so on—which could never have evolved among sparse populations of hunter gatherers. We didn't like it, so after some delay but then in three generations we have learned how to protect ourselves from them.

We are now doing better for our descendants than our ancestors did for us; we are leaving our descendants all the knowledge and the techniques required. Radioactivity is far easier to detect than are the chemical poisons that make thousands of miles of British rivers undrinkable. As already mentioned, Los Angeles already puts drinking water

through ion-exchange filters to remove undesirable chemical and biological contaminations. Ion-exchange methods can also remove radioactive contamination. So long as they have any kind of nuclear industry to
keep the instrumentation going our descendants, as now, will be more
threatened by the evolution of novel viruses than by the small and
unchanging group of radioactive poisons.

I hope I have shown in the last few pages that it is not necessary to add
anything contributed by the long-term handling of radioactive wastes to
the death rates per GWye expected to arise from the operation of nuclear
power plants recorded in table 10.4.

THE NATURAL REACTOR AT OKLO

About 1800 million years ago in Gabon (IAEA 1975), a country on the
equatorial west coast of Africa, a rich ore of uranium dioxide (uraninite)
was being laid down from a shallow sea. At that time the proportion of
U-235 in uranium was about 3%, four times the present proportion; the
proportion has fallen since then because the half-life of U-235 is much
shorter than that of the U-238 which is the major constituent of ordinary
uranium.

With oxygen as a moderator, helped by the water with which the ore
was necessarily soaked during its deposition, the ore eventually reached
the critical density at which a chain reaction could start, in spite of the
absorption of neutrons by some of the other elements present in the ore.
The reaction would have gone very slowly at first, because as the ore got
warmer the neutron moderation would become less efficient; but as the
isotopes of the elements such as boron and cadmium which absorb
neutrons very efficiently were reduced in amount, the chain reaction
would speed up until it became fast enough to boil off the water content
and so slow itself down again. The reactor seems to have switched itself
on and off periodically for the better part of a million years, in doing so
using up nearly half of its U-235. For the rest of the 1800 million years it
has been soaked and dried many times, and over the whole of that time
even the technetium-99, the most rapidly moving element according to
the experiments behind the NRPB scenario, seems to have moved only a
short distance in spite of its long half-life of 200 000 years. This is shown
by the fact that the non-radioactive ruthenium-99 into which it decayed
is depleted in the richer parts of the ore, compared with the abundance
of the other stable isotopes of ruthenium, but enhanced in the outer
parts a few metres away. Many elements, which in their original
chemical form could have dispersed rapidly, formed highly insoluble
sulphides and silicates which did not disperse. The uranium and its
descendant radium, whose transport by water was one of the chief

dangers in the theoretical study, stayed put. A number of other radioactive isotopes escaped—we have of course no information about how fast or how far they went—but it is clear that at least for some of the elements presenting apparently large long-term risks the actual risks could be very small.

DECOMMISSIONING

It is important that when nuclear reactors come to the end of their useful lives they should be treated responsibly and that the CEGB should not follow the example of many earlier industries in leaving large areas of land derelict and permanently poisoned with a variety of chemicals and toxic metals. The total areas involved are of course tiny compared with the areas left contaminated by disused gas works for example. However, worldwide 43 reactors had been taken out of service by the end of 1988, and by 2000 a further 63 will be retired or nearing retirement.

Recent planning by the CEGB is to re-use reactor sites by building a new power station as each existing station closes. This will make it possible to leave the highly radioactive structures to cool off where they are for at least another 40 years with little positive action other than sealing off the reactor to keep people out. In the USA, however, the Shippingport 72 MWe PWR is now being dismantled to greenfield status. This reactor was shut down in 1982, and the central project, after removal of all fuel units, is a one-piece lift-out of the reactor pressure vessel and its subsequent transfer to Hanford in Washington State for land burial. The vessel, together with the irradiated core components, was filled with concrete grout, after which it was possible to stand within 10 ft of the pressure vessel assembly without risk. The neutron shield tanks will then also be grouted and lifting equipment installed to enable the 900 tonne package to be shipped down the Ohio and Mississippi rivers to the Gulf of Mexico, where it can be taken to Washington State via the Panama Canal. Larger reactors might need to be cut into pieces by remote-controlled equipment and carried away in large concrete filled tanks, but there is no great hurry and it will be worth waiting for several half-lives of cobalt-60 (half-life 5.3 years) which is the main active constituent of the steel parts to be removed.

The UKAEA is proceeding with the decommissioning, down to the point allowing unrestricted use of the site, of the 33 MWe pilot plant for the British AGRs at Windscale. The fuel was removed some years ago and the plant left idle for some time, but now the equipment for handling the various parts is being developed, and this should give valuable information for the later decommissioning of full-scale plants.

Following the Russian treatment of the wrecked reactor at Chernobyl, the Chairman of the UKAEA has suggested an alternative method, after defuelling, which would be to entomb the whole reactor in concrete, perhaps after helping with another problem by stuffing it with packaged low-level waste to attenuate the remaining intense gamma radiation (*Atom* Jan. 1989, p. 17). Monitoring, and convection-driven ventilation systems, would be needed; but this plan should be a great deal cheaper than disassembly, and the area covered by all the reactors in the country would be smaller than the area covered by one large coal waste dump.

SUMMARY

Essentially all of the cancer deaths occurring at present from the wastes of the nuclear power industry arise from the liquid effluent discharged from Sellafield into the Irish Sea. The consumption of contaminated fish and other organisms might cause one hypothetical death in Britain and one in the rest of Europe for each four years of operation at the present level. Very elaborate methods have been worked out for disposal of low-level and intermediate-level solid wastes, either on the ocean bottom or underground on land, and neither would add any appreciable extra risk. No long-term method of disposal of the very large quantities of high-level radioactive wastes now stored at Sellafield has yet been agreed, but work is well under way for the conversion of the present high-level liquid wastes into solid form in sealed containers, and for the air-cooled storage of these, probably for four or five decades.

Detailed plans have been described for future burial, to depths of 500–1000 m, of all the wastes expected to be available by the year 2000, in one large repository. It is likely that these would remain securely isolated from the surface environment until even the longest-lived radioisotopes have decayed; but there is a real if small chance that underground water, which could penetrate the repository long before this, will dissolve some of the stored material and carry it back to the surface. A very thorough quantitative study of the possibility of this is described in this chapter. It is shown, at least to the author's satisfaction, that the most probable number of future deaths resulting will be close to zero.

Chapter 12

Wastes from other Power Sources

WASTES FROM COAL

The wastes from the mining of coal have already been mentioned (Chapter 8). Despite the disaster at Aberfan, spoil heaps usually have more effect on the quality of life in their neighbourhood than on the length of life of their neighbours. The risks to the public of ill-health or death arise from the burning of coal rather than from the winning of it. We are at present burning about 120 million tonnes of coal every year, two-thirds in electric power stations and the rest in a great variety of smaller industrial and domestic units.

The direct effects produced by inhalation of smoke and acid fumes from fires or furnaces have been discussed in Chapter 10. In this chapter I shall be concerned with the longer-term effects on people of the wastes produced. In contrast to the nuclear pattern, the main chemical carcinogens and mutagens produced by the fossil fuels do not last long; they are soon deposited directly from the air or washed down by rain, and being fairly reactive such organic compounds are rapidly destroyed by soil bacteria. Important ecological consequences are beginning to appear as a result of sulphuric and nitric acids washed down by rain at long distances from their point of origin. It is unlikely that this will have any appreciable effect on human health, so that discussion of acid rain will be deferred until Chapter 14.

Carbon dioxide and climate

Potentially the largest and most important in its effects, and unfortunately the least quantifiably dangerous waste from any form of energy production, is carbon dioxide.

Carbon is the element chiefly responsible for the basic structure of all living creatures, forms nearly all of the burnable material in coal, and is the major constituent of oil and natural gas. Plants use the energy of sunlight to combine carbon dioxide with water and other materials, liberating oxygen in the process, while animals gain their energy in the same way as does a fire—by using the oxygen of the air to oxidise plant compounds back to carbon dioxide and water. A rough balance between absorption by plants and exhalation by animals has been maintained for at least 2500 million years, the turnover rate now being roughly 70 000 million tonnes a year, of which the breathing of human beings is responsible for around 1000 million tonnes. The balance has of course fluctuated in the past, but it is probably now changing faster than ever before.

Table 12.1 Estimates of atmospheric CO_2 content and prediction of the corresponding temperature rise (over the preindustrial temperature) (Rycroft 1982).

Year	CO_2 from fuel† (10^{10} tonnes)	CO_2 content‡ (ppm)	Temperature rise§ (°C)
1900	3.8	295	0.02
1910	6.3	297	0.03
1920	9.7	299	0.07
1930	13.6	302	0.09
1940	17.9	305	0.11
1950	23.3	309	0.15
1960	31.2	314¶	0.21
1970	44.0	322¶	0.29
1980	63	335¶	0.42
1990	88	351	0.58
2000	121	373	0.80
2010	167	403	1.10

† Annual growth of 3% is assumed after 1972
‡ The pre-industrial value of 293 ppm is assumed; and 50% of CO_2 produced from fuel burning is assumed to remain in the atmosphere
§ Assumes a 0.3 °C increase for each 10% rise in the CO_2 content
¶ The observed values in Hawaii (see figure 12.1)

As a result of burning first coal, then coal and oil, and now coal and oil and large forests, the proportion of carbon dioxide in the atmosphere has been increasing ever since the beginning of the industrial revolution, when it was at most 290 ppm. We are now producing about 10 000 million tonnes of carbon dioxide a year from coal and oil, plus 30% more (Brown and Lugo 1984) by burning the forests which might have taken some of it up. By 1958 the concentration had already increased by 8% to 313 ppm. Table 12.1 shows the measured changes, with predictions for the next thirty years, and figure 12.1 shows the variation of carbon dioxide concentration at Mauna Loa in Hawaii, far from any large industrial area, since 1958 (Rycroft 1982). In 22 years it has risen a further 8%. In 1988 the average concentration of carbon dioxide as measured at Mauna Loa had risen to 351.2 ppm. At the expected rate of increase of burning of fossil fuels, the concentration of carbon dioxide should double over the next 50 years.

In one way this could be advantageous. In warm, wet, sunny conditions the rate of plant growth can be limited by the amount of carbon dioxide available for photosynthesis, so that larger concentrations of carbon dioxide could lead to a greater rate of production of food (Rogers *et al* 1983). Over most of the planet however, plant growth is limited by availability of water and sunlight or by temperature, rather than by the availability of carbon dioxide. The possible disadvantages are overwhelmingly more important as carbon dioxide in the atmosphere has a controlling effect on climate.

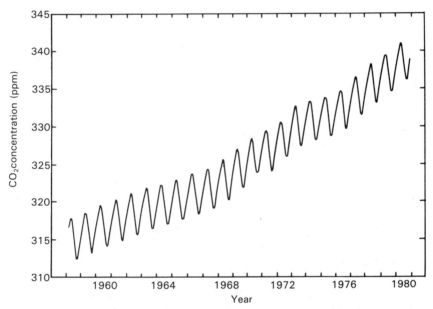

Figure 12.1 Atmospheric CO_2 concentrations measured at Mauna Loa, Hawaii, from 1958 to 1980.

The average temperature of the earth, as of every other body in the solar system except the sun itself, is determined by a balance between the heat absorbed from the sun and the heat radiated away. If the heat absorbed is greater than that radiated, the temperature rises; if less, it falls. On a sunny day we can feel directly the heat that we are absorbing from the sun; we may at the same time be absorbing nearly a quarter as much heat as a result of infrared radiation from our surroundings, but we do not notice this because we are radiating as much or more ourselves, also in the infrared. Our local surroundings are radiating heat all round the clock, while they are receiving direct sunlight for only part of this time.

The main constituents of our atmosphere, nitrogen, oxygen and argon, are practically transparent both to the visible light from the sun and to the infrared radiation from the earth. Carbon dioxide and water vapour, however, are transparent to the visible light from the sun coming in, but absorb an important part of the infrared going out, and reradiate much of it back to earth. As a result, the earth is some $10\,^{\circ}C$ warmer than it would be if it had no carbon dioxide in the atmosphere, and every increase in carbon dioxide makes it a little warmer still. This effect is often called the 'greenhouse' effect because the glass of a greenhouse keeps in the sun's heat in just the same way. The steady increase in carbon dioxide must be steadily changing the radiation balance. More heat energy in the atmosphere means more evaporation of water from the oceans. Water vapour is less dense than air and the moist air rises, expands, and cools until the water vapour begins to condense again, releasing in the process the latent heat of evaporation that it absorbed when it left the ocean. This release of heat helps the column of air to expand and rise faster than it otherwise would. Also, it is this release of heat that supplies the energy for the winds and storms that make our weather so interesting, as well as carrying water over the land to supply the rain that keeps alive both ourselves and the plants that feed us.

It is tempting to think that the frequency with which extremes of weather have been occurring over the last few years is due to the increased level of carbon dioxide, but the apparently random variations in British weather are quite large enough to account for these, and there would be no justification for blaming the fossil fuel industry for the freak waves that recently drowned a man and his dog and three police officers who tried to rescue them at Blackpool, or even for the abnormally fierce storms that hit the south of England in the autumn of 1987. Nevertheless, although specific effects cannot be recognised, more energy retained by the atmosphere must do something. More storms and bigger storms must surely be among the somethings, but we do not know nearly enough yet even to guess how many extra deaths (or even perhaps lives saved) result per million tonnes of carbon dioxide released. In any case, this is not the main source of the concern aroused by carbon dioxide. The extra carbon dioxide will not be affecting us *directly* in any way, although each of us is taking in with every breath something like a million million million (10^{18}) more molecules of carbon dioxide than our grandparents did.

The two possible effects that would matter are the melting of the Antarctic ice-cap and changes in the world's climate. Neither of these can be predicted at all easily. More evaporation of water from the oceans will enhance the greenhouse effect, because water vapour itself absorbs some of the relevant infrared radiation. On the other hand the increased amount of cloud will increase the amount of sunlight that is reflected away without ever being absorbed by the earth or the atmosphere round it at all. Again, the increased evaporation from the oceans may lead to more precipitation in the form of fresh snow near the poles, thus simultaneously reflecting more of the sunlight in the summer which would have been absorbed by the older compacted ice of the ice-cap, while adding more frozen water to be melted.

The importance of the Antarctic ice-cap as opposed to that of Arctic is that a large part of the former is resting solidly on the ground while the latter, apart from the Greenland ice-sheet, is floating. The weight and the volume of the sea displaced would be unchanged by the melting of the floating Arctic ice-cap, so this would have no effect on the global sea-level. Ice is less dense than water. The water from each melted iceberg just fills the space originally occupied by the underwater part of the ice when floating, without causing either a rise or fall of the surrounding water level. If the land-based ice melts however, it will run off into the sea and the global sea-level will rise. There was in fact a rise in mean global temperature of 0.3–0.6 $^\circ$C, and a rise in mean global sea-level of 4.5 cm between 1890 and 1940 (Etkins and Epstein 1982). At least half of the rise in sea-level may have been due to thermal expansion of the upper layers of the ocean as they got warmer, but the rest must have been due to the melting of Antarctic ice. In the following 40 years the global mean sea-level has risen at an average rate of at least 3 mm per year. During this period the mean global temperature has if anything fallen by 0.2 $^\circ$C, so the rise cannot be attributed to warming and expansion of the ocean. The only plausible explanation for the recent rise therefore seems to be the melting of 50 000 km^3 of ice during the last 40 years. The fall of mean temperature is not necessarily inconsistent with the increased melting rate of ice, which depends on a rise of summer temperature in a fairly small area. The melting absorbs an enormous amount of heat; to melt 50 000 km^3 of ice would require 400 million GWy, and would produce 46 000 km^3 of ice-cold fresh-water which, being less dense than sea-water, could have spread out over huge areas of sea. This would do little to warm the rest of the world.

While predictions are still dangerous, we can at least say with certainty that *if* the West Antarctic ice-sheet has already begun to melt because of the greenhouse effect, the process will accelerate until in 50 to 100 years' time that part of the ice-cap will have vanished, with a 5 m rise in global sea-level that will flood an important part of Europe, and low-lying parts of many of the great cities of the world such as Calcutta and Bombay.

The higher, colder and thicker East Antarctic ice-sheet will take longer. If this goes too, the sea-level could rise by a further 50 m or more—but might take 1000 years to do so (Mercer 1978).

Climatic changes are even more difficult to predict than the future of the ice-caps, but it is difficult to believe that they will not occur—and easy to believe that they are already beginning. It is very unlikely that any area will become uninhabitably hot; what is likely is that there will be shifts of the present climatic zones towards or more probably away from the equator. A shift of some hundreds of miles north by the Sahara and other great deserts of the world would be a major catastrophe; and a very small reduction of rainfall in the grain-growing areas of the USA—many of which already have to be left fallow every second year to allow the ground-water to build up for the crop-growing years—would destroy the world's largest producer of wheat. At present we cannot rule out anything; and although we hope that the more disastrous possibilities mentioned below are the less likely ones, there would be nothing whatever that we could do about them quickly with our present technology. If the world population were only 1000 million it would be possible for the unfortunate ones to migrate to the new areas in which efficient food production was possible, and to teach the rice farmers to grow wheat or the wheat farmers to grow rice or whatever may suit the new climatic pattern; but the period between 50 and 80 years' time when action might become urgent will be the time when world population reaches its peak, 10 000–20 000 million, before—if we are lucky—beginning to decline.

The mild climate of Britain is due to the Gulf Stream. This forms part of the clockwise circulation of water in the North Atlantic. Every second 90 km^3 of warm equatorial water is driven through the gap between Florida and the Bahamas by easterly winds. This runs up the American coast and spreads into several branches, bringing warm water and mild air to Britain. If this wind pattern in the Gulf of Mexico changed, the warm water might go elsewhere, and the British climate would change to that of Labrador, which, with an area much the same as that of Britain, supports a population of about 20 000.

An important and disturbing point about chaotic systems like that of world weather behaviour, is that one broadly stable pattern may be switched suddenly and discontinuously by quite small factors to a second broadly stable pattern which is very different from the first, and that the intermediate states are entirely unstable so that the changeover is unpredictable and rapid. It is still most probable that recent extremes of weather as far apart as Britain and Australia are due merely to the apparently random chances which control the details of weather behaviour. Nevertheless, if they do represent a gradual destabilisation of present climatic patterns, it is possible that this may not remain gradual and may not require us to wait for 50 to 100 years. If such a change displaced the easterly winds in the Gulf of Mexico, Western Europeans would have to move.

Having described the possible risks, it is only fair to report a small gain from carbon dioxide production. Atmospheric carbon dioxide contains the radioactive isotope carbon-14, naturally produced by cosmic rays, which is absorbed by plants and passed on to animals and ourselves, giving each of us a radiation dose of 1.5 mrem per year. There is no carbon-14 in oil, coal or natural gas. The 12 million tonnes of non-radioactive carbon dioxide produced each year by a coal-fired power station in producing 1 GWy of electricity will dilute slightly the carbon-14 in the atmosphere, and hence will eventually reduce the annual dose to each of us by nearly 0.005 μrem. This, on the linear theory, should save from dying of cancer about one person in the present world population of 4800 million every 500 years. This is hardly enough to balance the risks.

It is impossible to guess how many people will actually and recognisably die of the indirect effects of carbon dioxide wastes from fossil fuel. Perhaps none. But perhaps future historians will blame us for our irresponsible failure to foresee the climatic change and to invest enormously more effort in building up the known alternatives.

Radioactive waste from coal

In the last chapter it was pointed out that uranium is widely distributed in many types of rock. It is also present in coal, although usually at a rather lower concentration than in most rocks. Some American coals have as much as 10 ppm; but in Britain the average is about 1.5 ppm (1.5 grams per tonne) (Eadie 1978).

According to Lord Marshall, Chairman of the CEGB, speaking in 1986: 'Earlier this year British Nuclear Fuels released into the Irish Sea some 400 kilograms of uranium, with the full knowledge of the regulators. This attracted considerable media attention and, I believe, some 14 Parliamentary questions. I have to inform you that yesterday the CEGB released about 300 kilograms of radioactive uranium, together with all its radioactive decay products, into the environment. Furthermore, we released some 300 kilograms of uranium the day before that. We shall be releasing the same amount of uranium today, and we plan to do the same tomorrow. In fact, we do it every day of the year so long as we burn coal in our power stations. And we do not call that "radioactive waste"; we call it coal ash.'

About 3.5 million tonnes of coal are burnt in a coal power station to produce 1 GWy of electricity†. In this coal will be 5.25 tonnes of uranium, with an activity of nearly 2 Ci. All of the radioactive descendants of uranium will also be present; in particular there will be nearly 2 Ci of radium-226. Almost all of these will be present in the 500 000 tonnes or so of powdered fly ash, most of which will be captured by the filters and electrostatic precipitators in the tall smoke stacks, but a few thousand tonnes will escape into the atmosphere carrying with it a proportional number of millicuries of uranium. The very fine particles that escape the filters may carry rather more radium activity, some compounds of which are volatile at furnace temperatures; the fraction that escapes will be bound to solid particles which, at least from inland sites in Britain, are likely to reach the ground somewhere in Britain.

According to the CEGB (Manning 1980) the maximum radioactive dose to a hypothetical individual living near a large coal-fired power station would be 6 mrem per year, mainly delivered through the food chain—presumably carried by locally grown vegetables.

The main bulk of the fly-ash will be used for land-fill, perhaps in areas in which houses are to be built, in making breeze blocks for building, or for a variety of minor purposes. None of these will be deep underground, and most of the ash will be where it can easily be leached by rain—often acid rain from the current sources of burning fossil fuels.

In table 11.2 the uranium content of the high-level nuclear wastes expected by the year 2000 was given as 18 Ci—less than one year's output from our existing coal-fired power stations. In the last chapter it was shown that this would be unlikely to reach the surface from a deep repository in which it was stored in sealed containers. There is however no reason to expect this in the case of the coal wastes, in which most of the uranium and radium is directly placed in a finely particulate form immediately accessible to rain. This does not mean that they will immediately appear in our drinking water. The transport of uranium and radium through the surface soils of river or well catchment areas might be just as slow as in the nuclear waste case, in which the uranium was expected to take 300 000 years to travel 10 km—only about 3 cm a year. On the other hand, material in the form of fine particles washed off roofs and roads straight into drains will obviously travel a great deal faster than this, since small particles are not adsorbed in the way that individual atoms are. It is difficult therefore to guess how much of the radium from coal-ash uranium will get into how much of our drinking water. But some of it should be a great deal quicker than the million years calculated by Hill and Grimwood in their preliminary study discussed in Chapter 11.

† This is greater than the figure given earlier for the coal consumption of 1 GWe station in a year, because shut-downs for maintenance mean that a 1 GWe station takes more than a year to produce 1 GWye.

The arguments I presented in the last chapter for the unimportance of uranium in the wastes, based on the almost complete absence in our drinking water of radium from the 600 million Ci of uranium and radium at depths in the ground shallower than the proposed repository for nuclear wastes, may not apply to uranium and radium distributed directly over the surface. I do not think that the danger even from this is very great. But there can be no question that the 40 Ci of uranium or radium disgorged from coal-fired power stations each year (for about 20 GWye) must be a far greater danger than the 18 Ci in buried wastes from the production of 330 GWye by nuclear power stations over some decades.

The NRPB estimates that the annual collective radiation doses to people in the UK are:

	Nuclear	Coal
To workers	10 000 man-rems	9 500 man-rems
To the public	5 000 man rems	500 man-rems

This neglects the radiation doses received from the coal wastes accumulated since the industrial revolution.

Apart from the deaths which may arise from the consumption of radium in our food or drink, there is an additional risk from the radon to which the radium gives rise. The concentration of radon in the air of big cities is several times greater than in uncontaminated country air in the west of Britain, and this extra radon must surely be derived from the past deposition of uranium and radium from coal smoke (Haque *et al* 1965). A typical city-excess of perhaps 0.0003 μCi of radon per cubic metre of air is likely to cause an extra two or three deaths from lung cancer per year in Britain. Only a small fraction of this excess is due to the recent wastes from coal-fired power stations with tall chimney stacks, and little to the recent output from smaller coal-fired units. It is no good crying over milk split before the smoke-control acts of the 1950s; there is little we can do about it and nothing that would be worthwhile to reduce so small a risk. Already, in Chapter 10, much more cogent reasons, concerned with chemical pollution, have been given for retiring the small-scale coal burners as soon as it is practicable to do so.

However, the future output of radon for which the current power stations are responsible needs consideration. For each GWye of electricity produced, the 2 Ci of uranium and radium in the coal will maintain 2 Ci of radon. In Chapter 10 it was shown that if all the radon escaped from uncovered tailings from the 80 Ci of uranium mined in Canada for the production of 1 GWye, one death from lung cancer due to its inhalation could be produced about every 300 years. The British population density is over 100 times that of Canada, so that 2 Ci here, on the same assumptions, should kill perhaps a little faster.

More important for the calculation of cancer deaths is the fact that the rate of killing by uncovered tailings would fall off with the half-life of thorium-230 (80 000 years), and the total number killed could amount to 400 by the time the whole of the activity has decayed. The rate of killing by radon from the coal used to produce 1 GWye on the other hand will fall off with the half-life of uranium-238 itself (4500 million years). The total could reach about 40 million deaths, although we should worry only about the 32 million who would be killed in the 10 000 million years before the sun is expected to expand into a red giant and engulf us.

Having now demonstrated the overwhelmingly greater responsibility of coal-fired power stations over that of nuclear stations for deaths in the far future, all of which our successors are likely to prevent, this may be a good place to explain why I think that this difference should be totally disregarded in deciding which types of power supply we should employ. It seems absurd to worry about a risk equivalent to that of living a foot nearer to the top of the atmosphere (with the increased risk from cosmic rays). There is an even greater absurdity in supposing that our descendants will be unable to control a so obviously undesirable part of the environment as the long-lived alpha emitters if they think it worth the trouble. What seems to me really immoral is the diversion of money, thought and youthful idealism into reducing such utterly trivial risks to people of the future and away from the well-recognised problems of ignorance, hunger and disease affecting 2000 million of our fellows on this planet today.

It is clearly possible, though I think unlikely, that a nuclear war a generation hence, or the genocidal wars for living space which must necessarily occur within the next hundred years or so if we do not control our multiplication, might wreck an excessively artificial civilisation completely, and that our descendants will decline into small groups of hunter–gatherers fighting each other for food-bearing territory, and limited in numbers by internecine warfare and disease. In this case, the average death rate over long periods in a constant environment must exactly balance the birth rate, and anyone who dies of radiation-induced cancer or anything else will leave room for someone else to grow up. We are not in a position to say whether in that far-off time it would be better, *for them*, to have more people with slightly shorter lives or fewer people with slightly longer ones.

WASTES FROM OIL AND GAS

The major long-term risk from the oil-fired stations is the same as the major long-term risk from coal: carbon dioxide. The total contribution of oil burnt in Britain is running at about 250 million tonnes of carbon dioxide a year; but the proportion of this which comes from electricity production is very much less than the proportion from coal, most of the oil burnt being used for heating and transport. One car doing 30 mpg (10.5 km/litre) and 7000 miles (11 300 km) per year produces three tonnes of carbon dioxide every year. Transatlantic flights are releasing some 16 million tonnes in a year.

The other long-term wastes from oil are similar in kind to, but smaller in quantity, than those from coal, the ash content of crude oil being a much smaller proportion of the combustible part.

In the case of gas, the main risk again arises from the uncertain danger from carbon dioxide. Gas is used mainly for heating and for a number of specialised industrial processes, and the carbon dioxide yield per GWy of heat varies with the origin of the gas. In total the gas being burnt in Britain is at present adding about 60 million tonnes of carbon dioxide per year to the atmosphere. No important amounts of wastes offering any other long-term hazards are produced, the sulphur content of natural gas being small.

DECOMMISSIONING OF COAL, OIL AND GAS

The decommissioning of coal-fired power stations will be expensive, but little pollution of ground by coal stocks will have occurred.

The decommissioning of oil refineries and oil-fired power stations at the end of their lives will be much simpler, cheaper and quicker than that of nuclear power stations, but it will still be expensive to render the contaminated ground on which they stand safe and usable for agriculture. There is little more risk to the labour force than in the demolition of any other group of large chemical plants (other than places like Seveso) and it does not seem to merit any special discussion. However, decommissioning of North Sea oil rigs could be very expensive, in far more dangerous conditions, and needs a lot more attention.

When supplies of natural gas, distributed through a national grid of high-pressure pipes, displaced the coal gas previously employed, a very large number of local gas-works became redundant. Most of the sites are contaminated with a great variety of toxic materials, including phenol, cancer-causing coal tars, and metals such as lead and arsenic (Pearce 1981).

The London Borough of Greenwich spent £500 000 decontaminating three small housing estates that had been built on old gas-works tips, finding 2% concentrations of lead and nearly 0.1% of cadmium (which of course, being non-radioactive, will last forever) in children's playing areas. It would be a great deal better and safer to decontaminate the areas *before* building houses and play-areas on them, but this could be expensive if all of the many hundreds of sites are to be treated. A consultant's report for Gateshead Council recommends the removal of huge amounts of poisoned soil, with apparently no recommendation as to where it should be removed to; the excavation of underground tanks, reservoirs, and ducts; and covering the whole area with one or two metres of inert material such as fly ash from power stations. (Fly ash itself contains arsenic and mercury, not to mention the radioactive elements described above, but doubtless in far lower concentrations.)

It is of course most unlikely that these sites will cause any direct deaths, but if not properly cleansed they could hardly fail to cause some substandard health, leading to lowered resistance to more serious conditions.

WASTES FROM RENEWABLE ALTERNATIVES

As wind turbines get old, fatigue of metal parts subject to periodic strains and to corrosion will eventually make them dangerous. The removal of any tall deteriorating structure is a risky job, but the risks are well understood and short-lived. The cost of removal should be more than covered by the value of the scrap metal.

From wave machines there are again no long-term risks. Returning the huge structures to docks for disassembly at the ends of their lives will be a ticklish job if an unexpected storm should blow up; but again the materials are recoverable and there are no long-lived poisonous wastes. Until they are in use we cannot forecast the risks from other dangers such as breaking up and leaving wreckage drifting about.

Biomass will produce carbon dioxide but is recycling rather than originating this; and solar power leaves no wastes while in use. Therefore difficulties and long-term risks do not arise from the decommissioning of solar or biomass power producers.

SUMMARY

Coal produces not only chemical wastes but also radioactive ones, because it contains uranium and its decay products. There are 40 tonnes of uranium in the annual output of fly ash from British power stations. The collective dose from this single year's output could kill many more people during the life of our planet than could be killed by the buried wastes of nuclear power stations producing 15 times more electricity. It would not be desirable or sensible to take these small effects into account when deciding between coal and nuclear power.

The main objective of this book has been to compare the risks of electricity production by the means now available. The result is that the present risks to individuals of oil, coal and nuclear production differ by less than the uncertainty of any of them, and are all so small as to deserve little attention when making decisions as to how much of each to use. It is most important to remember that there are other decisions to be made. It is worth repeating again and again that an urban excess of 10 000 cancer deaths occurs each year in Britain; and I find it impossible to believe that a large proportion of these are not due to airborne carcinogens produced from vehicle exhausts and coal burnt in small units. Replacement of either by electricity from any source will save lives.

As a result of the burning of the fossil fuels, coal, oil and gas, the proportion of carbon dioxide in the atmosphere has gone up by 16% since the beginning of the industrial revolution, and may double over the next 50 years. We do not yet know whether this could be quite harmless to us or whether it could have disastrous effects on our climate. It is important to recognise that this risk is qualitatively different from any of the individual risks discussed. The global warming and rise in sea level for which carbon dioxide is expected to be responsible would directly or indirectly affect the whole human race: those who were forced to move, those who were expected to receive those moving, and those who would have to learn new kinds of agriculture.

Part 4

Environmental Effects, the Opposition to Nuclear Power, Conclusion

Chapter 13

Terrorism and Proliferation

It should be clear from Chapter 10 that the nuclear, oil and liquefied combustible gas industries can be subject to major accidents leading to large losses of life, but that in none of these industries is such an accident likely. Many people who accept that such accidental disasters are sufficiently unlikely to give rise to concern are still seriously afraid that disasters might be made to happen by terrorists.

Terrorists can be divided into two groups: the political terrorists who use terrorism to further political ends and the criminal terrorists who use violence or threats of violence to extort money. The two groups may overlap; political terrorists not infrequently rob banks to buy arms to be used for political objectives. For either group too small a threat is ineffective, but too large a threat may be counterproductive and dangerous. Any major action will involve a lot of preparation, which must inevitably mean several active and a lot of peripheral supporters who can be relied on to accept minor actions for monetary gain but many of whom could not stomach the killing of very large numbers of innocent people and could not be trusted to keep quiet if large enough rewards were offered. The Provisional IRA usually get away with the murders of individuals who could at least be thought to be their active enemies; but when 27 young people were killed in a Birmingham pub the perpetrators were caught in less than 24 hours. Somebody who might well have kept quiet about the killing of a soldier must have given a tip-off to the police. If the Provisionals attacked a nuclear power station or blew up one of the ships carrying liquefied natural gas as it came into port, they would be likely to lose nearly all their support in Ireland and gain the active opposition of any sympathisers among Irish immigrants in England, who might with some justification fear that after such an atrocity anyone with an Irish accent was going to have a hard time for a while. More importantly still, they would lose a very large number of their supporters in the USA who provide most of their financial backing.

Theory about what terrorists may think worthwhile is of course unconvincing, but there is a lot of experimental evidence that terrorists do not want to kill too many people. It is so extraordinarily easy to kill a lot of people and they haven't done it. I used to be very careful not to disclose ways in which this could be done, obvious though they must be to any thinking terrorist, but chlorine has so often been mentioned by others that the simplicity of hijacking one of the tankers full of liquid chlorine that regularly run down the M6 can hardly be news to them. Repainted as a milk tanker and run—or perhaps I should say crawled—into any big city in the dark of a misty winter afternoon rush hour, when any kind of evacuation would be impossible, it could hardly fail to kill a thousand or so. The perpetrators would have a fair chance of getting away if they simulated a breakdown and left a bomb with a five minute fuse to release the gas. Such an act would not require specialised knowledge, or employment of qualified engineers, which some other possibilities cannot do without. A great deal could be done with a little ingenuity and a few petrol tankers, which could presumably be hired without alerting the authorities as hijacking a chlorine tanker would do.

The fact that terrorists so far have not wanted to do too much damage, however, does not prove that this must always be so, and it is worthwhile to discuss what is possible, though unlikely, as well as what is probable.

TERRORISTS AND NUCLEAR POWER

As in the case of accident, people are more afraid of what terrorists could do in the nuclear field than of what they could do in the traditional fields of energy production. The two chief fears are that terrorists might cause the explosion of a reactor, or that they might steal enough plutonium to make a bomb, having been encouraged in print by technically inexperienced people who assure them it would be easy to do so. It is recognised that the terrorists might kill themselves in either case, but some political terrorists are prepared to die for their cause—the number may be exaggerated, but great numbers might not be needed.

It must be accepted that armed guards at the gates, while a useful deterrent for people who just want money, will never be able to stop a determined group of men who are prepared to accept casualties and who have acquired detailed knowledge of the establishment to be attacked and of the routine procedures of the guards. Such a group could certainly take over the control room of a power station for a considerable time before adequate deployment of security units could take place. With full information they could withdraw the control rods used to limit the power output. The effect of this would be to start an unlimited rise in the reactor output, but this would merely bring into action the automatic systems for shutting down the reactor when the output rises too far. Enormously expensive damage could be done to the external pipework and cooling system with explosives, which might lead to a melt-down; this would take many hours to develop in a PWR, a day or more in a Magnox reactor, and could not happen in an AGR. This would leave plenty of time for counter measures and for evacuation if this proved desirable. It is difficult to see how terrorists could break open the 3.5 m reinforced concrete biological shield. A large shaped explosive charge might blow a hole in it, but in real life all of these activities would take a lot of time especially in view of the variety of disabling booby-traps that the terrorists could expect to be activated.

All in all, the chance of success of such an attack before the terrorists had to divert their attention to holding off the forces brought up to remove them must be very small. Terrorists may well be prepared to die for a successful coup, but they must see an appreciable chance of its being successful. The risk that such an attempt might succeed in killing many people may look serious to us, but the risk of failure must look very much more serious to any group of terrorists who have the technical understanding that would be essential to such an enterprise. Such an attack would be far less cost-effective (in the terms of the terorists—lives destroyed among the public per terrorist killed) than an attack on an oil refinery or liquefied fuel-gas depot.

Stealing reactor-grade plutonium from Sellafield may not be so difficult, given detailed information as to its whereabouts and a quite moderate amount of plastic explosive to deal with locked doors; and provided the store is not guarded, as it could easily be, by the gamma rays from a canister of high-level waste dropping into the store if the door were tampered with. To avoid having a critical mass the plutonium may be stored in a number of fairly small containers, easy to carry, and a light van or a small motor-boat would be perfectly adequate to take the load away. But getting away with it afterwards would be a problem. The roads to and from Sellafield are few and easily blocked, and a boat or helicopter could be kept under surveillance by radar while appropriate forces were brought up. Finding boats or helicopters in the dark cannot always be certain and the thieves might escape, but would-be thieves will be more impressed by the large chance that they would be caught than by a small chance that they would not be caught.

Stealing the plutonium is only the beginning. The next stage is what to do with it. The plutonium stored at Sellafield will be in the form of oxide, suitable for use in the fuel of a fast-neutron (breeder) reactor. It must therefore be reduced to the metal in small enough batches to avoid the risk that a chain reaction might start if carbon or other material that could act as a moderator were involved in the reduction. The pure metal must then be cast and machined into accurately controlled forms, the machining being done in an oxygen-free atmosphere to avoid spontaneous ignition in air. A major difficulty is that the reactor-grade plutonium stored at Sellafield will contain an unknown proportion of the heavier isotopes, plutonium-240, 241 and 242, as well as the plutonium-239 that is wanted for bombs. The heavier isotopes are fissionable by fast neutrons as is needed in a bomb, and are efficiently used in a fast-neutron reactor, but their critical masses will be different and so the optimum dimensions of the bomb cannot be calculated without knowledge of the proportion present. More seriously still, plutonium-240 and plutonium-242 both undergo spontaneous fission as well as alpha-particle decay, so that there are always a few neutrons wandering around in any large quantity of them. It is necessary for an efficient explosion that a good source of neutrons should be available at the instant when the separate subcritical parts of the bomb come together into a supercritical mass; it is however important that there should not be neutrons present while the parts are being brought together. Successive generations of neutrons take only a thousand-millionth of a second, in which time the parts could move only a thousandth of a millimetre at the speed of a rifle bullet; so that if fissions start too soon the parts may be evaporated by a relatively slow build-up of fission rate before they are properly together. To put it more briefly, the bomb will misfire, giving a lethal local cloud of evaporated plutonium but a very much reduced explosion.

Using highly sophisticated methods the Americans have recently succeeded in obtaining effective explosions from reactor-grade plutonium, but even for them this is none too satisfactory, so that they are still producing the much more expensive military grade of nearly pure plutonium-239 and are putting huge sums of money into the improvement of methods of isotope separation to remove the heavier plutonium isotopes from the cheaper power-reactor-produced supplies.

Such sophisticated methods as the Americans used for a reactor-grade bomb will not be available to amateurs, however skilled chemists and mechanics they may be. Instead of measuring the proportion of heavier isotopes they will have to use the method known as 'tickling the dragon's tail'. This typically consists in bringing the parts of a proposed bomb very slowly together with a neutron detector near by, and measuring carefully the distance apart at which the neutron output goes up by an accurately measured amount. The output goes up extremely rapidly as the distance is reduced, and in the best possible conditions people in bomb-production plants have been killed by the heavy burst of neutrons and gamma-radiation that follows a tiny accidental movement. Although the operations are easy on paper, several spare terrorists are likely to be needed to be sure of making an effective bomb.

If they got enough plutonium there would be no great difficulty in making a minor explosion, but the energy required to evaporate 10 kg of plutonium vigorously enough to blow it apart is less than one millionth of the energy liberated by the Hiroshima bomb, with a correspondingly smaller amount of radioactive fission products. A well placed tonne of TNT mixed with thallium might do just as well. That of course is a minimum possibility; a home-made bomb *might* do a lot better than that (without approaching the Hiroshima level). The point is that there is no way for terrorists using reactor-grade plutonium to know whether it would do better or not and the scale of effort and skill required would produce far more effective results in other ways with far less risk of being caught during the necessarily lengthy period between the stealing of the plutonium and the completion of the bomb.

If terrorists really wish to have a nuclear bomb, by far their best policy would be to steal one. Then not only would they have at once a bomb that would work, but it could be much better used as a threat because everyone would know that it would work. I neither know nor wish to know what precautions against armed theft our armed forces take when storing bombs or transferring them from place to place; but if the same precautions are taken for stocks of plutonium in store or in transit, it is as easy—or as difficult—to steal one as to steal the other, and there is no risk of attempts being made to steal anything but finished bombs.

Some of the people who are concerned about the risks that plutonium may be stolen are also much concerned about the danger to democratic freedom of having armed guards at Sellafield or other nuclear establishments. I can see no way in which my democratic freedom is reduced by my being unable to force my way into other people's factories, whether to steal plutonium or secret plans for a new car. It does not therefore concern me whether the guards that all factories have at their gates are armed with whistles or machine guns. We live now in a world in which armed gangs exist, and I can only hope that stores of dangerous chemicals and oil refineries are also effectively guarded. Military establishments have always had armed guards on their gates to enable us to defend democratic freedoms. Defences must be planned to defend against the dangers that there are, not the dangers that we would like there to be.

The other option for terrorists, with similar counter-productive effects, would be to attack one of the big tanks each holding around 100 tonnes of high-level wastes in solution. The activity of these will be 1000 times less than it was just after removal from reactors, but would still be around 100 million curies per tank. The tanks are well protected with an outer concrete wall up to a metre thick and a massive lid.

The problem facing the terrorists is how to get a large part of this activity effectively dispersed. None of it is volatile, so that blowing a hole in a tank would merely let it run out—locally lethal but unlikely to spread much beyond the underground part of the outer containment tank. As was pointed out in connection with the Kyshtym disaster, to disperse bulk material into fine particles or droplets efficiently requires an intimate mixture of material and explosive.

The terrorists have first to choose a time when the weather forecast has promised a wind blowing towards a well-populated area. Then, having shot their way in, they have to achieve this dispersion before security units alerted by the initial shooting arrive to interfere. During the hour that they could expect, they will have to distribute a tonne or so of explosive through the tank without exposing themselves to its contents—which would render them incapable of intelligent action in a matter of seconds. The nearest they could get to an effective distribution would be to cut the water-cooling tubes at the bottom and poke explosives in round the series of bends inside the metre of concrete needed for screening—these bends being needed to prevent radiation emerging if the water pipes were empty. It is difficult to see how they could get much more than a tonne of explosive in place inside an hour, even when the logistical problems of bringing it to the site had been solved. Finally, almost all of the energy of the explosion would go into a shock wave destroying the outer structure and hurling large volumes of liquid for tens of metres, rather than breaking it up into the micron-sized droplets necessary for efficient dispersion.

Surrounding buildings would be unapproachable until 60–100 cm of sand or earth had been spread by a remote-controlled earth-mover or air jet. Doubtless people for some way around would be evacuated until the distribution of activity had been found; but the criminals could not rely on giving anyone outside the works a lethal dose—even a gram of the liquid (with a curie of activity) falling directly on to someone would not give a dose of more than a few tens of rems if washed off within a few minutes.

This is how a competent project manager could be expected to report to his terrorist HQ. There are of course enormous uncertainties, and the results could be a great deal worse, with tens of square kilometres being made uninhabitable for decades. But the amount of activity in a tank is 50 or 100 times less, though with more long-lived material, than would be present in a big reactor undergoing the improbable worst-case breakdown described earlier. If a large area were contaminated but evacuated within 24 hours, a lot of people might have doses of tens of rems, but the number killed would be small compared with the number promptly killed in a well-planned attack on an oil refinery or liquefied gas store.

The damage done in the works by just blowing a hole in a tank and shield could of course be appalling and perhaps irreparable, and I would like to see BNF get on with the glassification of wastes—in which form they would be immune to this kind of attack.

TERRORISTS AND FOSSIL FUELS

I can see no way whatever in which terrorists can exploit either coal stocks or coal-fired power stations to kill a lot of people outside the works. On the other hand, I can see no way whatever in which we can reliably protect an oil refinery, or a liquefied gas store, from an unexpected commando-type attack with rocket launchers. I hope that the enterprises concerned have armed guards adequate to prevent armed intrusion by one or two individuals; but it is most likely that people sufficiently competent to organise a really effective attack are also bright enough to see how counter-productive to their causes such a major attack would prove.

There are of course people who want only a lot of money and who have no prejudice against killing a lot of other people to get it; however, it is the *threat* to kill a lot of people that would bring in the money, not the actual murders. You actually kill someone only to back up a new threat when the first threat doesn't work, so you have to convince people that you have a second hazardous exploit organised which will work in spite of the warning given by the first, for which you need at least two bombs in stock. No local authority is going to give away millions—even if they had millions to give—on the basis of a written threat without evidence of ability to carry it out; and this threat must necessarily give both a warning and clues to the kinds of skills possessed by the people involved. People interested in cash rather than glory are unlikely to accept the risks of death or capture in an inevitably time-consuming operation at a large oil refinery or at Sellafield, and will not accept the very real risks of handling plutonium. Furthermore, in an operation inevitably involving 20 or more people held together only by hope of gain, the risk that one of those getting only a minor share in the proceeds will inform on the rest for the safe and large reward that would obviously be offered, would be discouragingly large.

PROLIFERATION

Proliferation of the use of oil, gas and coal on the huge scale needed by many parts of the Third World for any large improvement in their abysmally low standard of living will cause a large increase in carbon dioxide output over the next four or five decades. Solar power may be competitive on a very large scale for small-scale uses, but the construction of photovoltaic cells needs high technology, and the replacement of wood and dung for cooking even by mirrors and boilers is beyond the capacity of village technology, and leaves the villagers at the mercy of foreign or indigenous suppliers as much as would a supply of electricity based on something other than wood or fossil fuel.

Proliferation of nuclear technology is another matter. The fear here is, and very naturally is, that proliferation of nuclear power stations may lead to a proliferation of nuclear weapons.

The first point to make is that the possession of a nuclear power station is not essential to a country which decides to build a nuclear bomb. Every country which now has nuclear bombs was testing its first bomb five years or more before it had even a small nuclear power station. A simple air-cooled reactor, of the form of the Windscale military reactor which later caught fire, fuelled with natural uranium and run at 150 °C, could be designed and built from widely published information in three or four years if money was no object, and would be producing plutonium at least five years before a power-producing plant could be ready to work. It is also perfectly possible to make a U-235 bomb, like the one dropped on Hiroshima, without a reactor at all, using an isotope-separating plant.

The difficult part of bomb production using plutonium is the processing plant, which has to be capable of handling more work than would the processing plant for a power reactor of similar size. This is because in order to get uncontaminated military-grade plutonium the uranium fuel units must be removed for processing very much sooner and more often than they would be if the reactor was run as long as possible, which leads to reactor-grade plutonium. If the units are removed five times sooner, there will be five times as many to be processed each year.

It is more important to discourage the proliferation of processing plants than of power reactors. There are many countries, Japan being the most important, with little coal or oil, for which a large contribution from nuclear power is essential. It was ill-advised for the opponents of proliferation to oppose the reprocessing of Japanese nuclear fuel in Britain or France. Japan has of course the technological skill required to build a reprocessing plant for itself; and once built it would be difficult even for the Japanese legislature to make it impossible for the armed forces to slip in a few batches of short-stay fuel rods, extracted early on some excuse, or fuel rods from the less reactive outside of the core, against the expressed wishes and intentions of the vast majority of Japanese. There must be a chance that some Japanese general who does not share the majority view would like to have enough military-grade plutonium at least to establish the techniques for making bombs.

Now that after a dangerous delay the USSR and the USA have agreed to cut their nuclear missile stocks as the Non-Proliferation Treaty (NPT) required them to do, the hopes of renewal of the treaty by other countries are more likely. The cuts are small, but the medium-range weapons concerned would be important for fighting in Europe. Neither side would have agreed if it had any intention of launching an attack in the foreseeable future; and the consequent relief of tension is out of proportion to the actual reductions.

The NPT, besides requiring the signatory countries which already have nuclear weapons to reduce them, requires the signatories which do not possess them to agree not to make nuclear explosives either for military or peaceful purposes, and requires them to accept regular inspection of their nuclear establishments by the IAEA which is supported and financed by all of the bomb countries.

In the same treaty, the existing nuclear bomb countries undertook to supply nuclear power plants and their technology to other signatories. Again it is essential that they should do so when asked, if we are to hope that *any* countries will renew the treaty.

Regular inspections by the IAEA can check that imported nuclear fuel—especially enriched nuclear fuel—goes to the places authorised to receive it; that spent fuel units are not removed early and are safely stored until sent to a recognised destination, and so forth. It is of course true that the number of IAEA inspectors is insufficient to check more than a fraction of the various processes, and that many improper practices could escape notice. It would however be difficult for any great number of fuel units to disappear altogether—which they would have to do if they were sent to a secret processing plant, especially as the production of military-grade plutonium requires each fuel unit to be removed before it has made very much plutonium. When it is so relatively easy to build a reactor specially designed for military production, no military establishment is going to accept the unpredictably effective reactor-grade material. Furthermore, while a large percentage of infringements of the treaty requirements may be missed by the limited number of inspectors, sooner or later an infringement will not be missed. This could lead to various kinds of sanctions including a ban on the supply of fuel which, if an important proportion of the electrical energy of the country concerned was derived from nuclear sources, could be disastrous to it.

A large number of reactors *have* run for a long time in many countries without any important infringements being detected—which doesn't mean that none *could* have occurred, but does mean that infringements cannot have been common. A great many individual infringements have to occur successively before you have stolen, processed, purified, cast, machined and assembled the material for even a single bomb. The Indians exploded a bomb, but they had not signed the NPT, and there is no evidence that any signatory other than the countries which already had some bombs has disobeyed its provisions.

The only important way in which the establishment of nuclear power plants can aid a country wishing to build bombs in secret is that it will enable native engineers and technicians to gain practical experience of running a reactor and of handling the intensely radioactive spent fuel. It would seem to be a valuable additional precaution for the IAEA to require a register of the managerial and technical staff of all reactors under their supervision. If a series of highly qualified people disappeared—after allowing for any who may have reason to believe they are wanted by the police—nothing could be done about it, but it would be an effective warning that closer inspection was needed.

The mere presence, at least for several years, of the foreign experts needed to set up reactors bought from abroad, is itself valuable in getting to know personally a lot of native experts and to hear relevant gossip.

I certainly think that reactors, processing technology and know-how should not be exported to countries which have not signed the treaty, and postgraduate students for nuclear engineering courses should be accepted only from countries which have signed it.

Apart from detailed arguments, it must be recognised that the basic information required has now all been published and there is no way of preventing any country with reasonably good engineering resources from building nuclear bombs if it wishes to do so. However much we regret this we cannot destroy the knowledge. Any kind of sanctions, short of military occupation, can only delay, not prevent, eventual success in constructing bombs. It is not clear even that delay can be relied on. Any serious sanctions are likely to stir up a nationalist reaction that will make it easy to spend a great deal more money on the project than would otherwise have been politically practicable.

Building nuclear power stations for a country which is short of fossil fuel and cannot build its own does make it slightly easier for that country to find qualified staff to build nuclear weapons in secret, and we cannot have 100% certainty that this will be detected before they are built. *Not* building nuclear power stations for such a country makes if far more vulnerable to the ultimately inevitable shortage of oil, and hence more afraid of involvement in increasingly bellicose competition for diminishing supplies. This would provide a powerful incentive to 'develop a bomb capability' as military jargon describes it, in secret, whether as a deterrent to somebody stronger or as a threat to somebody weaker.

The incentive to build nuclear weapons is thus far greater if the nuclear powers refuse to carry out their undertaking to supply nuclear plants for civilian use; and if they do refuse, no external deterrent to building dual-purpose reactors remains at all. It is obviously a matter of opinion, but it seems to me that proliferation is far more probable if we lead other countries to want to build nuclear weapons in spite of some difficulty and the high cost than if we make it a very little easier to do but remove an incentive to do it.

To summarise this section, I believe that the proliferation of nuclear weapons to new countries is extremely undesirable, and that IAEA inspection of authorised power stations in the treaty countries cannot with certainty prevent it. Nevertheless, if we refuse nuclear power to the treaty countries we would ourselves be breaking the treaty, which requires the bomb countries to help the non-bomb countries to build up nuclear power for civilian use. Then there would be no remaining incentive to adhere to the NPT, which for the reasons given above is clearly better than nothing. The resulting free-for-all pattern would actually increase the incentive to build a weapon when other surrounding countries were also free from inspection. And if all of the spent fuel is processed or stored by the bomb countries, these will be in a position at least to make a rough check on the amounts that might be being held back.

I believe that the steady proliferation of American and Soviet weapons has been a more serious risk than the spread to fresh areas. The danger of the Great Power proliferation could increase very seriously when the supplies of economically extractable oil begin to approach their end. Then there is going to be an enormous temptation to the USA and the USSR to try to control what remains. In the absence of at least a prospect of adequate alternative supplies, military control of the last ten years of oil production could mean economic control of the world.

It is widely believed that neither the USA nor USSR will intentionally start a nuclear war; but if the Middle East ever became a vital interest of both, the level of strain could rise to the point where each really begins to think that the other *might* start it. Then computer or early-warning errors, which now are disbelieved even though it takes time to find out what went wrong, might be taken seriously and a war could start with both sides certain that the other had started it.

None of the risks of power production described so far seem to me as serious as the risk of having too little non-oil production in a few decades time. Finally, I can see no way in which the building of further nuclear power stations in Britain can make the probability of proliferation of weapons elsewhere either greater or smaller.

Chapter 14

The Environmental Effects of Power Production

The rise of humans to pre-eminence in the animal world has been founded on our ability to make tools and weapons followed by the control of fire and the use of animal skins to allow us to live in inhospitable climates. These activities had little effect on the main features of our surroundings, although the disappearance of a number of species of large animals may have been hastened or caused by early human hunters.

Serious changes in our environment began with the development of agriculture. This led both to a huge increase in our numbers and our impact on other species competitive with or dangerous to ourselves, as well as to widespread replacement of the complex natural system of interacting flora and fauna by crops of value to us. Artificial irrigation enabled us to spread our food crops and animals into areas in which neither could have survived originally. Until the industrial revolution the changes that we were making were essentially local. The changes in each area from the natural to an artificially maintained and much simpler ecology were due almost entirely to the activities of the local people living on the proceeds of these changes.

Since the development of steam power for industry and transport this situation has changed. Many pleasant valleys and hillsides became covered with the machinery and wastes of coal mining. The cutting down of whole forests in the northern hemisphere, for newspaper production and for the timber requirements of the industrial towns of Europe and America, was made possible only by the application of the energy derived from fossil fuels. Now the effects of our numbers, our industry and our power supplies are affecting the environment of our entire planet, and it is more important to think of what we are doing to the future than it has ever been before.

The destruction of hundreds of thousands of square kilometres of rain forest in Brazil for the short-term production of crops or cattle, and made possible by the power-driven chainsaw, is far more serious than any of the effects of power generation in use in the northern hemisphere. But it is the latter for which Europeans and North Americans are mainly responsible, and so this will be the main concern of this chapter.

The smoke and grime of the large industrial towns at the end of last century were less unhealthy than the appalling effects of ill-managed human excrement in the mediaeval cities, but they were far more damaging to the environment and the scenery. The gas works that were such a striking architectural feature of even the smaller towns were far more widespread and certainly no more pleasing to the eye than the relatively few featureless blocks and cooling towers of modern power stations. The little gasometer in Chard in Somerset, on top of which the manager would sit on fine Sunday mornings reading his paper to increase the pressure for the housewives cooking the Sunday joint, had a human touch about it which is not shown by Drax or Sizewell, but it really wasn't any prettier, and smelt much worse. However, power supplies have more important effects on our environment than arguable features of their appearance, and these I shall now go on to discuss.

ENVIRONMENTAL EFFECTS OF BURNING OIL AND COAL

The most conspicuous environmental effects of fossil-fuel burning in Britain have arisen from the mining of coal, and in particular from the spoil tips which dominate the scenery in the older coal mining areas. Modern mechanised mining greatly increases the volume of spoil but has made possible very much better means of distributing the 40 million tonnes a year that go on to the tips. These are lower and less intrusive than the older tips, but take up more land. It is estimated that 220 to 250 hectares a year will be needed over the next 20 years (*Coal and the Environment* 1981). This is a great deal more than would be needed for decommissioned reactors over the same period. The appearance and ecology of the land covered will be greatly altered, although not necessarily for the worse; in some cases agricultural production will eventually be possible, in others new leisure parks. Disused mines in low-lying areas can eventually cause pollution problems; the mine may fill with water heavily contaminated with various minerals and this may overflow into local water systems. These of course are strictly local problems.

More extensive trouble is caused by the products of combustion. The effects of burning raw coal, in open fires or to heat small steam boilers, were obvious and drastic. From the mid-1800s up to the 1940s every large city had a blackened wasteland around it, and the damage extended for tens of kilometres in the direction of the prevailing wind—in Britain to the north east. A 'lichen desert', an area within which the more attractive lichens were unable to grow, would extend for 80 km (50 miles) or more in this direction; vegetation was limited to the few hardy plants which could stand not only the effects of large concentrations of sulphur dioxide but could also stand the choking of their pores by soot. Insects and their predators were likewise limited to the species whose ecology could be based on this limited variety of plants.

A striking effect was shown by a large night-flying insect known as the peppered moth. Before 1850 the only form of this which was generally known was white with a fine speckling of black, which made it very difficult for insect-eating birds to see when it was asleep in the daytime sitting on a silver birch (one of several trees on which the caterpillar feeds) or on any other tree with patches of pale lichens. A very rare all-black variety had been found by a Mr Doubleday, and called *var. doubledayaria*, but few entomologists of the time could ever have seen one. By the end of the 1930s this black variety, which was invisible on smoke-blackened trees, had become the common form in the Black Country north of Birmingham, and near other big cities, where the white form would be conspicuous and quickly eaten. In the country and in other far from urban areas the white form was still the normal type. There have been few more striking examples of the effects of natural selection working over so short a time.

Thirty years ago, half a litre of rainwater collected in a clean vessel out in the open in Edgbaston, one of the cleaner bits of Birmingham, would still yield a teaspoonful of black sludge when boiled down. And I can remember the shiny coal-black knees of our baby daughter after she had been crawling round an apparently clean lawn on summer afternoons. The large reduction in the burning of raw coal since the smoke-control laws came into force has greatly reduced these effects, just as the replacement of steam engines by diesel and electric engines on the railways has been followed by a vast improvement of richness and variety in the herbage on either side of the railway. There is still however widespread smoke pollution. For confirmation of this statement I am indebted to the activity of aphids. These live by sucking plant juices and excreting the excess carbohydrate as a sweet sticky liquid, which captures particles of soot, and these provide a visible indication of air pollution. In figure 14.1 are shown two examples of such blackened leaves picked ten miles or more from any major industrial area.

The switch of demand from coal of any kind to natural gas, electricity, and oil for both domestic and industrial use has helped enormously, although fuel oil produces nearly as much sulphur dioxide as does coal. Unfortunately, smoke control does not reduce the output of sulphur dioxide. Although the points of release are changed when such smoke-less fuels as Coalite are used, a lot of the sulphur from the raw coal will have escaped during the process of manufacture. Furthermore, the higher temperature at which coal is burnt in the more efficient systems now available produces more oxides of nitrogen, which are also produced by motor transport, and which are converted to nitrous and nitric acids by the moisture in the atmosphere. A mixture of the oxides of nitrogen and sulphur will reduce the growth rate of several grasses at concentrations less than half of those needed to produce similar growth-rate reductions by either acid acting alone (Ashenden and Mansfield 1978).

Figure 14.1 A sallow leaf from Derbyshire (from Dr Diana Franklin) and a poplar leaf from the west side of the Malvern Hills (from Dr M C Scott), showing how aphid honey dew is a pollution detector. Part of each leaf has been wiped with a wet cloth. (Photograph Mr J E James.)

The important point, however, is that the more efficient burning of coal and the fact that two-thirds of British coal is now burnt in electric power stations with high smoke stacks does not at all reduce the total amount of acid gases released, although it does reduce greatly the local fall-out. As a result it necessarily increases the more distant fall-out. The oxides of neither nitrogen nor sulphur remain for long in the atmosphere; both are highly hydrophilic and are washed down with rain or snow. The result is that acid rain and snow are falling over much of Europe and North America in concentrations high enough to do considerable damage. The effects of this are not all bad; plants in some areas are deficient in sulphur and grow faster with a small addition. But to be useful on balance the addition must be small.

Acid rain

Rainwater is naturally slightly acid because the carbon dioxide normally present in the atmosphere dissolves in water to form carbonic acid. This is a very weak acid, easily neutralised by very small amounts of the alkaline minerals normally present in soil. The acidity of rain and snow falling before the industrial revolution has been preserved in glaciers and in polar ice-sheets. Measurements of 180 year old ice in the Greenland ice-sheets have shown a spread between a very weak degree of acidity and an even weaker alkalinity, probably produced by small amounts of mineral dust also in the atmosphere (Likens *et al* 1979). In large areas of North America and Europe, rain storms can now have an acidity 100 times greater than that produced by atmospheric carbon dioxide; a storm at Pitlochry in Scotland in 1974 was over 1000 times more acid.

Both sulphur dioxide and oxides of nitrogen are produced naturally, in volcanic eruptions and by natural biological processes. The effect of burning coal and oil on the present enormous scale is to increase very greatly the amount of these, not to introduce something which has never been in the global environment before. As a result of this, most living things can tolerate acidities up to five or ten times greater than that produced by carbon dioxide alone, but only a few species can stand 100 times more.

A review of work on acid fall-out (*Nature* 1984) showed that it is dry deposition of acid that does the main damage to plants, rather than acid rain in which the acid is diluted. But the effects are not completely separate and I shall use the familiar term *acid rain* to cover all types of acid deposition in the following discussion.

The greatest visible and reliably measurable damage is done to northern hill lakes, in areas where the land is covered with snow for several months each year. Here the steady fall-out of acid snow accumulates throughout the winter, when the production of acid fumes from coal-burning is at a maximum. In the spring, when the snow melts in the mountains and runs down quickly into the mountain lakes, the accumulation of many months of acid fall-out may all run into a lake in a week or two. The natural neutralisation by minerals in the lake bottom is slow and cannot keep up with this rush of acid, so that the acidity of the whole lake rises to a level which can be lethal to fish, and especially to their newly-hatched young. As a result, many lakes in the Adirondacks in the USA, in Canada and in Scandinavia, have lost their entire population of fish. Such lakes while not devoid of all life as some exaggerated reports have suggested, can support only an impoverished ecology based on the few plants and invertebrates that can survive in highly acid water. A feature which appears attractive at first sight is that such lakes look cleaner and more transparent and hence more beautiful than ever before—without the cloudiness of water richly populated by tiny living organisms.

A very detailed study of the evidence available has been made by a working group sponsored by the International Electric Research Exchange (IERE, a group comprising electrical utilities of 16 countries) (IERE 1981). Such a group can hardly be expected to exaggerate the effects of their power station effluents. This study leaves no doubt that a large number of lakes in Europe and in North America have become very much more acid over the last two decades. They also show, however, that acidity is not the only factor affecting water organisms; known factors such as the concentration of calcium and other elements can modify the effects of moderate excesses of acid. A lake with an impervious granite bottom will be less able to release minerals that can neutralise the acid, and may therefore remain acid for a longer time. Unknown factors might include the time of year of maximum acidity, or the rate of build-up towards the maximum, which could affect the ability of plants or animals to adapt themselves to the change.

Fifty-two lakes in four countries, all of them initially less acid even than normal rainfall, far from an industrial area, are listed as having increases of acidity of from 20 to nearly 200 times over a period of 10 to 30 years. Nine Canadian lakes showed a nearly 200 times increase over an average period of only nine years. About 3000 lakes, in Norway, USA and Canada, were examined for the presence or absence of fish; half of these lakes having more than four times the acidity of rainwater with only the carbonic acid derived from normal atmospheric carbon dioxide, and half of them less than four times this acidity. Of the less acid group 14% were barren, against 57% of the more acid group. It is not stated why so high an acidity was taken as the discriminating level, rather than the 20 times lower figure which would have been typical 30 or 40 years ago, when even the slight acidity of natural rain was normally neutralised by minerals in the lakes. It is possible that no appreciable number of lakes could be found with so low an acidity in the later period. The result of the high dividing line is that the lakes of which only 14% were barren had up to 20 times the acidity typical of the earlier period. The high dividing line, although perhaps practically convenient, must lead to an underestimate of the ecological effect.

On the other hand the term 'barren' used in this report may suggest too large an ecological effect; the 'barren lakes' were fishless, not altogether lifeless, although the surviving species other than fish must have been much reduced in number, and may have been different from any of the species normally present.

The effects of acid rain on forests

A possibility far more serious than the effect on lakes was highlighted at a United Nations Environment Conference at Stockholm in 1972, and considered further at later meetings (Becker 1982), that direct or indirect damage might be done by acids to forests and other natural vegetation. Damage had been found in forests in the eastern part of West Germany and in Scandinavia, and the effects of acid rain were thought likely to be responsible.

It seemed unlikely that the trees could have been affected directly by the acidity of rain, since laboratory experiments with acid mists and artificial rain needed much higher concentrations to show adverse effects, and the sulphur content so far from the industrial areas responsible was much lower than in some areas nearer to the sources where trees showed no effect. It is however very difficult to tell how well the laboratory conditions compared with natural ones. If leaves are thoroughly wetted by rain with a quite low acid content and then dried by the sun, the concentration of acid on the leaf surface will increase as the water evaporates but the less volatile sulphuric acid does not; and the surfaces of leaves could easily be subjected for a time to very much higher concentrations than existed in the rain. It is now known that much of the damage due to acid 'rain' from coal-fired power stations is due to dry deposition of sulphur dioxide gas. It has also been shown experimentally that serious damage to spruce trees done by oxides of nitrogen from vehicle exhausts can be eliminated by the use of catalytic converters (Kammerbauer *et al* 1986). The effect of reductions of nitrogen oxides on the concentration and distribution of ozone is however still uncertain.

A suggestion has been made that the acid may be affecting the soil rather than directly poisoning the trees. Whichever is the case, the effect has aroused very serious concern in Germany and the Scandinavian countries, all possessing extensive coniferous forests of major economic importance, both for their tourist and their timber industries. The effects of continued deposition of acid could be cumulative and, on the more pessimistic assumptions, it seems possible that entire forests might disappear within a few decades.

Fifteen years of intensive investigation since the Stockholm conference have recorded very serious amounts of damage, but have still not yet solved all of the problems involved. Experimental studies on plants in various soils treated with different concentrations of nitric and sulphuric acids show severe toxic effects, but only at concentrations very much larger than those which can be occurring in the forests suspected of being endangered. It is however not clear that laboratory experiments lasting only a few years can adequately represent the effects on mature trees of exposure to weaker concentrations for 30 years, and though the case for early and serious damage has been much weakened it is not disproved. A slight weakening due to acid rain of resistance to virus or fungal disease might not have been important in experimental studies in areas where such diseases happened to be rare.

A very detailed study of the available evidence by the IERE working group already referred to has found no evidence of any effects that could be proved to be attributable to acid rain. The 'Greens', the politically important ecology party in West Germany, on the other hand, find the evidence completely convincing, and accuse the electrical industries of bias. Since it would raise the cost of electricity from coal by up to 15% if all coal-fired power stations had to install equipment to remove acid—apart from the problem to be mentioned later of getting rid of the resulting sludge—the electrical industries could hardly help being biased. The Greens, who are accepting suggestive and strongly circum-stantial evidence as absolute proof, must also be biased. This does not mean that they are wrong. There is no evidence that the technical experts on the industrial working parties allowed any bias they might have started with to affect their report. But that does not mean that they are right. It might be thought that the effects of acid rain in cultivated areas would be more easily observed, but it is unlikely that any effects will be observed on land cultivated for cash crops; the farmer may find that it is useful to lime his land a bit more than previously, but will not query why he needs to do so.

Such fresh evidence as we have describes symptoms rather than explaining causes. According to a report from Bonn, published in *The Times* on 20 October 1983:

> Some 2 500 000 hectares of forest are now affected in various degrees, especially in Southern Germany. In Baden-Württemberg, where the famous Black Forest grows, almost half of the trees are sick, and emergency programmes to combat the destruction have been rushed through the local parliament as well as that in Bavaria.
>
> Equally worrying however, is the spread of damage elsewhere, especially in the Harz mountains on the East German border and in the forests west of the Ruhr Valley.
>
> Some 80 per cent of all Germany's fir trees are affected, and the latest survey shows that spruce and pine are increasingly turning yellow and sickening. Deciduous trees are more resistant.
>
> A hot dry summer accelerated the destruction, and a helicopter ride over central Germany gives a dramatic view of the yellowing

woods, where the needles start falling from the trees until they are almost bare. The situation is now regarded as a national catastrophe.

The Government has launched a multi-million Deutschmark programme to halt the damage, and earlier this year set up a foundation to coordinate private and public efforts to save the forests, but many fear it is too late.

Atmospheric pollution leading to rainfall contaminated with sulphur dioxide, heavy metals, nitrogen and photo-oxidants, is held to blame, and strict new laws have been passed tightening up the already strict control on industrial emissions.

However, with prevailing winds from the west, much of the pollution comes from France, which has been tardy to recognise the problem. Bonn wants to make acid rain a top concern of the European Community. East Germany and Czechoslovakia have also reported considerable damage to their forests, and have both agreed to start talks with Bonn.

Clearly the Germany government must be interested in votes as well as in evidence, but it is extraordinarily unlikely that it would be spending millions on measures to reduce acid production—which will not only cost money now but will increase permanently the cost of electricity—unless the scientific evidence were very strong indeed.

In the USA as well as in Europe the tree damage is most striking in high-elevation coniferous forests where environmental stresses are already high. In many sites more than 850 m above sea level in the Adirondacks, the Green Mountains of Vermont and the White Mountains of New Hampshire, more than 50% of the red spruce have died in the last 25 years. Tree ring records from such forests show sharply reduced annual growth beginning in the early 1960s (Mohnen 1988).

Obviously research should be increased, and financed on the basis normal in wartime: that any promising study should be supported rather than that some defined sum of money should be allocated to be competed for by experienced writers of applications. (Such a competition may be won by those who have most experience and who spend most time in writing applications, rather than by those who spend all of their time in thinking about worthwhile things to do.)

There seems to me to be no question that the damage done to lakes is due directly to acid falling in rain or snow, but there is much evidence that damage to forests is more complicated and that ozone may be the chief agent directly responsible. Ozone is produced by the action of sunlight on oxides of nitrogen. More of these may be derived from vehicle exhausts than from power stations, and, rather inconveniently, might be increased by the use of lead-free petrol. Further research is urgently needed, together with a lot of thought on what we do about it when we know the answer.

Meanwhile it is important to extend the research from diagnosis to cure. For example, a large number of areas chosen to include both badly affected and unaffected trees could be sprayed by aircraft with a suspension of slaked lime or other mild alkali, with or without fungicides or nutrients. If each area were a well-defined square of half a kilometre each way, regular aerial photographs round the year should show visible effects in a lot less time than will be taken before the cause can be absolutely certain.

Finally, work on prevention of damage should be started as a matter of urgency. Again it is cheaper to be wrong in doing too much than in doing too little. Windpower enthusiasts and the nuclear industry will approve the replacement of fossil fuels. Hardly anyone will like the cost and trouble of reducing the acid output of existing sources, and only the non-motorist will like the cost of elimination of pollution from vehicle exhausts. In the short term the reduction of acid output from existing sources *could* be quickly implemented, and it is to be hoped that the EEC will make a firm decision soon. A detailed study of the problems of acid rain is given in a special issue of *CEGB Research*, August 1987, No.20.

Reducing acid output is expensive. In a letter to *The Times* (2 April 1983) the Secretary of the CEGB estimated the capital cost to be £75 million per GWe for large coal-fired power stations, which would appreciably reduce the capital-cost advantage of coal or oil stations over nuclear ones. It would also add about £12.5 million per GWye to the annual maintenance and waste-disposal costs. Applying this to the 20 GWe of coal-fired power stations in Britain would reduce our total sulphur dioxide output by only 25%.

Most of the scrubbers (devices in which the waste gas is passed through a medium in which it has a high solubility) used in USA, which already has these in some large plants, use a slurry of lime or limestone in water as the scrubbing agent. Both are throw-away processes that consume hundreds of thousands of tonnes of scrubbing material per power plant per year, and produce huge quantities—around a million tonnes a year—of wet sludge, chiefly calcium sulphate, which must be disposed of. For a new power station at Shippingport, Pennsylvania, a valley is being drained and is expected to be filled, through an 11 km long pipeline, to a depth of 120 m over the 25 years of operation of the station (*Science* 1976). The sludge is not toxic and should offer no risk to life, though it would be messy to fall into.

Processes are being investigated that will recover the sulphur and the regenerate the scrubbing agent, but there is as yet no economic process that has operated on the scale required.

The first plant scheduled to be modified in the UK is the new Drax power station at Selby in North Yorkshire, for which the obvious source of the 350 000 tonnes of limestone needed each year is Ribblehead in the

Yorkshire Dales. As much as 1.2 million tonnes will be needed if the other two stations to be modified are included. The Council for the Protection of Rural England is naturally violently opposed to this, and presumably also to the resulting millions of cubic metres a year of the resulting sludge which have to be kept out of the local lakes and the River Ouse.

Carcinogens

As shown in Chapter 12, the carcinogens produced by burning oil and coal have some importance to us, but they have no importance to wildlife or to natural ecosystems. Very few wild creatures live long enough to have any appreciable risk of dying of malignant diseases; and the death of an elderly animal, past the age of efficient reproduction, is hardly to be regretted in biological terms as more resources are available for reproductively active animals.

The only way in which carcinogens might affect the ecology of an area arises from the increased mutation rates that carcinogens usually produce. The rapid changes of all kinds that humans have made in the world have made rapid evolution an advantage rather than a disadvantage to wildlife. But small increases in the mutation rate, although they may certainly do no harm, contribute very little to the rapid changes needed to re-adapt to a changing environment in a few decades. All successful—and hence numerous—species already carry a huge number of minor recessive mutations, which in the past have had small or undesirable effects on viability. When conditions change, new combinations of these mutations may be selected for without waiting for a rare and improbably novel mutation. The results of selection from existing mutations are demonstrated in the large variety of shapes and sizes of dogs which can be seen nowadays.

Carbon dioxide

So far the direct ecological effects of extra carbon dioxide have been few and positive in that the growth rate of some plants is increased. It is possible that this effect is already large enough to be detectable. It can be seen in figure 12.1 that the spring drop in carbon dioxide recorded at Mauna Loa, which results from the rapid growth of plants in spring and early summer, is significantly greater in the last few years recorded than in the earlier years.

At first sight this seems all to the good, but the ecological effects are not simple. The rich variety of plant life in the wild, however stable it may seem, results from an intense and continuous competition for living space and nutrients and against insect enemies. A slight advantage for the species whose growth rates are limited by the available carbon dioxide necessarily worsens the competitive strength of plants whose growth rates are limited by light, water or mineral nutrients, rather than by carbon dioxide. The apparently stable mixture of species will alter if any major factor is permanently altered even a little. Some plants, with their associated fauna, will become commoner; some will become rarer; some may become extinct. The more rapid growth indicated by the results from Mauna Loa may divert attention from an eventual impoverishment of variety. Such processes are slow, and it is unlikely that many extinctions have yet occurred, but a doubling of the concentration of carbon dioxide in the next 50–100 years could produce a very important reduction in ecological richness and variety even without any climatic change.

This delay does not mean that climatic changes will be ecologically unimportant. Even if these changes should represent only a shift in the location of existing climatic zones—southwards in Europe if the Gulf Stream changed its course, and away from the equator if the world were more or less uniformly warmed—they would still be important. Plants well adapted to the British climate might be thought to be equally well adapted to the same climate if this occurred 1000 miles further south, but if the changes occurred in a few decades such plants would be wiped out in Britain, and only by chance—or active intervention—could their seeds be transferred across the seas to suitable new locations. Again, the shifted climatic zones cannot repeat exactly their original conditions; distribution round the year of temperatures, rain and snow might be close enough to Britain's but the winter days would be longer and the summer days would be shorter. For many plants day-length is the controlling factor in initiating the production of flowers and fruit. Even small climatic shifts will lead to the extinction of fauna and flora in areas near the limits of their ranges, and large and rapid climatic shifts would lead to massive extinctions. These would be numerically few when compared to the numbers of extinctions occurring now as the tropical forests are destroyed, but they would not be trivial in the northern areas involved.

ENVIRONMENTAL EFFECTS OF NUCLEAR POWER

Under normal running conditions there are no environmental effects of nuclear power outside the sites of the establishments. The mutagenic effects of the escaping radioactive materials on the land areas of the world are far less than 1% of those due to the natural radiation background; and the effects of the natural radiation background are less than 1% of the effects of the chemical carcinogens produced by the burning of fossil fuels. As pointed out already, the increased mutation rate produced by the fossil fuels can hardly be doing any harm to organisms trying to survive in a changing world.

The very much larger amounts of radiation due to radioactive effluents routinely poured into the sea at Sellafield need more consideration. The body-burden of artificial radioactive substances in many fish and sea plants, and some invertebrates, is very much larger than are the body-burdens of the natural radioactive elements apart from polonium-210. But as mentioned earlier increased cancer rates, radiation-induced or otherwise, if they occur, are unlikely to have any ecological importance.

By far the biggest effect on the radioactive fishes themselves is likely to be unambiguously favourable: if the humans get worried about their radioactivity and stop buying fish it could reduce over-fishing which is worrying both conservationists and fishermen.

ENVIRONMENTAL EFFECTS OF RENEWABLE ALTERNATIVES

These must surely be small for the wind turbines, and zero for the roof-top solar heaters, although the former will have a dominant effect on the scenery over considerable areas. A series of wave machines would materially change the amount of wave energy now dissipated in disturbing the shores and eroding the coasts in stormier seas. This cannot fail to affect the shore distribution of shellfish and other invertebrates which are adapted to the difficult conditions in the area between low and high tides. It is, however, unlikely to cause extermination of any species, and is likely to prove an advantage to as many species as those for which it proves a disadvantage. It is not a factor which is likely to be taken into account in making decisions on the use of wave power.

HYDROELECTRIC POWER AND THE ENVIRONMENT

The really important renewable supply from the point of view of the environment is hydroelectric power. The small dams and ponds needed by the traditional water-mill certainly changed their immediate environment and the distribution of water-plants and animals over the sections of the stream that they affected, but in general they enriched the ecological variety of the area by providing suitable extra sites for uncommon species where agriculture had impoverished it. Large-scale hydroelectric systems however make drastic changes over immense areas.

The Parana dam in South America has replaced virgin forest by a lake

151 km long. The Kariba dam on the Zambesi covers over 5500 square kilometres, 40% more than the County of Kent, and required the explusion of 35 to 60 thousand Tonga tribesmen—who had to forget about a free life hunting and food gathering and learn to grow mealies 'and pay taxes instead' (Kenmuir 1978). In both cases electric power alone was the objective. The 110 m high Aswan high dam in Egypt will provide rather more than 2 GW of cheap electricity, and requires a reservoir holding 100 000 million tonnes of water occupying perhaps 10 000 km^2 of land. Associated works to ensure a steady supply of water call for storage of an extra 200 000 million tonnes of water in Lake Victoria and Lake Albert, raising their average water level by three metres and affecting drastically the annual pattern of changes along the lake margins. The associated works also envisage the by-passing of the swamps of the Sudd, which cover an area of 7000 to 15 000 km^2 according to the rainfall, and in which at least half of the water entering evaporates and is lost to the supply of the Aswan reservoir. By-passed, the swamps will dry out and one of the largest wetland areas of the world, with its unique ecology and large numbers of unstudied species which live nowhere else, will vanish for ever. Somewhere or other, too, the rainfall fed by evaporation from the Sudd must be reduced.

The scale of damage done by these African and South American dams is so vast that the relatively tiny dam that the Tasmanian government proposed to build on the Gordon River to produce 0.25 GW seems hardly worth mentioning. It is however worth a brief comment because for the first time the damage to an irreplaceable ecology gained not only local but worldwide attention—assisted by the much publicised trip of the naturalist David Bellamy from Britain to Tasmania for the express purpose of getting himself arrested for interfering with the preparatory work. Such publicity may help to discourage future similarly destructive projects. The proposed reservoir in Tasmania would not only have destroyed a large area of one of the few temperate rain forests still surviving in the world, but would have flooded a cave recently discovered to contain what appear to be the records of 100 000 years or more of continuous human occupation, covering a period of pre-agricultural human history for which the discontinuous records in the northern hemisphere have unbridgeable gaps.

BIOMASS AND THE ENVIRONMENT

The production of methane or other burnable fuels from combustible wastes, cow dung or human sewage, is valuable in two ways: in removing safely material which could present hazards to health and in conserving non-renewable fossil fuels. On the other hand, as pointed out earlier, the growing of crops such as sugar cane or fast-growing trees necessarily uses up land that could have grown food or timber for construction purposes, and thus directly or indirectly leads to the destruction of a part of our shrinking 'natural' environment. The figures are necessarily speculative, and could be wrong by a factor of ten in either direction, but the area that Brazil is proposing to use for the production of fuel alcohol for transport may represent more than a hectare (2.5 acres) of forest cut down for each car that is kept running for an average number of kilometres per year on the fuel produced. To describe this as using a renewable resource is misleading. Sugar cane is indeed renewable. Clear-felled tropical forests are not.

DIRECT SOLAR POWER AND THE ENVIRONMENT

The photovoltaic cells used to produce electricity directly from sunlight have important local ecological effects. If solar electricity is to be produced on a large scale in tropical areas not already covered with buildings, the area used will in biological terms be close to a desert. As was pointed out in Chapter 5, the true sand deserts or any area liable to regular sandstorms will not be usable for this purpose. The area to be used will therefore have had something growing on it. Where the solar panels are established nothing at all can be permitted to grow; tropical plants grow fast and quickly get out of control. Even in the sunnier parts of the tropics something like 30 km^2 would be needed to replace one fossil fuel or nuclear station producing 1 GWe, assuming a 20% efficiency of the solar cells, and allowing for reflection and absorption in the protecting cover. Such an efficiency has not yet been achieved on a large scale, but probably will be in the not too distant future.

Even the people most seriously concerned about the effects of acid rain would not expect a 1 GW fossil-fuel-fired power station to wipe out all life over 30 km^2. It is likely that many smaller solar-powered stations will be set up in areas of little ecological interest. The area sterilised is 20 times or so less than the area covered by fuel crops for the same energy output, but the ecological effect within such an area is enormous.

Well away from the equator it will pay to mount panels at an angle to the horizontal, in rows spread apart, to use sunlight more efficiently; and this will allow a useful amount of sky light to reach the ground between the rows. A solar power station sponsored by the European Community and built on these lines was put into operation in July 1983 on the German North Sea island of Pellworm. This occupies 1.6 hectares, gives a peak output of 300 kW (with a year-round output of perhaps 40 kW) and has panels set at a height which allows sheep to graze below them (IEE 1983b).

To most people, changes in environment matter very little. They are mostly a long way off and, over most of human history, the greater part of the globe has been practically empty of people anyway and unaffected by anything that we were capable of doing. In the last century and half all of this has changed; the combination of modern medicine and modern methods of power production made possible by applied science and technology has enabled us to escape from most natural limitations and our numbers are currently doubling every 40 years. Scientific developments have given us the capacity to change the ecological conditions of every portion of our planet.

The science that enables us to tell what will be the approximate proportion of iodine-129 in the iodine of the biosphere in two million years' time can also tell us a lot about the more direct consequences of our present actions in 50 years' time. It is important that we should give a higher proportion of our attention to the latter.

Chapter 15

The Opposition to Large-scale Power Production

They make all these emotive allegations, but when you nail them down they don't stand up.

Punch, 3 August 1983

In this chapter I shall present some serious criticisms of many of the opponents of coal, oil and nuclear energy production. I would like to begin however by expressing my conviction that without their vociferous and highly publicised criticisms the big power industries would be far less safe than they are. The threats of sulphur-bearing smoke, of acid rain and of carbon dioxide, would have continued to be merely prolific sources of highly technical controversy in the scientific journals, unknown to the public, without the often exaggerated voices of conservationists and others concerned with the long-term welfare of mankind and the preservation of wildlife.

The Western nuclear industry in particular, whatever its prospects for the future, could hardly have achieved its unparalleled safety record without such opposition. If less of the opposition had been ill-informed abuse, the well-informed part would have had more attention still.

THE SIMPLE LIFE?

Opposition to large-scale industry of all kinds comes from a great variety of sources. Some of it is very general; for example the more extreme of the misinterpreters of Schumacher's dictum that small is beautiful would like to renounce all large-scale systems and return to a simpler life style, in which small social groups could feed, clothe and warm themselves in their own chosen ways. It is doubtful whether any large number of people have ever actually wanted to do this in a climate like Britain's, when faced with the drudgery of fuel and water collection and of agriculture without powered mechanical aids, of spinning and weaving, of housebuilding and maintenance. Very few people in Britain wish even to grow all of the vegetable part of their food in their own allotments or gardens.

Whether or not some people would prefer such a life, which could indeed be happier and more satisfying than many modern life styles, it is no longer possible for the majority. There are too many of us. Whenever anyone, including me, says that there is too little of any resource, what is meant is that there are too many people wanting to share it; and in Britain there is neither the wood for fuel nor the land for cultivation to enable everyone to live the simple life. If one uses horses instead of machinery, each horse needs about as much land to grow grass and hay for its own food as it can cultivate to grow food for its owners. It would be nice to live in a world with a small enough population of constant size for everyone to have the choice of living in this way, obtaining the technical necessities such as saws and cooking pots, solar panels and lamp bulbs, made by robots in underground factories, at the cost of a few hours a week of community service. But the world is not like this, and we shall need even *more* technology if we are to feed and clothe the world population which, in the absence of war, we are certain to have throughout the next few generations.

SMALL-SCALE POWER UNITS

A larger and more practical group, including the proponents of intermediate technology, would like, on conservation grounds, to replace by renewable sources all of the power derived from the destruction of the world's limited capital of coal and oil. As we have seen, not all of the sources described as renewable can in fact be indefinitely renewed with a growing population, but much more could be done than is being done. Probably the most articulate and thorough proponent of this group is Amory Lovins, who has explained this viewpoint at length in his book *Soft Energy Paths* (1977) already quoted in Chapter 7. He makes a valuable contribution by showing the advantages of concentrating on what people actually want—such as warmth—rather than on economic 'demands' for example for electricity. In stressing the advantage of small units for house heating etc, he is either unaware of or unconcerned with the very much larger local concentrations of pollutants such as sulphur dioxide and carcinogens produced by smaller and less efficient combustion units; although he does say that fluidised-bed coal burners using lime to remove sulphur compounds could be developed for domestic use. How such a system could be made to run entirely automatically for months on end at a reasonable cost he does not explain; nor on the other hand does he show how an ordinary household could be taught and persuaded to maintain it. This however is a side issue—considerable energy gains could be made without household fluidised beds for everyone.

Lovins' desire for more effort and money to be put into development of renewable systems I share wholeheartedly, although I don't think they will soon, or perhaps ever, produce more than a third of our needs. But his optimistic approach to what is practicable leads him to expect too much. For example, he recognises that there will be an interim period during which oil is no longer available, and the renewables have not yet expanded enough to replace it. He therefore proposes (p. 49) that 'Coal can fill the real gaps in the fuel economy with only a temporary and modest (less than two-fold at peak) expansion of mining...'. Building an extra coal mining industry in the USA comparable in size to the present industry—modest, he says.

I cannot illustrate Amory Lovins' uncritical approach better than by quoting his own words (p. 23):

> Taking the initial steps towards a soft energy path, and following up to be sure they work, will not be easy, only easier than not taking them...a soft path simultaneously offers jobs for the unemployed, capital for business people, environmental protection for conserva-tionists, enhanced national security for the military, opportunities for small business to innovate and for big business to recycle itself, exciting technologies for the secular, a rebirth of spiritual values for the religious, traditional virtues for the old, radical reforms for the young, world order and equity for globalists, energy independence for isolationists, civil rights for liberals, states' rights for conserva-tives.

Many of his practical proposals seem both possible and desirable, as I have mentioned above, but if he claimed less he might exert a larger influence on the people who would actually have to do and pay for the things he would like to have done.

He rules out the use of nuclear power because the USA has exported a great deal of very highly enriched uranium, usable for bombs, for research reactors. I too object very strongly to this; but he suggests no reason why the building of power reactors in the USA or Britain should increase the risk of proliferation of bombs elsewhere. As I have said before, I cannot myself see how we can hope to cope with a world population of nearly 5000 million increasing at 100 million a year without using all the energy-producing industries that we have, including the oil industry. I have also a strong feeling that some effective competition for the latter would be very good for it.

THE FEAR OF NUCLEAR POWER

Most of the opposition to nuclear power is expressed in terms of the greater dangers believed to exist from its use when compared with the dangers from other sources of power (although some opponents of the nuclear industry fail to mention *any* dangers from other sources). Friends of the Earth (FoE) and some members of Greenpeace are now seriously concerned about acid rain and the increasing concentration of carbon dioxide; indeed there are erstwhile opponents of nuclear power in California who regard these as even more serious. Some opponents have based their continued opposition on the fact that many people are afraid of nuclear power, independently of whether or not the fears are justified. Clearly a democratic government must take account of people's fears, whether the fears are justified or not; but it is immoral to use the fact that some people are frightened to frighten others without producing a clear case that the fears are justified or not.

One of the most serious effects of Chernobyl in Western Europe has been the intensification of fear, in spite of the smaller-than-expected numbers of 'prompt' deaths and although the accident was of a type which could not have occurred in any Western reactor. An important corollary to excessive concentration on exaggerated fears is that it necessarily leads to underestimates of real but less well advertised risks, and hence to unnecessary deaths. This should not be acceptable.

Nearly all the opponents of nuclear power want to convert the majority to their way of thinking, and do genuinely believe that nuclear power will harm people more than do the alternatives. It is with the arguments of these, who have a genuine concern for people and who have mostly taken the trouble to learn at least some of the facts, that I shall be occupied for the rest of this chapter.

UNION OF CONCERNED SCIENTISTS

Events began with, and were perhaps triggered by, a letter to the President of the USA at the end of 1974 from 32 leading American scientists, including eleven Nobel Prize winners, pressing for further development of nuclear power. They included nine who were former directors of major AEC laboratories or members of the AEC Commission or its Central Advisory Committee, and between them had experts in most nuclear fields. In April 1975 the *Bulletin of the Atomic Scientists* published a letter sent by eight Nobel Prize winners of the Union of Concerned Scientists (UCS), none of them radiobiologists or nuclear engineers, to President Ford, expressing concern about the dangers of American nuclear reactors, and giving very large figures for the possible numbers of casualties should a reactor accident occur or nuclear wastes escape into the environment. A number of other points were made, including some justified criticisms of the American Atomic Energy Agency. This had indeed made a number of *ex cathedra* statements on nuclear safety matters that it first would not, and later in several cases could not, justify. As a specific example, after 17 years of study and the expenditure of $25 million, the AEC concluded that a proposed storage repository for high-level wastes at Lyons, Kansas, was totally satisfactory. Later however the AEC abandoned the proposal.

Too much of this controversy was concerned with personalities rather than with technical facts. Professor Charles Schwartz, in a document initially sent by him to the *San Francisco Chronicle* but not published by that paper, listed the 32 pro-nuclear signatories and discounted the value of their letter on the grounds that fully two-thirds of them had personal ties with big business as consultants, or as directors of major US corporations—including such giants as Exxon, IBM, Xerox, TRW and Owens-Illinois. Why directorships of IBM, Xerox, the oil company Exxon, etc, should bias people in favour of nuclear power was not made clear, and neither was the reason why Professor Schwartz made no similar analysis of the associations of the UCS members who were not in favour of nuclear power.

I am not denying the existence of bias. Everyone has some biases. Ignorance is however a more reliable source of bias than is knowledge. People who feel strongly about something of which they know nothing may by accident be right or may by accident be wrong, but they must by definition be biased.

Unfortunately, even the greatest concern for facts does not immunise one against prejudice. It is impossible to see the relevance of or to remember every fact that one meets, and if you have assimilated a large number of facts pointing in a particular direction it is far easier to see the relevance of a new fact pointing in the same direction than it is to see the relevance of, and hence to remember, a new fact pointing in the opposite direction—as exemplified in the general neglect of the effect of moderate irradiation in increasing the longevity of rodents, discussed in Chapter 4. Reading the words of those opposed to nuclear power has led me to look for the factual bases of many things that I might have taken for granted, and, for example, to do the arithmetic which confirmed the undesirability of simply sinking all of the high-level wastes to the bottom of the deep ocean. Without the exaggerated concern of Greenpeace with the trivial effects of intermediate-level wastes I might never have bothered to do this. This exaggerated concern did not, however, lead me to look instead into the antecedents of the members of Greenpeace to find out how many of them had rich uncles with investments in the oil industry.

To return to the UCS, who had been worried about the very large numbers of deaths that might be produced from a nuclear accident as a result of plutonium or other active materials being evaporated, or from the deep burial of radioactive wastes, I wrote separately to each of the signatories of the letter to President Ford. Among other things, I asked in each case for a quantitative analysis of any way by which the high-level wastes could kill a large number of people if put deeply underground in ancient rock containing no obviously useful ores that might attract excavators.

Seven out of the eight replied at some length to my general comments, but not one answered the specific questions, although Professor Urey did say that he thought it possible that I was right (in thinking the wastes not a significant danger and in thinking even the most serious accidents unlikely to kill more than many chemical accidents might do). This was not good enough. However eminent you are in one technical field you should not write to a President about risks in another technical field without for yourself confirming the quantitative accuracy of your statements.

I should perhaps finish this section by saying that the UCS later published a book on radioactive wastes in which the conclusion was reached that acceptable waste disposal should be achievable, although not within the optimistic time schedule set by the federal government (Lipschutz 1981).

FEARS OF LOW DOSES OF RADIATION

I shall now go on to discuss the arguments of a series of people who oppose the use of nuclear power on the grounds that small doses of radiation are very much more dangerous than is supposed by either the ICRP or the BEIR Committee.

Well ahead of the rest of the field is a proposition, the origin of which I have failed to discover, that 1 kilogram of plutonium would be sufficient to kill off all life in the world. This came to my notice in a naturally worried letter, dated 14 June 1976, from C A B Smith, Professor of Biometry at University College, London, to the Cumbrian County Council; he claimed that the proposition could be found in reputable journals. He presumably realised his error later, as he did not reply to a request from me for a reference to any reputable journal reporting this statement. The natural uranium, radium and polonium in the ocean give off about 400 alpha particles per gram of sea-water per year. One kilogram of plutonium distributed in the oceans of the world would be able to provide only 1 atom in each gram of water, which would give about 1 alpha particle per gram in 50 000 years. The uranium does not appear to affect the life in the ocean. Nor would the plutonium.

At the end of the Second World War plutonium contamination was a matter of concern in the laboratories engaged in developing nuclear weapons. The health authorities were therefore anxious to determine the excretion rate of plutonium once it had been absorbed. I do not know who authorised the experiment, or what inducements were offered to the subjects, but plutonium in solution was injected into 17 people who were believed to have only a short time to live. Eight however survived for more than eight years, and four were still alive 30 years later; and although most of the plutonium injected was still in their bodies not one had died of either bone or liver cancer (the main part of plutonium in the body is divided almost equally between the liver and the skeleton) (Mays 1982). If the ICRP figures had been seriously wrong such cancers would surely have appeared.

It is likely that the long-term cancer-producing effects had not yet shown themselves in Japan at the time of the injections, and that the risks of the experiment were not appreciated. Injections of Thorotrast (radioactive thorium oxide) to improve the contrast of x-ray photographs were still being given in the early 1950s, and these led to a number of deaths much later. At that time it was still believed that radiation doses below a quite high threshold had a zero effect.

The Hanford study

A suggestion for a more modest reduction of permissible levels by twenty times, which is much more important because it depends in part at least on actual observations, and not on pure theory, and also because it is still very frequently quoted, was put forward by Mancuso, Stewart and Kneale. They studied the death rates from cancer at the American nuclear establishment at Hanford, in a near-desert area of Washington State. The conclusion of this investigation was that the ICRP figures were underestimating the risk of radiation by twenty times. The effective member of the triumvirate was Dr Alice Stewart, an experienced epidemiologist, who had earlier, when working at Oxford, been the first to notice an increasing number of deaths from leukaemia among young children. She had suspected the responsibility of x-ray examinations in utero, and showed that these were indeed responsible for a large fraction of the leukaemias observed. This observation, which must since have saved the lives of thousands of children owing to the reduction of x-ray examinations of pregnant women, both led her to be very much on the look-out for other unexpected effects of radiation and led the radiobiologists of the nuclear industry and many others to take the Hanford publication much more seriously than they would otherwise have done.

At the time Stewart became involved, Mancuso had already been collecting health data for many years but had not combined his results into a report. Dr Stewart, after a careful study of the data, could see that the cancer rates among employees who had received some irradiation in the course of their work were rather higher than among those who had not. Joined by Mr G W Kneale, a statistician, the three have presented arguments to show that about 5% of the 832 cancer deaths recorded at Hanford were due to radiation, and that this implies a 20-fold greater danger from low levels of radiation than is estimated by the ICRP.

The method employed was the unusual one of comparing the irradiation histories of the Hanford employees who died of cancer with those of Hanford employees who died from other causes. Details of the data obtained from the very complete health records for the Hanford employees, and of the statistical methods employed to examine them, were presented at an international symposium in Vienna in 1978 (Kneale *et al* 1978). During the period of study from 1944 to 1977, 832 cancer deaths and 3201 non-cancer deaths were recorded. The 5% of deaths from cancer believed to be due to radiation would then amount to 42, the other 790 cancer deaths being due to natural causes. Of course the number of cancer deaths from natural causes out of 4000 total deaths will vary somewhat, and there is about a 1 in 20 chance that natural causes might on this occasion have accounted for 42 more than expected. Nevertheless, the number of deaths from cancer was definitely greater for the irradiated group than for the un-irradiated group; and although this could have been due to chance alone the fact that the probability of this is only 5% makes it sensible to look for a real carcinogenic factor.

Unfortunately, the detailed figures given in a series of tables in the paper quoted do not prove that this factor was the extra radiation received at Hanford. To start with, the total cancer death rate at Hanford was unusually low compared with the non-cancer death rate. In the defined age range between 40 and 59 this ratio in the whole male population of England and Wales in 1981 was 37% higher than it was at Hanford (Mortality Statistics 1981). (The male population only is taken because 92% of all Hanford deaths were males.)

The figures given by Kneale *et al* show that the ratio of cancer deaths to non-cancer deaths increased by 33% as the radiation dose accumulated at work increased from less than 80 mrem to over 5 rems. No indication is given of the average age at death of people who had experienced different levels of irradiation. It would be expected that on the average those who had received the largest cumulative doses of radiation had worked for longer and were older than those who had received the smaller doses. In England and Wales an increase in age of

11 years would be sufficient to give the 33% increase in the ratio of cancer deaths to non-cancer deaths observed at Hanford.

The second point of criticism is concerned with two factors not controlled, namely cigarette smoking, and drinking, which were not recorded at all during the survey. It is stated in the summary of the paper that the 5% of extra cancers among the irradiated groups were probably concentrated among cancers of the bone marrow (25), lung cancer (215) and pancreas (52), the figures in brackets being the numbers of deaths concerned as given in Table 5 of the paper. The number of deaths from lung cancer is more than a quarter of all deaths from cancer, and 20% more cigarettes smoked per man among the irradiated groups (who would have been more skilled and probably better paid) would have accounted for the entire effect observed.

Thirdly, no record was kept of the origin of the workers. The great majority of cancers recorded were the commoner kinds with long latent periods, so that any radiation-induced or non-radiation-induced deaths from cancer occurring in the first 20 to 30 years of a man's work at Hanford would almost certainly have been due to radiation or chemical carcinogens received before he arrived at Hanford. The Hanford works was purposely established a long way from any inhabited area, and the entire staff must therefore have been recruited from elsewhere. It is not only possible but likely that it would have been the technical staff who would have received the largest doses of radiation at Hanford, and the technical staff were presumably recruited from industrial regions of the USA. These regions include just those which have the highest cancer rates as shown in figure 4.3, while the more numerous manual staff would be more likely to have been recruited from the nearer states with higher natural radiation backgrounds but lower cancer rates. The low average cancer rate among Hanford employees is consistent with the majority of the workers having come from 'low cancer' states.

This lack of data on employee origin meant that no attempt could be made to find out even what radiation doses had been received from the natural background before arrival, although these must have been greater for over 80% of those dying of cancer than were the doses received after their arrival. (At the average US background rate of 120 mrem per year, by 20 years old a dose of 2.4 rems has been accumulated; while at the rate of 240 mrem per year characteristic of Wyoming and Colorado, 4.8 rems would have been accumulated.)

Leukaemia (15 cases) which has a short latent period showed no increase. Multiple myeloma however (10 cases) increased enough to imply that the addition of 3.6 rem to a lifetime background of 5–10 rem would double the natural rate. This is much less likely than that the increase occurred by chance.

To summarise, there is a 5% probability that the slightly enhanced cancer ratios at the higher radiation dose levels were in fact purely a matter of chance. If they do show the effect of an extra causative factor in operation at Hanford, this could be due to the greater age at death of the more heavily irradiated; it could be due to a coincidental larger cigarette consumption by the more heavily irradiated; it could be due to differences in pollution experience or radiation received by the victims before their arrival at Hanford, or it could be a mixture of all of these. There is no reason at all for choosing the radiation doses received after starting work at Hanford as the operative factor.

The really important positive argument against radiation as the relevant factor at Hanford lies in the evidence mentioned in Chapter 4 from the Argonne Laboratory and illustrated in figure 4.3. By the time they were 50 years of age the entire populations of Wyoming and Colorado had on the average received 5 rems more than the mass of Americans in the lowland States—and had a death rate from cancer 17% less than the American average.

In view of the overwhelming evidence from the Argonne study, it may seem unnecessary to have discussed the Kneale *et al* work in such detail. I have done so for three reasons. Firstly, if it were true it would have serious consequences. It would mean that the radon in British houses must have condemned many thousands to death already, and would make the draught-proofing needed for a serious heat conservation programme quite seriously unwise. Secondly, there are still many people so impressed with the amount of effort that has gone into the work, and the statistical sophistication employed to analyse it, that they feel it must still be taken seriously. The third reason is the moral to be drawn for future investigations of the effects of industrial irradiation at establishments such as Sellafield or Aldermaston. The dose rates received by many workers at Sellafield are as high as or higher than those recorded at Hanford, and an extensive study of cancer rates and health records is proceeding there†. It is important that the factors omitted from the pioneer Hanford study should not be neglected again.

THE EXAGGERATION OF RISK OF ACCIDENTS

The majority of opponents of nuclear power are less concerned with the exact doses of radiation needed to kill people than with the supposed susceptibility of the nuclear power stations to lethal accidents. Several books have been written, and sold in large numbers, to stress the risks of such accidents, but many of these seem to show that their authors understand very little of the matter of which they are writing. Most of

† The results have shown no relation between cancer rates and radiation doses.

the books are good in parts, the authors having adequate knowledge about at least some of the relevant subjects; but erroneous categorical statements about subjects they know less well throw serious doubts on their judgement in general. I shall discuss a few examples.

Poisoned Power by J W Gofman and A R Tamplin, published in 1973 but still being quoted, demonstrates this pattern very well. The authors begin with an excellent chapter on how nuclear reactors work, and a simplified chapter on the effects of radiation on people—marred by their overestimation, by a very large factor, of the dangers of plutonium—and follow these by presenting a horrifying but convincing report of cover-ups and distortions by the American AEC. Some of the attacks on the AEC are however unfair and dependent on using different meanings for the word 'safe'. For example, the AEC and a number of propagandists for nuclear power quite fairly, if injudiciously, said that nuclear power was safe—using the word as I would if I said my children would be safe if they didn't have to cross the road on the way to school. Gofman and Tamplin used the word to mean no risk at all—a concept applying to no human activity whatever, past, present or future. In fact they make this a point of principle. On pp. 268–9 they present a set of prerequisites for sound decisions on technological dangers. I quote three of them.

(1) Abolition of 'experts' or 'standard-setters' as decision makers.

(4) Open forum debate, followed by decision either by public vote or vote of *public* representatives.

(6) Recognition of the principle that the appropriate dose of a man-made poison is *zero*. Deviation from zero allowable pollution must be allowed only by public decision to be polluted in exchange for some benefit it chooses to receive.

(1) and the first half of (4), if they mean anything, mean that decisions should be made only by people the majority of whom do not know the facts. Alternatively, the second half of (4) could mean that in a democracy the government should decide. Since the US government accepted the decision to go ahead with nuclear power and most of the book is disapproving of the result, it presumably doesn't mean this. Certainly the decisions should not be made by experts alone, though they too have children and grandchildren and are likely to take trouble to use their expertise to reduce the risks to an acceptable level. Equally certainly, decisions should not be made without the experts; no safe or useful decision can be made by those ignorant of the facts. Both experts and the public—or their representatives—must be involved.

Point (6) for practical purposes means that nothing should ever be done for the first time. No 'public' could ever decide to be polluted without practical experience of the benefits—which could not be gained without some pollution. Presumably it would now be impossible to get a majority vote banning motor vehicles; but if when the internal combustion engine was first invented it had been possible to forecast the degree of pollution by petrol and diesel engines, the vote would surely have been 'no'. And this pollution is immensely greater than anything nuclear power stations can do, carefully monitored as they are (by the experts).

Rather injudiciously Gofman and Tamplin claim (p. 97) that an average extra dose of 0.17 rem a year to the population of the USA would lead to an extra 32 000 deaths from leukaemia and cancer each year (assuming the effects of radiation to be ten times the ICRP estimate). Experimentally, it is the large number of US states that have an average annual radiation dose 0.12 rem *less* than Colorado or Wyoming that have the extra 30 000 cancer deaths per year.

The main fault of the book, however, is not so much the few blatant errors as the complete disregard of the risks of the fossil-fuel alternatives—which are cheerfully dismissed with a confident statement (pp. 222–3) that fossil-fuel plants could be built free of pollution and producing cheaper power for a few hundred million dollars a year. Why the fossil-fuel companies, with a turnover of tens of thousands of millions of dollars a year do not bother to do this is not explained. Although the book mentions the tiny and well-understood risks from carbon-14, it does not even mention the possible risks from air pollution or carbon dioxide.

Race to the Finish, the Nuclear Stakes by Dervla Murphy is concerned to a large extent with the responses the author got from the people she met on an extensive fact (and opinion) finding tour in USA and Britain, and she quotes on a fair and not too unequal basis statements from both sides. Her reactions to the two sides however were different. She saw the film *The China Syndrome* during the early part of the Three Mile Island breakdown, and says (p. 57):

> Officials at Three Mile Island now say that the likelihood of a meltdown is 'extremely remote'. Which means still possible.... A strange frisson went through the audience when one of the characters explained that if the meltdown happened it would 'render uninhabitable an area the size of Pennsylvania'.... This has been the most heartening world-wide consequence of TMI [Three Mile Island]—what seemed almost impossible to nuclear experts and general public alike is now known to be possible. And the stupidity, dishonesty and ruthlessness of the nuclear bosses and their allies, which would have overstrained the average cinema audience's credulity before TMI has since appeared only too plausible.

Her only evidence for dishonesty was (*a*) that the fuel did not in fact melt and a catastrophe didn't happen, and (*b*) a character in a fictional film said with gross exaggeration that if it did it would be catastrophic. And the 'heartening world-wide consequences' of these bits of non-evidence are that the stupidity, ruthlessness and dishonesty of the nuclear energy experts have now become plausible, apparently because they were right when Dervla Murphy thought they were wrong.

I am willing to believe that some American (or indeed British) bosses in any big industry are prepared to mislead or to withhold information from the public; but the evidence given in the case quoted proves nothing but the bias of the author.

As a further example she is prepared to add the whole group of international experts comprising the ICRP to her list of dishonest or incompetent characters. On p. 122 she says:

> In a report known as ICRP-26, they allocated different risk rates to different organs, if *irradiated on their own* [her italics].... However, even a layman can see that the concept of different risk rates for different organs is absurd—a bit of abstract nonsense dreamed up by men who experiment on animals in laboratories. How does one irradiate a liver or a kidney or a thyroid gland *on its own* while it is still in daily use?... As is often the case with high-powered special-ists, some ICRP researchers concentrate on the use of abstruse equations and computer models to *estimate* the effects of radiation and seem indifferent to the results when their conclusions are tested in the crucible of real life.

The arrogance that enables the author to make, apparently without checking, such a confident assertion of the superiority of her instinctive judgement over the summation of four decades of expert study would be unbelievable if it had not been demonstrated in print. Practically all the dangerous radioactive materials are seriously hazardous only when they are absorbed by the body, and when they are absorbed by the body they are nearly all concentrated in particular organs. Thus ingested plutonium is mainly concentrated in liver and bone, and these are therefore the organs which are irradiated by the short-range alpha particles produced. Iodine-131, which could be responsible for most of the prompt deaths following a catastrophic reactor accident, is concent-rated in the thyroid gland, and the consequent beta irradiation of this gland is vitally important, while the associated gamma rays mostly escape without doing appreciable harm to the rest of the body. Cad-mium isotopes are concentrated by the kidneys. All diagnostic and therapeutic x-ray irradiations are concentrated on specific organs, and the permissible exposures are limited by the doses to these organs.

Although Dervla Murphy exaggerates nuclear risks, she is probably right in believing that large numbers of ordinary people exaggerate the risks at least as much as she does, and her book would be valuable reading for many nuclear engineers who want to understand the magnitude of the misapprehensions with which they ought to be dealing; and the fault that I think most seriously reduces the value of the book is again her lack of comparison with non-nuclear risks. We may not need all the energy we are using, but we must have some. It is about time that people such as Dervla Murphy, with a gift for expressive writing, should begin to deal with the things which are now damaging future generations in a major way, rather than with the ones which *might*, very rarely, do damage on a hundred times smaller scale.

Many of the books I have read show how tragically ill-informed are some of those who have written in opposition to nuclear power. There is no discredit in knowing very little about so complex a mixture of physics, engineering and biology as is required for making useful judgements about the risks of any kind of power production. If you do not know much there is credit, not discredit, in voting according to the little bit you know. But if you want to write a book it is desirable to learn about the things of which you want to write. Honest ignorance is fair enough—all of us are ignorant about most things—but to lead others to think that you are not ignorant when you are is not honest.

To illustrate this I would like to quote the suggestion, in quite another context, that the emission of CO_2 from coal power stations could be reduced by using slaked lime to absorb it. This shows ignorance of the fact that to produce slaked lime from chalk by heating it releases as much CO_2 as it is then later able to absorb, if each stage is 100% efficient. In practice it will release more.

Two highly inaccurate but apparently best selling books are important: *No Immediate Danger* by Dr Rosalie Bertell and *Red Alert* by Judith Cook. Judith Cook is an experienced journalist specially interested in social and environmental issues. She claims no technical knowledge herself but describes Dr Bertell as an international radiation expert. Dr Bertell has a PhD in mathematics and is a member of the Order of Grey Nuns. As the following comments will show however, she is no more a radiation expert than I am a nun.

Among the list of 35 errors that I sent to her after reading her book were the following few. On page 17, referring to the Oklo natural reactor, she calls it 'a small fission reaction in South Africa... hundreds of thousands of years ago'. In fact it was a major reactor in Gabon that operated on and off about 1700 million years ago, and nearly all its fission products have remained where they were ever since.

On page 19 she states that 'About 300 different radioactive chemicals are created with each fissioning.' They are not; two, or rarely, three are produced, each decaying with a few radioactive descendants. When millions of uranium atoms undergo fission, then there may be 300 different radioactive atoms produced, two or three at a time. Her page 106 contains the statement that 'The fission process is violent, causing normal molecules ... in near proximity to become radioactive.' It is not the violence of the fission process but the absorption of the neutrons it liberates that causes this. Page 121 gives us two examples. 'It was low level waste, not reactor fuel, which apparently exploded at Chelyabinsk and almost exploded in Hanford'. As I have shown on page 199 (Kyshtym) this was not possible. 'The dream of sending nuclear waste into space is utopian because of its extremely heavy weight, and the volume required to prevent the formation of a critical mass.' Almost all of the fissionable material has been removed from the wastes, an infinite amount of which could not produce a nuclear explosion. The space proposal is impracticable because rockets sometimes fail and may burn up, distributing the wastes in the atmosphere.

On page 252 she says with regret 'Nor is there a UN agency entrusted with the task of distributing serious anti-nuclear research.' Anti-nuclear propaganda is understandable and, if based on facts, fair. Research is done to answer a question the answer to which is unknown, and is usually more reliable if you do not start with the intention of proving that the answer you would like is right.

I am puzzled to know how Dr Bertell acquired so much incorrect information. A year or so after reading her book I heard her give a lecture on the plight of the natives of the Pacific island used by the Americans for an underwater bomb test, illustrated by her own photographs. She was passionately moved, but had shown considerable ability both as an observer and as a concerned reporter of her observations. It is to be hoped that she will direct her energy and ability into the far more serious though less publicised dangers from pollution of the air we breathe, and also the long-term dangers produced by the effluents of burning fossil fuels. I would also like conservationists in general to take much more interest in population problems and to realise that what is damaging our planet most is too many people.

Judith Cook has written a shorter book with fewer errors, some of these being due to her belief in the reliability of Dr Bertell. The errors she makes are however striking. Following the proposition that I discussed on page 289 above, she claims (page 26) that '1 lb of plutonium, universally disposed, would be adequate to kill every man, woman and child on earth.' It would not. On page 8 she says that enriched fuel contains plutonium. No, it is extra uranium-235.

In a number of places she refers to the dangers of the 'China Syndrome', and in her Glossary defines it as 'The theoretical consequence of a core melt-down, when the heavy molten mass of highly radioactive material actually goes straight through the vessel in which it has been contained and down through the earth's core. From the USA it would appear in China.' It is surprising what educated non-scientists can believe. Presumably it arrives at the earth's core with enough kinetic energy to go all the way *up* again to the surface—in the middle of the Indian Ocean.

Nowhere does Judith Cook mention the risks of any other source of energy or any other dangerous form of pollution.

The thing that worries me most about this book—and about almost all of the 13 books recommended in the Selected Biography under the heading 'Nuclear Energy'—is that they represent a growing pseudo science based on emotion rather than reason. For this, years of study and experience of practical research are not required, and indeed are attacked as leading to bias. Instead, the gurus demand the funding of *anti*-nuclear research. I would love to see a grant application for this—and offer free the suggestion that they might get support from the American oil industry which has profited heavily by their activities.

I do not think I can conclude better than by quoting for the benefit of her readers the advice of Paddy Ashdown MP in talking to Judith Cook: 'I believe that questioning what we are told is a fundamental step in reaching good decisions.'

WHY PICK ON NUCLEAR POWER?

However this may be, the really difficult question is why so many of the energetic and well-meaning uninformed concentrate their efforts on this particular field. The mass media have a serious degree of responsibility. More readers and viewers are attracted if you produce headlines and feature programmes prophesying doom than if you keep reiterating the obvious fact that we in the countries based on applied science are warmer, more comfortable, healthier, safer and longer-lived than any past population at any time and any place on our planet.

The uniformity of propaganda—and especially of the more exaggerated propaganda—against nuclear power in particular, rather than all large sources of power in general, suggests to some social observers an organised orchestration by groups behind the scenes who have something material to gain for themselves. There are three groups who might be interested.

The most powerful and the most obvious is the American oil industry. So long as there is no alternative, it is not necessarily harmful to the industry if the oil supplies become scarcer; if every time that the available reserves are halved the price in real terms is doubled, they can do well indefinitely. Nuclear power does not yet offer much competition, but the delay due to the opposition, here and in the USA, in building nuclear plants has given nearly an extra ten years before it can offer much competition.

Although it has frequently been accused, and rumours have been spread that it had planted people well beforehand in areas suitable for nuclear stations to lead local opposition should construction be proposed, I know of no evidence whatever that the oil industry, in America or anywhere else, has been paying for anti-nuclear propaganda. This would anyway be foolish and if proved would be counterproductive. If some rising young executive in the oil industry thinks he could gain credit with the boss by a bit of private enterprise in producing selected data in a scientific format for the use of the anti-nuclear organisations, one could hardly blame the industry as a whole. Energetic careerists in the nuclear industry must surely be interested in collecting data on acid rain and chemical carcinogens. Where either group is collecting and disseminating facts, even selected facts, there can be no objection; between them they could find and quantify risks in both industries that less well-informed outsiders would probably miss.

The coal industry in Britain is more directly threatened than the oil industry, since so much higher a proportion of its output goes into the production of electricity. The mineworker's leader Mr Scargill has been producing a great deal of anti-nuclear propaganda. He did say, at the Windscale Inquiry, that he could be reconciled to nuclear power if it could be proved that not one person would ever be killed by the use of nuclear power. For someone representing an industry which was killing a miner a week in an accident and which is responsible for a significant but unknown proportion of the deaths from cancer occurring annually in Britain, this was disingenuous.

The second group to be considered consists of the confidence tricksters, who are to be found in every sphere of human life. It has been well said by operators in the field that a mug is born every minute. This formulation was coined to describe characters who would buy gold bricks from strangers, or universal remedies and cancer cures from unqualified quacks. To qualify as a mug you had only to grow up with a comprehensive ignorance of such people; any technical knowledge that a mug might possess would neither help nor hinder. In the scientific field it would be more appropriate to say that a sucker is born every second, since a comprehensive ignorance of science is commoner than a similar ignorance of people; and it is impossible that there should be no con-men around to take advantage of so obvious an ecological niche in society.

If you wish for a scientific career in either industry or academic life, you must be prepared to work pretty long hours for a decade or more on a specialised subject, and among the things that you do not have time for is explaining the basic facts of your field to people who know nothing about it whatever (except of course to a registered class of specialist students who are themselves wanting to learn from the beginning). If on the other hand you have no interest at all in facts, and spend a similar period in learning how to sound convincing, and how to use the expert's inadequacy in communication to throw doubts on his reliability, you can win friends and influence people.

The scientific con-man is unlikely to make a vast fortune, although he may get a pretty good living out of lecture fees and books; and if he does well he may get a good deal of foreign travel in carrying his message to susceptible groups around the world. What he does gain is fame and an enthusiastic following of true believers, which for many of those already provided with the necessaries of life may well be more attractive than cash.

While I cannot doubt the existence of persons who for various motives are taking advantage of the scientific gullibility of a large number of well-meaning people, I may be overestimating the importance of their influence. It is perfectly possible that most of them are merely opportunists who see a fuss going on and can gain a boost to their egos by joining in and running shouting out in front. They may even begin to believe their own ill-informed propaganda.

I would like to add that I do not suspect of insincerity the writers of any of the books that I have discussed above. If I had suspected them to be insincere I would have considered it far less useful to criticise their errors; and in all these cases, if they ever read this book, I am hoping that they may learn from their mistakes and take more trouble to check their facts in future. This would have been a forlorn hope with the insincere. The unmasked con-man doesn't change or argue—he merely waits a bit or moves to somewhere where he isn't known and carries on as before.

The third group of opponents of nuclear power consists of those who firmly believe that proliferation of nuclear bombs in new countries is to a large extent a result of spread of nuclear power, and that this is more seriously dangerous to the world than is the proliferation of American and Soviet weapons. This is a perfectly fair opinion, held by several well-known scientists. I have already explained why I do not myself accept that building nuclear power stations in Britain will increase the risk of proliferation, and will not repeat my reasons here.

It is true that the development of nuclear power was accelerated by the money and effort spent on research for the early military applications—so was the development of flint tools, bronze, steel, aircraft, communications satellites, and indeed the sophisticated aerofoils used for wind turbines. Nevertheless, there is no reason for not taking as much advantage as possible, for peaceful purposes, of the military research. It appalled me to see the Conservation Society, the Friends of the Earth, Greenpeace and, of all things, the Town and Country Planning Association, using money subscribed or allocated to them for conservation of the earth's resources and protection of amenities, appealing for additional government money to help them to slow down the retirement of fossil-fuel stations by opposing the building of a PWR at Sizewell. They have never made such a large effort to gain government money for research on wind power or the really valuable work they have done in opposing the draining of marshland and other special types of habitat in our rapidly disappearing countryside.

I have been pleased to notice some improvements. Recent display advertisements by Greenpeace in the media have been concerned entirely with the climatic risks of the increase of atmospheric carbon dioxide, with no mention at all of the risks of nuclear power. It would be unreasonable to expect Greenpeace yet to give positive support to nuclear power, but their recognition of it as a lesser evil would enable a lot of people to begin rethinking.

THE DANGER OF EXAGGERATION

The really serious energy problems that we face, as well as our problems in many other fields, are hardly due at all to the actual risks we run, but to the appallingly large gap, which I hope this book may do something to reduce, between the actual risks and what the public believe are the risks. It is the perceived and not the real risk that controls policy, and if enough of the public believe in myths it must be myths upon which the government of a democratic country has to act.

The exaggerated public perception of the risks of nuclear power is not only, or even mainly, the responsibility of the publications discussed in this chapter. It has been built up and fed by the newspapers and television—especially, because of its impact, by television. Mostly, as mentioned before, this has been due to the greater news value of exciting and frightening suspicions as compared with humdrum facts, amplified perhaps by the spice of jealousy that leads one to enjoy damaging the reputations of people better informed than oneself. There have however been some quite intentional and systematic distortions. For example, the BBC arranged a discussion (BBC 1, 23 April 1979) on the reliability of the ICRP estimate of the dangers of low levels of radiation, in which Dr Alice Stewart, who believes that the risks are twenty times greater than does the ICRP, and the late Dr John Reissland of the National Radiological Protection Board, which has for many years had a team of scientists studying the question, took part. The interviews were recorded on tape, and Dr Stewart gave a clear and full account of her work on the death rates at Hanford which was described above. Dr Reissland, who had recently published a very detailed critique of the Hanford work, explained in detail the numerous reasons—some of which have already been given—why he considered Dr Stewart's arguments to be wrong. At the end of this the interviewer asked him what would be the consequences if Dr Stewart should prove to be right. Dr Reissland demurred, on the ground that he had just explained in some detail why she could not be right. Well, said the interviewer, but just supposing she *were* right, what should we have to do? So Dr Reissland admitted that if she *were* right the regulations concerning radiation would have to be tightened up, to reduce the dose allowed. Dr Stewart's contribution was presented in detail; the only part of Dr Reissland's contribution presented was his last sentence stating that if she were right the doses permitted by the regulations would have to be reduced.

An even more blatant distortion was presented by Yorkshire Television on 1 November 1983 and repeated by ITV on 3 April 1984. In 32 years from 1953 there have been eight cases of leukaemia among children at Seascale, a small town with 2200 inhabitants on the coast $2\frac{1}{2}$ km from Sellafield. This is about ten times the national rate and clearly cannot be due to chance. These cases were presented in the broadcasts as being obviously due to contamination of the beaches and of household dust by radioactive materials from Sellafield.

This, together with the findings of the *Black Report*, was discussed in Chapter 11 (pp. 216–17). It was pointed out that the radiation doses received were hundreds of times too small to account for the leukaemias observed. No other possible cause was suggested in the film, and in particular no mention was made of possible infective agents carried by untreated sewage on the Seascale beach.

The director of the YTV film was quite aware of the relation between radiation doses and cancer. He had taped a long interview with me a few months before, in which I described the Argonne results (figure 4.3) showing that small increases in radiation may on balance do no harm at all, which I had thought would be interesting to a lot of people. Like three other taped interviews I had had with the BBC and ITV on topical radiation events over the last few years, this was not broadcast. Being myself biased by this treatment, I suspect that if I had always prophesied doom and destruction instead of presenting quantitative data in a simple form all four would have been broadcast.

Some of the arguments presented at the Sizewell inquiry and now being repeated at the Hinkley Point inquiry appear very strange coming from bodies concerned with conservation. The argument that there is no present need for a new power station, for example, must assume that we have no interest in the conservation of the extra 60 million tonnes or so of coal that it would replace during its working life. To say that we do not need the output now, when the world will have twice the number to feed and fuel before the station completes its working life, suggests an immoral lack of concern for either our own descendants or for the rest of the world which does not seem compatible with the various aims of any of the societies mentioned. To reduce current unemployment by building for the future helped to pull the USA out of the slump of the 1930s, and I can see no reason why it should not help Britain now.

Signs of second thoughts are at last beginning to appear in all the societies claiming to be concerned with conservation; but there is a long way to go and nothing can restore the ten-year gap in nuclear construction. One new nuclear station started in each of those years would have cost a lot, but between them they could have saved over 500 million tonnes of fossil fuel during their lifetimes. If the societies mentioned had concentrated the money and effort spent *against* nuclear power on pressure *for* government aid to wind and wave research and to promising conservation schemes, which would have produced even more jobs per pound spent, we might have had more of both wind and nuclear power—which would be better for conservation of oil and coal than more of either alone.

For anyone with technical knowledge of the facts, it is difficult to understand why so many people fear the effects of nuclear power so much more than they fear the effects of the well-known risks of traditional sources, and shudder at the very word 'nuclear'. Professor Petr Beckmann has given a very plausible explanation (*The Times* 28 March 1989). He says: 'First and foremost, the underlying thrust of the anti-nuclear campaign is not about energy: it is about enmity to technology, industry, profits and free enterprise. Nuclear power is the ideal surrogate target sought out by the coercive collectivists of the intellectual establishment. It is vulnerable to scare campaigns presenting it as mysterious and menacing, and opposing it makes people appear moral—above all to themselves.'

He is speaking for the USA, but it is clearly relevant to Britain, where the militant left of the Labour Party fits his description well.

In spite of the flood of noisy expression of opposition to nuclear power, well supported by the media, it appears that the British public has a majority who believe in the value of nuclear power. A poll taken in October 1988 showed that 51% believed that the share of nuclear power should be increased, or at least stay the same; and 70% think that Britain's need for nuclear energy will increase in the years ahead. UKAEA has been making considerable efforts to inform the public more fully about nuclear power, and in particular has produced useful packs of teaching material and information for use in senior schools. BNF has made a success of its open doors at Sellafield, with large numbers of visitors.

SUMMARY

Collectively the opponents of nuclear power have done good by impressing the industry with the need for ever increasing safety measures as the industry grows. Nearly all have done good by pressing for more development of such sources as wind generators which are non-polluting and little danger to the public. Individually some, such as Amory Lovins, have made valuable contributions to discussions on the relative usefulness of the different forms of energy in different applications. Too many have displayed such ignorance of the basic facts which determine both the kind and the magnitude of the hazards of nuclear energy to humans that their stated opinions can be given little weight. Hardly any have considered quantitatively the risks associated with any source of energy other than nuclear.

I am sure that most readers of this book will have felt that I am prejudiced in favour of nuclear power. Even if I have succeeded in giving a perfectly balanced account this feeling could persist, owing to the consistent selective bias of the media in the opposite direction. If you have been for some time in a car with the windows tinted green, when you wind down the windows the world looks pink.

Chapter 16

Finale

Comparisons are odorous

Dogberry in *Much Ado About Nothing*

I am much aware that my extensive use of arithmetic, and in particular the arithmetic of large numbers, is distasteful to most people. Unfortunately most people also want information that needs both a lot of knowledge of basic but unfamiliar features of the world and a lot of quantitative comparisons which cannot be obtained without arithmetic. There is nothing unfamiliar in wanting the goods but not wanting to pay for them—and nothing necessarily immoral or undesirable either; the combination has led to a lot of honest original thought and invention. At the same time, when it comes to the pinch, you have to choose between paying for the goods and doing without.

The importance of this goes far beyond its relevance to nuclear power. All of the decisions which have to be made by a democratically elected government are made on the basis of the cost of implementation and on the value of the effects. If both cost and value depend to an important extent on technology it is important that at least some of the decision makers should have access to and should understand the quantitative information involved.

It is a great strength of representative government in a democracy that people who understand the ends but not the means can elect representatives who agree with them on ends and who are capable of understanding the means; but it is essential that at least some of the electors do themselves understand the means. Just as a countrywide democracy could not be effective in the past while the great majority were illiterate, it could become ineffective in the future if too many are innumerate.

If the nuclear power controversy can teach the protagonists to follow and appreciate quantitative technical arguments it will have much more lasting importance than any effect that it may have on the proportions of different sources of electricity that we shall be using in the next few decades.

As a result of the enormously rapid increase in our understanding of the world and of the technological advances made possible by this, most of the traditional health hazards up to middle age have gone, and the remaining health risks are small and unfamiliar. The effect is a widespread feeling that it should also be possible to eliminate the remaining health hazards; and risks which would before have seemed trivial have now gained unprecedented importance. Where the risks of a particular group of hazards are small, there is naturally little information as to just how small they are; and to put them in order of importance it is imperative to use relevant and established technical knowledge and accurate numerical assessments. This means a little science and a lot of figures. The cost of acquiring the necessary skills is considerable, and many people will not feel that they can afford it, having already enough other, more pressing, interests to occupy them. This is perfectly fair. All that I would add is that if you are offered all the answers for negligible intellectual cost, the value of the answers offered is exactly the same —negligible.

WHAT OF THE FUTURE?

I may unintentionally have given the impression that I am unconcerned about the future. This is not so; concern for the future is my main motive for writing this book. But our descendants will know what they want better than we do, and will be far better equipped with knowledge of how to improve things for themselves. I have no idea to what extent radiation will contribute to their advantage or to their detriment; but it is more important to leave to them the knowledge of how to detect and control natural radiations that they must anyway face, than to worry about tiny additions from man-made sources. During the next 10 000 years the total natural background dose delivered at an average place on our planet will be 2000 rems, so that a constant population of a million people over this period anywhere in the world will between them have received a total collective dose of 2000 million man-rems. Nothing that we can do in this generation can add much to either the carcinogenic or the mutagenic effect of natural radiation over even 10 000 years. What we *can* do is to leave them the knowledge that will enable future generations to cope with them.

HOW TO AVOID IRRADIATION IN THE FUTURE

Even with our present knowledge, we can see ways in which people in the future could reduce the natural background if they wish to do so. For example, they would surely build radon-free accommodation and could decide to reduce the carbon-14 and potassium-40 content of plant food with the use of CO_2 derived from fossil fuels and of isotope separators respectively. I doubt whether the latter possibility will ever be considered worthwhile, even if it turns out that the natural levels of beta and gamma radiation do kill more than they save. It is important to realise however that the engineering systems needed by the nuclear industry for its power reactors could be used to reduce the radiation doses received by people by very much more than the effluents of the nuclear industry are increasing them.

It is probable that in a couple of generations' time, if we can divert our main attention from the minor carcinogenic and mutagenic effects of the nuclear industry to the major effects of our chemical environment, we shall have reduced the initiation rate of both cancers and mutations to a small fraction of the present level. It is also probable that we shall understand and be able to control the later stages which must occur before cancers actually develop. What is really unlikely is that in even 100 years we shall still be seriously concerned with the things that most seriously concern us now. This is effectively illustrated by our present uninterestedness in the shortage of flints, which would have been of such serious concern to our neolithic ancestors, or in the fear expressed last century that within a further hundred years the streets of London would all be impassable because they would be knee-deep in horse droppings. Other concerns of the not so distant past were smallpox, typhoid and cholera, the first of which has been eliminated, and the other two of which are of no importance to countries with even a moderately advanced technology and adequate supplies of energy for the production of chlorine and large-scale water supplies. When we think that it is only about 100 years since Lister led the movement towards antiseptic surgery (before that operations frequently led to death from 'hospital gangrene'); that it is only about 100 years since streets and houses were first lit by electricity; and less than 50 years since the first antibiotic, we surely should be able to expect the next generation or two to make accelerated advances in the fields which matter most—even if these are not always those which get the best financial support.

It is incumbent upon us to provide the best conditions we can for the present world and for 50 to 100 years hence; it is sensible to worry about large effects on the further future such as unchecked population growth but it is not common sense to stop a useful action now because there might be a small effect in 1000 years' time.

HOW TO HURT OR HELP OUR DECENDANTS

The real damage we can do to our descendants will arise in any or all of three ways. The first is from leaving too many of them too soon. There are no known limits to how many descendants we can have if they are spread over a long enough time to ensure that there are never too many at once; and given plenty of energy we can synthesise food for as many people as the world can carry. If, however, we continue to double our numbers every 40 years as we are doing at present, the world will reach its absolute limiting capacity in about 900 years' time, simply because it will not be possible to get rid of the heat liberated by our own bodies and by the equipment to keep ourselves going (Fremlin 1964). Practically all other life on the planet would have been wiped out long before that.

Assuming that we shall in fact limit our numbers, the second form of damage that we can do is to destroy the renewable resources of the planet represented by its still teeming richness and variety of wildlife. The third, and least important (which does not mean unimportant) is the using up of the non-renewable resources such as gas, coal and oil.

The early ways of limiting population—disease, famine and war—are all undesirable, and in current conditions the last might have catastrophic effects. The only acceptable means is a limitation of births. The first requirement of this means is that people should want to limit births. So far it is only the rich populations, with plenty of food, public provision for the old and good medical services, that wish to limit births and will do so effectively. In such a society, parents can confidently expect their children to grow up and will not be compelled to produce several 'spares' in case of accident to ensure that they themselves can rely on support in their old age.

To give the poorer part of the world the resources for this will need a great deal more energy than they now have. Although the large cities of Africa and Asia could use gas or electricity from large sources, including nuclear power stations, a big and rapidly multiplying part of the really poor in country villages cannot do so; they need cooking fuel, clean water supplies, practical education and time. Kerosine and coal are alternatives to the destruction of the forests, and the rich countries should be systematically planning to conserve these as effectively as possible and as soon as possible. Compared with the disastrous effects inevitable in an overcrowded world, none of the health hazards of any form of energy production are of any importance at all. Costs too, though not negligible, are of less importance in the rich countries in the long run than is the need to reduce dependence on non-renewable sources which are essential for the poor.

We shall ease the transition over the next 50 years or so, before geothermal energy could become available on a large scale, and reduce the climatic risks of too rapid a production of carbon dioxide, if the richer countries use as much wind and solar energy as they can, develop nuclear energy to replace all of the oil and coal used for production of electricity as fast as they can, and use nuclear electricity to replace oil, coal and gas in small-scale applications as fast as this can be afforded.

In the field of energy production, as in many other ways, the rich may in the long term do better for themselves as well as the poor if in the short term they pay more attention to the poor and less to themselves.

SUMMARY AND CONCLUSIONS

The main, and I believe inescapable, conclusion from the foregoing chapters is that whether from large-scale accident, from sabotage or from routine injury to the public, nuclear power is safer than power from fossil fuels. Oil refineries or liquefied natural gas will produce more frequent serious accidents, and coal kills more in normal operation. Fossil fuels burnt in power stations, however, cause far fewer deaths per unit of energy delivered to the public than do the fossil fuels used in transport and small-scale units.

Since the abandonment of open fires, which would not burn without excellent ventilation, the average radiation dose to the public from alpha-emitting radon and its descendants has risen to a level which may cause several hundred extra deaths from cancer every year, and this could easily be increased by two or three times by over-enthusiastic efforts to conserve heat and by reducing ventilation. The development of cheap and effective ways of reducing the seepage of radon through floors and walls would be hundreds of times more valuable than would the far more costly effort to reduce still further the already tiny radioactive effluent of the nuclear industry.

All forms of ionising radiation at high doses and high dose rates can induce cancer, but there is evidence that beta or gamma radiation at low dose rates in youth and early adulthood may more than compensate for this and may moreover extend the expectation of life. It is uncertain whether this is also true of alpha radiation, such as that from radon, radium or plutonium.

Chemical pollutants from fossil fuels cause more cancers than all the radiations together, including those from radon and its descendants, but all of the energy-producing industries combined are producing only a tiny fraction of the environmentally caused cancers now killing at least 70 000 people a year in Britain. Other causes must be things that we eat, drink and breathe, and may well include many natural and traditional foods. A large-scale epidemiological study of the relation between cancer and diet could pay much larger dividends than a study of any industrial hazards. It is likely that there are synergistic interactions between the effects due to chemicals in our diet, effects due to background radiation, and the effects of the carcinogens produced by burning fossil fuels. If so, a reduction of the first would automatically reduce the danger of the others.

In the long run major accidents kill far fewer people than do routine operations, except in the case of hydroelectric systems. Accidents causing the deaths of hundreds of people *can* occur at oil refineries, liquefied fuel gas stores or nuclear power stations, although more things have to go wrong together in the nuclear case. It is also likely that nuclear establishments can be less easily used by terrorists than could oil refineries or liquefied gas stores.

It was pointed out in Chapter 3 that the number of actual deaths caused by power production is less important than the number of years of active life that could otherwise have been expected before death occurred from some unrelated cause. Practically all of the deaths believed to be caused among the public by either fossil fuels or nuclear power are long delayed, often occurring 30 years or more after the initiation of cancer or of respiratory illness. Both are more likely to kill people after the age of 65 than to kill them earlier. The respiratory illnesses cause increasing discomfort and disability for many years before death; while cancers have no adverse effects on physical activity and the enjoyment of life until a relatively short time before the end—on the other hand this short time can be much more seriously painful. Although cancer deaths have increased rapidly throughout this century, the largest numbers of deaths are due to heart and circulation trouble, which like cancer are a consequence of our much increased life span and may be quite as difficult to prevent.

All of the major sources of energy have, by raising our material standard of living, lengthened far more lives than they have shortened, besides increasing our freedom to live life as we choose. The use of coal and oil enabled us to live long enough for cancer to become important. The knowledge and the radioactive tracers derived from nuclear reactors have continued the process and have been essential factors in showing us the way to the understanding of the complex and wonderful structure of our cells that may enable us to control cancer.

As Bhabha could have said: no energy supply is more dangerous than no energy supply. I would go a bit further and say that the differences in risk between the different large-scale sources of energy are so small, and the differences in cost so unpredictable, that neither risk nor cost should determine our decisions on the kinds of energy we should use. It is to me delightful to find that, perhaps for the first time, we can make some of the major decisions affecting the future with a vision no longer limited by extrapolation from the past.

If I have encouraged some of those who have worried about the tiny effects of nuclear wastes on our far distant descendants to worry instead about the thousands of our own children who are going to die in the world's big towns from the effects of pollution by burning coal and oil, the work of writing this book will have been worth while.

Appendix 1

A Mechanism of Cancer Initiation

The normal operation and reproduction of all cells is controlled by the deoxyribonucleic acid (DNA) in the nucleus. The arrangement of DNA in the nucleus is complicated. In human body cells it is contained in 23 pairs of thread-like structures, known as chromosomes, each of which consists of two immensely long molecular backbones or strands wound round each other in a double helix. Each of the strands carries, like ribs, a succession of sub-molecules known as bases. There are only four kinds of these in DNA: adenine (A), cytosine (C), guanine (G) and thymine (T). The bases on one strand interlock with the bases on the other, holding the double helix together, adenine always interlocking with thymine and guanine with cytosine, as in figure A1.1.

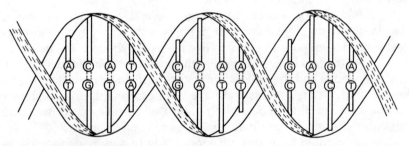

Figure A1.1 Simplified view of a section of DNA.

The four bases then form a kind of four-letter alphabet, in which is written the information required for the construction of proteins and the operation of the cells and hence of the complete organism.

It is very unlikely that all cancers arise in the same way, but it is generally accepted that in a large number of cases at least two stages are involved. The first may be a change within the nucleus of a single cell, while the second is the development of a colony of descendants of the changed cell, which grows out of control.

The first stage can be quick; at Hiroshima the radiation which initiated almost all of the resulting cancers was delivered in a few milliseconds. The second stage (which may involve several steps) rarely takes less than three or four years and may take several decades. The same pattern of initiation followed by a long delay seems also to apply to the cancers induced by chemical cancer-producers (carcinogens). There is no firm evidence that either radiation or chemical carcinogens affect the second stage, and I shall discuss only the first initiating stage.

The primary effect of radiation on a cell is simple. A fast-moving electrically charged particle, either an electron or an alpha particle, will ionise—eject an electron from—either the DNA or one of the protein or water molecules also in the cell. In all cases the electron ejected will have been responsible for holding together the atoms in the molecule from which it came (or will displace one of those that was responsible), and this may lead to irreversible chemical changes in the DNA. Such an irreversible change will change the information conveyed, and forms a mutation. If such a change takes place in an ovum or sperm it can lead only to changes—usually harmful—in the next generation. If it takes place in an ordinary body cell it constitutes what is known as a somatic mutation and cannot affect the next generation, but can produce effects, including cancer, in the individual of whom the cell forms part.

Even if only a protein or a water molecule in the nucleus is disrupted, without ionisation, the pieces (known as free radicals)—such as a superoxide (O_2^- or hydroxyl (OH)—can react with the DNA bases to produce similar irreversible changes. Many of these indirect changes will be identical with some of the irreversible changes in DNA produced by chemical carcinogens.

It is known that a specific type of bladder cancer is produced by a specific change, known as a point mutation, in just one of the pairs of bases that hold together the two strands that form the double helix of DNA (Reddy et al 1982). The single cell change is one that could be mediated either by radiation or by chemical action. There is however a difficulty: that such a change would be produced too easily by radiation—by which I mean that the amount of natural radiation we now receive would, if this change in a single chromosome were sufficient, produce far more cancers than have ever been observed. One fast electron traversing a cell nucleus will produce over 100 ionisations and two or three times as many free radicals (Fremlin 1980). This could lead to changes in hundreds of bases, but almost all of these would be repaired; permanent damage will be done only if the end result of the changes is a different but still complementary pair of bases, which will not be recognised as needing repair. In the case of the bladder cancer mentioned above this has happened. A G–C pair in a particular place in a particular gene has been changed into a T–A pair, which does not make nonsense of the information carried by the gene but evidently replaces it by erroneous and damaging information (as if a telegram intended to read 'do not attack' were transmitted as 'do now attack'). It is possible that this gene is one of those responsible for the receptors that every cell must have for the messages to tell it when to stop dividing. It happens that this particular change would involve little chemical difficulty, consisting mainly of a slight rearrangement of atoms already there. It is likely that as research proceeds similar changes in other genes will be found to be carcinogenic.

The probability of this particular change in one nucleus traversed by one electron is of course small; but there are over 10^{13} cells in the human body, and by the age of 20 every cell nucleus will on the average have been struck four times by electrons if the beta–gamma natural background throughout has been 100 mrem a year. Presumably a lot more damage still will have been done by the chemical carcinogens which produce 100 times as many cancers. There may be 6000 million base pairs in the nucleus of a human cell; but somewhere in the body any particular pair in a particular gene must have been seriously damaged some millions of times in 20 years. And yet, after all of this, the two rems received from the background in that time will, on ICRP figures, lead to only one extra chance in 5000 that the owner of all these cells will develop a cancer.

This difficulty disappears if we suppose that the change from (for example) G–C to T–A is a recessive defect which must appear independently in *each* member of the pair of chromosomes containing the gene concerned. The probabilities then become quite reasonable. If this is so, both defects may be produced by the same fast electron, when the chance of it happening will be proportional to dose; or they may be produced by two separate electrons, the chances of which will be proportional to the square of the dose.

This gives us an argument for a linear plus quadratic equation for the total dose, but does not apply to genetic effects, in which only one chromosome of each pair is passed on by either parent. The production of a critical base change in a successful germ cell therefore means that in the resulting offspring every single one of the 10^{13} cells would carry a faulty gene from the beginning, which at first sight could lead to a lot of cancers. The effect however would be that, in the offspring, the same critical injury by any carcinogenic agency to even one of the normal genes derived from the second parent could produce a cell capable of uncontrolled multiplication, which would be reliably lethal in a growing foetus from a very early stage, leading to one of the very common miscarriages for unknown reasons rather than to a recognisable cancer after birth. This might account for the babies who die of cancer before one year of age, who cannot have experienced the long delay between initiation and expression of cancer that is typical of adults. (The number of deaths from cancer for babies under one in the UK in 1986 was 19, only 10 of these classed as leukaemia.)

This discussion has all been concerned with cancers initiated by point mutations involving a permanent change of one base pair. The same arguments will apply if the backbone of the DNA—which is made up of an unchanging alternation of phosphate and ribose sugar groups like links of a chain—is broken by ionisation or chemical damage. Such a break in one strand will usually be quickly and correctly repaired, as the ends are kept close together by the second strand. If however a double break occurs in each of two neighbouring chromosomes, there will be four loose ends and there is considerable chance˙that the wrong pairs may be joined. This may cross-link two different chromosomes, and in each of these the gene in which the break had occurred would be effectively eliminated. This is a relatively common occurrence where large doses of radiation are received in a short time; cross-linked chromosomes and chromosomes between which sections have been exchanged can be seen under the microscope in suitable white blood cells and can be used to estimate the radiation dose.

As in the case of the point mutation, it is likely that at least two independent chemical or radiation-induced injuries will be needed, requiring a quadratic relation as before. The fact that many tumour cells show chromosome defects suggests that this is an important mechanism for cancer initiation.

This analysis has some features in common with the theory of dual radiation action by Kellerer and Rossi (1972), but uses a different approach.

EFFECTS OF ALPHA PARTICLES

Cancer initiation by alpha particles must differ in several respects from that by beta or gamma rays. Even with large whole-body doses very few cell nuclei will be struck more than once. An alpha particle would deposit about 300 times as much energy in the nucleus of a cell as would a fast electron. An alpha particle from one of the polonium decay products of radon for example could produce tens of thousands of ion pairs on its passage through the nucleus. A vast electric field will be set up briefly between the positively charged core of the track and the negatively charged surroundings containing the ejected electrons. The energy released by their rapid recombination with positive ions will products of radon for example could produce thousands of ion pairs on its passage through the nucleus. A vast electric field will be set nucleus is thus colossal and irremediable.

The proportion of the nuclear volume that is so comprehensively wrecked however is quite small, and although many cells must be killed some, although damaged, may still be capable of successful division and of providing a clone of malignant cells to develop into a cancer.

Appendix 2

Some Properties of Radioactive Nuclides

THE URANIUM-238 CHAIN

Nuclide	Emission†	Half-life	Activity of 1 g of the pure substance (Ci)
Plutonium-242‡	α	3.8×10^5 yr	3.9×10^{-3}
↓			
Uranium-238	α	4.5×10^9 yr	3.4×10^{-7}
↓			
Thorium-234	β, γ	24 d	2.3×10^4
↓			
Protactinium-234	β, γ	6.75 h	2.0×10^6
↓			
Uranium-234	α	2.5×10^5 yr	6×10^{-3}
↓			
Thorium-230	α	8×10^4 yr	0.02
↓			
Radium-226	α	1600 yr	1
↓			
Radon-222	α	3.8 d	1.5×10^5
↓			
Polonium-218	α	3 min	2.9×10^8
↓			
Lead-214	β, γ	27 min	3.3×10^7
↓			
Bismuth-214	β, γ	19.7 min	4.3×10^7
↓			
Polonium-214	α	0.5 s	1.0×10^{11}
↓			
Lead-210	β, γ	21 yr	82
↓			
Bismuth-210	β, γ	5.01 d	1.24×10^5
↓			
Polonium-210	α	138 d	4.5×10^3
↓			
Lead-206	Stable		

† α = alpha particle, β = beta particle, γ = gamma ray
‡ Not found in nature

THE URANIUM-235 CHAIN

Nuclide	Emission	Half-life	Activity of 1 g of the pure substance (Ci)
Plutonium-239† (F) ↓	α	2.44×10^4 yr	6.2×10^{-2}
Uranium-235 (F) ↓	α	7.1×10^8 yr	2.2×10^{-6}
Thorium-231 ↓	β, γ	25.5 h	5.4×10^5
Protactinium-231 ↓	α	3.25×10^4 yr	4.8×10^{-2}
Actinium-227 ↓·	β, γ	21.6 yr	74
Thorium-227 ↓	α	18.2 d	3.2×10^4
Radium-223 ↓	α	11.4 d	5.2×10^4
Radon-219 ↓	α	4 s	1.3×10^{10}
Polonium-215 ↓	α	1.8×10^{-3} s	3×10^{13}
Lead-211 ↓	β, γ	36 min	2.5×10^7
Bismuth-211 ↓	β, γ	2.2 min	4.2×10^8
Polonium-211 ↓	α	25 s	2.2×10^9
Lead-207	Stable		

† Not found in nature
(F) Fissionable with thermal neutrons

THE THORIUM-232 CHAIN

Nuclide	Emission	Half-life	Activity of 1 g of the pure substance (Ci)
Thorium-232 ↓	α	1.4×10^{10} yr	1.1×10^{-7}
Radium-228 ↓	β	6.7 yr	2.4×10^2
Actinium-228 ↓	β, γ	6.1 h	2.3×10^6
Thorium-228 ↓	α	1.9 h	8.3×10^2
Radium-224 ↓	α	3.6 d	1.6×10^5
Radon-220 ↓	α	55 s	9.4×10^8
Polonium-216 ↓	α	15 s	3.5×10^9
Lead-212 ↓	β, γ	10.6 h	1.4×10^6
Bismuth-212 ↓	β, γ	61 min	1.5×10^7
Polonium-212 ↓	α	3.7×10^{-7} s	1.4×10^{17}
Lead-208	Stable		

OTHER RADIOACTIVE NUCLIDES REFERRED TO IN THE TEXT

Nuclide	Emission	Half-life	Activity of 1 g of the pure substance (Ci)
Hydrogen-3 (tritium)	β	12 yr	9.8×10^3
Carbon-14	β	5700 yr	4.5
Sodium-24	β, γ	15 h	8.8×10^6
Argon-41	β, γ	1.8 h	4.2×10^7
Potassium-40	β, γ	1.3×10^9 yr	7.0×10^{-6}
Bromine-87	β, n	55 s	2.4×10^9
Krypton-85	β, γ	11 yr	3.9×10^2
Rubidium-87	β	5×10^{10} yr	8.3×10^{-8}
Strontium-90 $\downarrow$	β	28 yr	1.4×10^2
Yttrium-90	β	64 h	5.5×10^5
Zirconium-93	β	1.5×10^6 yr	2.6×10^{-3}
Technetium-99	β	1.5×10^6 yr	2.6×10^{-3}
Ruthenium-106 $\downarrow$	β	1 yr	3.4×10^3
Rhodium-106	β, γ	130 min	1.4×10^7
Iodine-129	β, γ	1.7×10^7 yr	1.6×10^{-4}
Iodine-131	β, γ	8 d	1.25×10^5
Iodine-135	β, γ	6.7 h	3.5×10^6
Xenon-133	β, γ	5.3 d	1.9×10^5
Xenon-135	β, γ	9.2 h	2.5×10^6
Caesium-134	β, γ	2 yr	1.3×10^3
Caesium-137	β, γ	30 yr	88
Plutonium-241 $\downarrow$	β, γ (F)	13 yr	1.1×10^2
Americium-241 $\downarrow$	α (F)	460 yr	3.3
Neptunium-237	α (F)	2.1×10^6 yr	7.1×10^{-4}

(F) Fissionable with slow neutrons

THE BUILD-UP OF HEAVIER NUCLEI FROM URANIUM

A heavy fissile nucleus struck by a slow neutron will not always undergo fission. The energy liberated when a neutron is captured may be emitted as a gamma ray before fission takes place. In a high-power reactor a nucleus may absorb several neutrons in succession before the fuel is removed, and the main processes leading to heavier nuclei are shown below. The relative fission and neutron capture probabilities are all given for thermal neutron absorption. With high energy neutrons all of the nuclei shown may undergo fission.

Where a nucleus has a half-life of only a few days, it is unlikely to absorb another neutron before decaying, so the effects of such absorption have been neglected. Where the half-life is 80 years or more, the products of radioactive decay have been omitted. Most of the nuclei for which no decay is shown will emit alpha particles; for example Pu-239 $\xrightarrow{\alpha}$ U-235.

The half-lives shown are of course for the spontaneous radioactive decay where *no* neutrons have entered the nucleus.

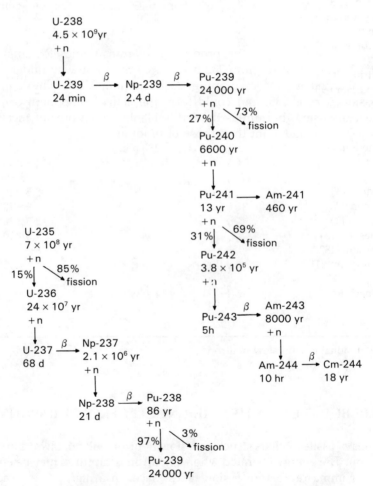

Figure A2.1 Production of transuranic elements.

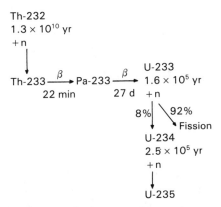

Figure A2.2 Behaviour of thorium-232 under neutron bombardment.

Figure A2.1 shows the production of transuranic elements. Figure A2.2 shows that thorium-232 behaves in the same way under neutron bombardment as does uranium-238. Uranium-233 is almost as easily fissioned as U-235, and could form the active component of breeder reactors using thorium as the fertile blanket. Rich ores of thorium are much more abundant than those of uranium.

Appendix 3

The Nuclear Physics of Reactors

The nuclei of atoms are composed of particles called nucleons. There are two kinds of these: protons which carry positive electric charge and repel each other, and neutrons which are uncharged and do not. Each particle has a mass close to 1.6×10^{-27} kg, the neutron being 0.1% heavier than the proton. All atoms of a given element have the same number of protons, this number being known as the atomic number of the element; but different isotopes of each element will have different numbers of neutrons. Thus natural uranium consists of three different isotopes, almost 99.3% being uranium-238 with 92 protons and 146 neutrons (making 238 nucleons), 0.7% being uranium-235 with 92 protons and 143 neutrons, and 0.006% being uranium-234. Most elements have one or more stable (non-radioactive) isotopes, and all elements have radioactive isotopes. The chemical properties of an atom depend on the number of electrons outside the nucleus, each of which has a negative electrical charge, and their number must be equal to the number of positively charged protons inside the nucleus for the atom as a whole to be electrically neutral. Different isotopes of the same element have the same chemical properties but quite different nuclear properties.

The protons in a nucleus repel each other with enormous force. Thus one single proton on the outside of a uranium nucleus is repelled by the other 91 protons with a force that would lift 23 kg off the floor, or give a rifle bullet (weighing as much as 10^{24} protons) the acceleration of $20\,000g$ that it would experience in the barrel of a high-velocity rifle. (g is the acceleration due to gravity at the earth's surface.)

The reason why all nuclei do not instantaneously fly to pieces is that the nucleons, whether protons or neutrons, are attracted to each other by a strong nuclear force. At the average distance between nucleons in the nucleus, about 10^{-15} m, this nuclear force is around 100 times as great as the repulsive force between protons; but it is effective, like glue, only at very short distances, while the electrical repulsion is still enormous at distances beyond the range of the attractive force. A neutron, as soon as it comes within range of the strong nuclear forces of for example a nucleus of U-235, will be pulled in with enormous acceleration, and will create a splash on its arrival that will set the whole resulting U-236 nucleus into oscillation, and may split it in two. This is the process of fission, and is the basis of nuclear power production.

By the time the two parts, which are usually unequal, have a clear space of little more than 10^{-15} m between them, the nuclear force has vanished and the electrostatic repulsion takes full control, causing an initial acceleration around 10^{27} g and final velocities in opposite directions of more than 100 000 km/s by the time the electrostatic repulsion has fallen to a negligible value.

It is misleading to say that the vast (for such small particles) kinetic energy that this represents was obtained from the loss of mass. It was obtained from the vast electric repulsive forces; and the mass of the fission fragments plus the mass of their kinetic energies (see p. 99) still adds up exactly to the mass of the short-lived U-236 formed by the original U-235 nucleus plus its absorbed neutron. *After* the fission fragments have come to rest in the surrounding material and have dissipated their kinetic energy as heat, the sum of their masses will be less than that of the U-236 by the mass of their initial kinetic energy, and the mass of the surrounding material will have increased by the mass of the same amount of heat energy. It is better to think of the loss of mass of the fission products as a result of the fission rather than as a cause. (In passing, it is perhaps as well to remind readers that when any material rises in temperature, this means that its molecules are moving faster—the temperature is a measure of the average kinetic energy of the molecules or atoms.)

Two vital aspects of fission have been neglected in this simplified description. The first is that either at the time of separation, or before the fission products have moved far, two or three neutrons are thrown out at much the same speed and are available to produce further fissions of other U-235 nuclei. It is important that in a few cases a neutron is emitted after a considerable delay; for example the fission product bromine-87 emits a beta-particle with a half-life of 55 s, and immediately afterwards emits a neutron, leaving krypton-86. The second is that the fission products are exceedingly radioactive. Most of the fission products are very short-lived, and decay in the reactor in which they are formed, usefully adding the energy of their radioactive decay to the kinetic energy of the fission products. Huge as is the activity of the wastes reaching Sellafield from the tonne of fission products made in producing 1 GWye, this activity is minute compared with the initial activity of a tonne of uranium undergoing fission in a large bomb explosion.

Uranium-238 will undergo fission if struck by a sufficiently energetic neutron—moving at over 15 000 km/s—but the energy liberated when a slow-moving neutron is captured by U-238 is not enough to cause fission and is emitted as a gamma ray, leaving a nucleus of U-239. This is radioactive with a half-life of 23.5 min, and emits a beta particle to become neptunium-239, which emits another with a half-life of 2.3 d to become plutonium-239 (see Appendix 2). This, like uranium-235, will undergo fission with the slowest neutrons, and as time goes on will contribute more and more of the fission products and energy in a reactor.

This process will take place in any uranium-using reactor, and even in a thermal reactor the plutonium produced and burnt may produce more energy and more fission products than has uranium-235 itself. By the time the accumulation of neutron-absorbing fission products has made it necessary to replace the fuel, in an efficient reactor much of the uranium-235 will have been destroyed, but the remaining unburnt plutonium may be equivalent to about 80% of the uranium-235 burnt. In a fast-neutron (breeder) reactor, with a blanket of 'depleted' uranium (i.e. U-238 with little remaining U-235) this process will produce plutonium-239 more efficiently, the amount of new fissile material produced being perhaps 10% more than that burnt in the core of the reactor.

As shown in Appendix 2, an alternative blanket material could be natural thorium-232, which like U-238 does not undergo fission when bombarded by slow neutrons but will absorb them, giving thorium-233 which goes through two beta decays to give the fissionable uranium-233.

Although the U-238 nucleus is not sufficiently distorted by slow neutrons to suffer fission, the 238 nucleons of which it is composed are in violent motion—as is the case in all nuclei—and there is an infinitesimal chance that these motions may momentarily leave most of the protons concentrated near opposite sides of the nucleus, when fission may occur spontaneously. The probability of spontaneous fission in U-238 is millions of times less than is alpha emission (and was discovered by Professor Flerov in the USSR only after neutron-induced fission had been known for several years). Nevertheless a tonne of uranium will produce 100 000 fissions a second, so that there is no need to supply an external neutron source to start up a power reactor.

To keep the fission process going at a constant rate in a reactor, the neutrons from each fission must produce on the average just one new fission. In a bomb, a sufficient mass of pure fissile material such as U-235 or Pu-239, suddenly put together at the last instant, will have on the average more than one neutron captured from each fission, giving an increase of 20% or more per generation, and the process will accelerate at a very rapid rate. In a thermal reactor the proportion of fissile material is small, and a moderator slowing and reflecting back neutrons is needed to keep the process going. Few neutrons escape altogether, so that nearly as many can be absorbed by U-238 to make Pu-239 as will cause fission of U-235, those remaining being absorbed by the moderator, supporting structures and control rods.

Even in this case, each generation of neutrons will take only a few hundred microseconds and if all the neutrons from a fission were emitted instantaneously an increase of only 1% per generation would double the output in less than a second. This would make control of output to close limits very difficult, and it is exceedingly fortunate that all the neutrons are not emitted instantaneously. Delayed neutron emitters such as the fission product bromine-87 introduce an automatic control. A small increase in the rate of capture of neutrons will not produce its full effect in increasing the output of neutrons for some minutes, giving a gradual rather than instantaneous rise in heat output. This makes the small adjustments needed for steady operation smooth and easy, and gives time for the operator to see the effects of the adjustments instead of requiring the reaction time of a racing driver.

Glossary

ACE Association for Conservation of Energy.

Adsorption Adsorbed atoms or molecules will be stuck to the surface of a solid, as distinct from atoms absorbed into the body of the solid.

AEC Atomic Energy Commission (USA).

Age-adjusted death rates *See* Death rates.

Age-specific death rates *See* Death rates.

AGR Advanced gas-cooled reactor.

ALARA As low as reasonably achievable. The ALARA principle requires nuclear installations to limit their discharges to levels ALARA.

Alpha particle A fast-moving helium nucleus emitted by nuclei of radioactive atoms such as radium or plutonium.

Ames test A test for the capacity of a substance to produce mutations in bacteria, which is often used to demonstrate the likelihood that the substance may also produce cancers.

Amorphous With atoms or molecules arranged randomly (in contrast to crystalline, when atoms are arranged in a regular pattern).

Becquerel (Bq) Unit of radioactivity. One becquerel is one disintegration per second. 1 curie = 3.7×10^{10} Bq.

BEIR (Committee) Committee on the Biological Effects of Ionising Radiation (USA).

Bentonite A clay mineral that swells with the absorption of water.

Benzo-a-pyrene (benzpyrene) A cancer-producing hydrocarbon ($C_{20}H_{12}$).

Beta–gamma radiation Any mixture of beta and gamma radiation; these have identical effects when absorbed in an organ.

Beta particles Fast-moving electrons emitted by the nuclei of radioactive substances, for example caesium-137 or carbon-14.

BNF(= BNFL) British Nuclear Fuels, which includes the processing plant at Sellafield and other major plants at Risley and Capenhurst.

Bq *See* Becquerel.

Carcinogen A substance which induces cancer.

CEGB Central Electricity Generating Board.

CHP Combined heat and power.

Ci *See* Curie.

Collective dose This is the product of the average dose in rems received by members of a population and the number of people in that population. Thus if 100 people each receive 100 rems, or a million people each receive 0.01 rems (10 mrems), the collective dose is 10 000 man-rems.

Critical mass/density The minimum quantity or density of a fissile material that will sustain a chain reaction. The critical mass depends on shape: a solid sphere has the smallest critical mass for any shape, and a thin sheet will never become critical however large its mass or area.

Curie (Ci) The unit used to measure the rate of radioactive decay. One curie is 3.7×10^{10} disintegrations per second, the decay rate of 1 gram of radium or almost three tonnes of uranium. Millicurie (mCi) and microcurie (μCi)—one thousandth and one millionth of a curie respectively—are also used.

Cytoplasm The part of a living cell surrounding the nucleus but within the surface membrane.

Death rate The number of deaths per year per hundred thousand (or sometimes per million). **Gross rate**: number per hundred thousand of the total population. **Age-specific rate**: the number of deaths per hundred thousand within a short age range, e.g. between 10 and 14 or between 60 and 64. **Age-adjusted death rate**: since death rates increase with age a high gross death rate may mean only an aging population. To compare two populations it is useful to calculate for each the death rate they would have had if they had had the same age distribution.

DNA Deoxyribonucleic acid (*see* Appendix 1).

Endemic (Usually of illness) regularly found in a population or an area.

Enrichment (of uranium). Increase of the proportion of the fissile isotope uranium-235.

Epidemic A disease that attacks large numbers in a particular area for a limited time.

Fertile (of uranium-238 and other heavy nuclei). Capable of absorbing neutrons to form a fissile product.

Fissile Capable of undergoing fission when struck by slow (low energy) neutrons.

Fission The break-up of a large nucleus into two smaller nuclei with the liberation of neutrons. **Spontaneous fission**: a rare form of radioactivity which occurs without the absorption of a neutron or other particle. **Induced fission**: fission produced by the absorption of a neutron or (rarely) another particle.

FRL Fisheries Research Laboratory at Lowestoft, Suffolk.

Gamma ray A penetrating radiation, like x-rays, given off from the nuclei of many radioactive substances such as caesium-137 or potassium-40. When absorbed by the body a gamma ray will produce a fast-moving electron.

Gigawatt (GW) A million kilowatts. **GWe**: a gigawatt of electricity. **GWy/e/h**: Gigawatt-year/of electricity/of heat. This is equal to 1 million kilowatt-years or 8760 million kilowatt-hours.

GCR Gas-cooled reactor.

GNP Gross national product.

Gray (Gy) 100 rads.

'Greens' Members of the (West German) Ecology Party.

Gy *See* Gray.

Half-life The time it takes for half the nuclei of a radioactive material to undergo radioactive decay, and hence for the activity of the material to be reduced to half.

Hormesis The stimulus given to an organism by non-toxic concentrations of a toxic substance.

HSE Health and Safety Executive.

IAEA International Atomic Energy Agency. With head office in Vienna, this has representatives of most countries, with active support from both USA and USSR.

ICRP International Commission on Radiological Protection. This has operated since 1928 and suggests permissible limits of radiation dose (using research in many countries), which in Britain have the force of law.

IERE International Electric Research Exchange.

Ion pair When an atom or molecule is ionised it is broken into two ions, one a negative free electron and the other a positively charged atom or molecule. These are known as an ion pair; they may recombine, or interact with surrounding molecules.

Joule (J) SI unit of work, heat or energy of any kind. Equal to work done when a force of 1 newton moves a mass through a distance of 1 metre. 1 joule = 0.000 948 BTU (British Thermal Units). 1 joule per second = 1 watt.

LET Linear energy transfer: the amount of energy transferred from an electron or alpha particle to its surroundings per unit length of track. Beta particles or the electrons produced by gamma rays have a low LET, alpha particles a high LET.

Leukaemia A malignant disease in which white blood cells multiply uncontrollably. It can be produced by irradiation of the bone marrow, but nearly all (over 90%) of the cases in Britain arise from other causes.

LFG Liquefied fuel gas.

LNG Liquefied natural gas.

LPG Liquefied petroleum gas.

MAFF Ministry of Agriculture, Fisheries and Food.

Magnox Earliest type of reactor built in Britain, in which the fuel consisted of natural uranium metal rods sheathed in a magnesium–aluminium alloy.

MEB Midlands electricity board.

Mega- A million times (denoted by the prefix M).

Megaton (of bombs). One producing an explosive energy equivalent to that from a million tons of TNT (trinitrotoluene, a high explosive). (1 American ton = 0.907 metric tonnes.) About 5×10^{15} joules.

Mesons Short-lived particles produced by cosmic rays in the upper atmosphere, some of which penetrate the atmosphere to ground level, and which cause damage similar to that caused by x-rays or fast electrons.

Micro- One millionth (denoted by the prefix μ, as in μg, μCi, μm etc).

Microgram (μg) One millionth of a gram.

Micron One millionth of a metre (μm). Used in preference to micrometre to avoid confusion with a measuring instrument called a micrometer.

Milli- One thousandth (denoted by the prefix m).

Moderator Material (e.g. graphite, water) surrounding the fuel rods or pellets in a reactor to slow down neutrons released in fission and hence to increase their ability to induce further fission.

Molecule The smallest part of a substance which has the properties of the substance; for example CO_2, the molecule of which consists of one atom of carbon plus two of oxygen.

mph   Miles per hour. 1 mph = 1.61 km/h.

Mutagen   A substance which induces mutations.

NCB   National Coal Board.

NII   Nuclear Installations Inspectorate of the HSE. An independent organisation set up by the British Government to control the design and operation of nuclear plants.

NPT   Non-Proliferation Treaty.

NRC   Nuclear Regulatory Commission (USA).

NRPB   National Radiological Protection Board.

Nuclide   Any specific isotope of any element.

OPEC   Organisation of Petroleum Exporting Countries.

PFR   Prototype fast reactor.

Photovoltaic cell   A thin slice of silicon or other material that will produce a small voltage (about 1 volt) when illuminated. The current produced in an external circuit will be proportional to the intensity of the light used and to the area of the cell.

ppm   Parts per million.

PWR   Pressurised water reactor.

Rad   The dose of radiation received when 0.01 joules of x-ray, gamma ray or beta-particle energy is absorbed in each kilogram of the organ or person receiving the dose.

RBMK   Transliteration of the Russian initials of a type of pressure tube reactor used only in the USSR.

Rem   The dose received when radiation of any kind is absorbed producing the same biological effect as 1 rad of x-rays.

Schistosomiasis (bilharzia)   A water-borne disease carried by water snails, easily avoided by filtering or boiling water but capable of infecting through the bare skin.

Sievert (Sv)   A large unit of dose received from radioactivity. 1 Sv = 100 rems.

Somatic cells   Cells forming all parts of the body, other than the germ cells in the reproductive organs which are passed on to the next generation.

Sterile male technique   A method of insect control which successfully exterminated screw-worm flies over large areas. Large numbers of male flies are sterilised by radiation and released; wild females mating with them produce no fertile eggs.

Sv   *See* Sievert.

Sv   *See* Sverdrup.

Sverdrup   An oceanic unit: 1 Sv = 1 cubic kilometre per second—roughly the flow rate of the Gulf Stream.

Synergistic   A synergistic substance is one that increases the effectiveness of another substance. (Used here in connection with increasing ill effects on health.)

Tailings   Accumulation of residues from the extraction of uranium or other materials from their ores.

TMI   Three Mile Island nuclear power station.

Tonne   1000 kilograms (=0.984 English ton or 1.102 US ton). In approximate measurements the difference between 1 tonne and 1 English ton is not important.

Tracheotomy   Cutting into the windpipe (to prevent choking to death).

UCS   Union of Concerned Scientists.

UKAEA   United Kingdom Atomic Energy Authority.

UNSCEAR   United Nations Scientific Committee on the Effects of Atomic Radiation, set up in 1955.

WLM   Working level month. One working level is that concentration of radioactive substances that will release a total of 2.1×10^{-8} joules of alpha-particle energy in each litre of air by the time all activity has decayed. A dose of 1 WLM would be received by someone breathing air at one working level for one working month of 170 hours. For a miner working hard this is a cancer risk roughly the same as 1 rem of whole-body irradiation; for normal indoor rates of breathing it is roughly equivalent to 0.5 rem.

References

Anderson A 1986 *Nature* **320** 475

Ashenden T W and Mansfield T A 1978 *Nature* **273** 142–3

Atkins D H F, Wiffen R D, Hardy C and Tarrant J B 1979 *AERE Report* R9426

Atom 1983 London dumping convention *Atom* no 318 88

Becker J 1982 *Nature* **298** 112

BEIR III 1980 Committee on the biological effects of ionising radiation *The Effects on Population of Exposure to Low Levels of Ionising Radiations* (Washington, DC: National Academy Press)

Bertell R 1985 *No Immediate Danger* (London: The Women's Press)

Black Sir Douglas 1984 Report of Independent Advisory Group chaired by Sir Douglas Black *Investigation of the Possible Increased Incidence of Cancer in West Cumbria* (London: HMSO)

Bodansky D, Robkin M A and Stadler D P 1987 *Indoor Radon and its Hazards* (Washington, DC: University of Washington Press)

British Safety Council 1980 *Report to Members of Parliament on Gas Explosions*

Brown S S and Lugo A E 1984 *Science* **223** 1292

Bungay H R 1982 *Science* **218** 643–6

Camplin W C, Grimwood P D and White I F 1980 *NRPB Report* R94

Carlson L D and Jackson B H 1959 *Radiat. Res.* **11** 509–19

Carlson L D, Scheyer W J and Jackson B A 1957 *Radiat. Res.* **7** 109–97

Cederlöf R, Doll R, Fowler B, Friberg L, Nelson N and Vouk V 1978 *Environ. Health Perspect.* **22** 1–12

Chan W H, Ro C U, Lusis M A and Vet R J 1982 *Atmos. Environ.* **16** 801–14

Clarke M J, Fleischman A B and Webb G A M 1981 *NRPB Report* R120

Clarke M J and Smith G B 1988 *Nature* **332** 245–6

Clarke R H and Southwood T R T 1989 *Nature* **338** 197

Coal and the Environment 1981 Commission on Energy and the Environment (London: HMSO)

Cohen A V and Pritchard D K 1980 *Health and Safety Executive Research Paper* 11 (London: HMSO)

Cohen B 1988 *Report at meeting of American Chemical Society*

Cohen B L 1976 *Bulletin of the Atomic Scientists* (February)

——1977 *Rev. Mod. Phys.* **49** 1–20

Cottrell A 1981 *How Safe is Nuclear Energy?* (London: Heinemann)

Collier J G and Davis L M 1987 *CEGB Report*

Cook J 1986 *Red Alert* (London: New English Library)

Crick N J and Linsley G S 1983 *NRPB Report* R135 and Addendum (London: HMSO)

Curran S C and Curran J S 1979 *Energy and Human Needs* (Edinburgh: Scottish Academic Press) p. 34 *et seq.*

Currie W M 1984 *Atom* no 327 (January)

Davies R V, Kennedy J, McIlroy R W, Spense R and Hill K M 1964 *Nature* **203** 1110–15

Department of Energy 1989 *Review: Renewable Energy* Issue 6

Doll R 1981 *Br. J. Radiol.* **54** 179–86

Doll R and Peto R 1981 *The Causes of Cancer* (New York: Oxford University Press)

Eadie A 1978 Answer to a written question *Parliamentary Debates, Official Report* House of Commons (*Hansard* 15/5/1978)

Eakins J D, Lally D E, Burton P J, Kilworth D R and Pratley P A 1982 *AERE Report* R10127

Electronics and Power 1987 June 365

Energy Paper 46 1979 (London: HMSO)

EPA 1983 *Regulatory Impact Analysis of Final Environmental Standards for Uranium Mill Tailings at Active Sites* EPA 520/1-63-010 (Washington, DC: US Environmental Protection Agency) pp. 4–8

Etkins R and Epstein E S 1982 *Science* **215** 287–9

Evans R D, Harley J H, Jacobi W, McLean W S, Mills W A and Stewart C G 1981 *Nature* **290** 98–100

Fetter S A and Tsipis K 1981 *Sci. Am.* **244** 33–9

Fleischmann M and Pons S 1989 *J. Electroanal. Chem.* **261** 301–8

Flowers Report 1976 *Nuclear Power and the Environment* (Royal Commission on Environmental Pollution, Sixth Report, Command 6618) (London: HMSO)

Fremlin J H 1964 *New Sci.* **24** 285

——1980 *Ambio* **9** 60–5

Frigerio N A, Eckerman K F and Stower R S 1973 The Argonne radiological impact programme *Environmental and Earth Sciences Report* ANI/ES-26 Part 1 (Illinois: Argonne National Laboratory)

Fryer L S and Griffiths R F 1979 *Report* SRD R49 (Safety and Reliability Directorate, UKEA)

Gittus J H 1982 *CEGB Proof of Evidence: on Degraded Core Analysis* 2 vols (London: CEGB) p. 16 and p. 30

Glaser P J 1970 *Microwave Power* 5(4) (Special issue on Satellite Power and Microwave Transmission to Earth)

Glasstone S and Dolan P J 1977 *The Effects of Nuclear Weapons* (UK: Castlehouse Publications) 3rd edn 1980

Gofman J W and Tamplin A R 1973 *Poisoned Power* (London: Chatto and Windus)

Grosch D S and Hopwood L E 1979 *Biological Effects of Radiation* (New York: Academic) 2nd edn

Grubb M 1988 *Windirections* (Newsletter of the British and European Wind Energy Association) **8** no 3 p. 6

Halliday J A, Lipman N H, Bossanyi E A and Musgrove P J 1983 *Paper presented at Wind Workshop 6, June 1983, Minneapolis, USA*

Hamilton L D 1983 *Testimony at Sizewell B Public Inquiry* (paper APG/P/4) (APG, 8 Ruvigny Mansions, Putney, London SW15)

Hammond R P 1962 *Nucleonics* **20** 45–9

Handbook of Chemistry and Physics 1981–2 ed R C Weast and M J Astle (Boca Raton, Florida: Chemical Rubber Company)

Hansard 1977 Parliamentary Reply given by Mr Benn 25 July 1977

——1988 Parliamentary Reply given by Mrs Currie 16 December 1988

Haque A K M M, Collinson A J L and Blyth-Brook C O S 1965 *Phys. Med. Biol.* **10** 505–14

HBRRG (High background radiation research group, China) 1980 *Science* **209** 877–80

Health and Safety Statistics 1989 *Employment Gazette* **97** no 2

Hesketh G E 1980 *Commission of the European Communities Document* V/2408/80 pp. 360–84

Hill M D and Grimwood P D 1978 *NRPB Report* R69

Hill M D, White I F and Fleishman A R 1980 *NRPB Report* R95

Hill M J, Hawkesworth G and Tattersall G 1973 *Br. J. Cancer* **28** 562–7

Hirayama T 1979 *Nutrition and Cancer* **1** 67–8

Hoffsten P E and Dixon F J 1974 *J. Immun.* **112** 564

Holmes L 1981 *Electron. Power* 810–52

IAEA 1975 *The Oklo Phenomenon* ST1/PUB/405 (Vienna: IAEA)

——1982 *Bulletin* **24** 31–2

IARC 1982 *Cancer Incidence in Five Continents* vol IV ed J Waterhouse, C Muir, K Shanmugaratnam and J Powell IARC Scientific Publication no 42 (Lyon: IARC)

ICRP 1977 Publication 27 *Ann. ICRP* **1**(4)

IEE 1983a *IEE News* no 90 (June)

——1983b *IEE News* no 93 (September)

——1984 *IEE News* (Jan) p. 3

IERE 1981 International Electric Research Exchange Working Group II *Effects of SO₂ and its Derivatives on Health and Ecology* vol 2 (Leatherhead, Surrey: CEGB Central Electricity Research Laboratories)

Inspector General 1981 *Report by Sir Richard West* (in *Birmingham Post* 25/3/81)

Izrael Y A 1987 *Ecological Consequences of Radioactive Contamination in the Chernobyl Emergency Zone* (Moscow)

Jones S E *et al* 1989 *Nature* **338** 737–40

Kammerbauer H, Selinger H, Römmelt R, Ziegler J A, Knoppik D and Hock B 1986 *Environ. Pollution* A **42** 133–42

Keeling C D and Ryecroft M J 1982 *Nature* **285** 180

Kellerer A M and Rossi H H 1972 *Curr. Top. Radiat. Res.* **8** 85

Kenmuir D 1978 *A Wilderness called Kariba: the Wild Life and Natural History of Kariba* (Zimbabwe)

Kinlen L 1988 *The Lancet* 10 December

Kletz T A 1979 Safety in numbers *Focus in Engineering* (ICI)

Kneale G W, Stewart A and Mancuso T F 1978 *Int. Symp. on the Late Effects of Ionising Radiation, Vienna, March 1978* IAEA-SM-224/510 (Vienna: IAEA)

Kneale G W, Stewart A and Wilson L M K 1986 *Cancer Immun. Immunother.* **21** 129–32

Kochupillai N, Verma I C, Grewell M S and Ramalingaswami V 1976 *Nature* **262** 60–1

Lamerton L F, Pontifex A H, Blackett N M and Adams K 1960 *Br. J. Radiol.* **33** 287–301

Lee B 1983 *New Sci.* **90** 227–9
Life Span Study Report 1986 TRI-86. Radiation Effects Research Foundation, Japan
Likens G E, Wright R F, Galloway J N and Butler T J 1979 *Sci. Am.* **241** 39–47
Lindop P 1965 *Sci. Basis Med. Ann. Rev.* 91–109
Lipschutz R D 1981 *Radioactive Waste; Politics, Technology and Risk* (Cambridge, Mass.: Ballinger)
Lorenz E 1950 *Am. J. Roentgenol. Nucl. Med.* **63** 176–85
Lovins A B 1977 *Soft Energy Paths* (Harmondsworth: Pelican)
MAFF 1975–88 *Aquatic Environment Monitoring Reports* (Annual Reports) (Lowestoft: Directorate of Fisheries Research)
Manning P T 1980 *CEGB Report* PL-GS/E/4/80
Mayneord W V and Clarke R H 1975 *Br. J. Radiol. Suppl.* **12** 58
Mays C W 1980 *A Symposium of the American Academy of Arts and Science* ed R G Sachs (Cambridge, Mass.: Ballinger)
——1982 *Proc. 17th Ann. Meeting NRCP, April 1981* pp. 182–200 (Bethesda, Maryland: NRCP)
Mercer J H 1978 *Nature* **271** 321–5
Meyers N 1984 *Nature* **307** 99
Mohnen V A 1988 *Sci. Am.* August 14
Morgan K Z 1975 *J. Am. Ind. Hyg. Assoc.* **36** 365
Mortality Statistics 1981 *Cause* (London: HMSO)
——1982 *Cause* (London: HMSO)
——1985 *Accidents and Violence* (London: HMSO)
——1986 *Cause* (London: HMSO)
Murphy D 1981 *Race to the Finish* (London: John Murray)
Nature 1975 **257** 525–6
——1983 **304** 301
——1984 **307** 87
New Scientist 1976 **72** 265
——1977 **74** 761
——1982 **94** 337
——1983a **99** 295
——1983b **98** 761
——1983c **99** 102
——1983d **99** 768
——1984a **101** no 1398 26
——1984b **101** no 1391 22
——1989 **122** no 1660 19
Niehaus R 1983 *IAEA Bull.* **25** 30–6
NII 1979 *The Leakage of Radioactive Liquor into the Ground, BNFL Windscale 15/3/1979* (London: Health and Safety Executive)
Nuclear Engineering International 1975 **20** 1015
Nuclear News 1988 February
Ockrent D 1981 *Proc. R. Soc.* A **376** 133
O'Riordan M C, James A C, Rae S and Wrixon A D 1983 *NRPB Report* R152
——1987 *NRPB Report* GS6
——1988 *NRPB Report* R190
Pearce F 1981 *New Sci.* **91** 643

Pentreath R J 1980 *Nuclear Power, Man and the Environment* (London: Taylor and Francis)

Pentreath R J, Lovell M B, Harvey B R and Ibbett R D 1979 in *Biological Implications of Radionuclides Released from Nuclear Industries* vol 2 (Vienna: IAEA)

PERG 1981 *Research Report* RR7 (Oxford: Political Ecology Research Group)

Pusch R and Jacobsson A 1980 *Symp. on Underground Disposal of Radioactive Waste* vol 1 (Vienna: IAEA) p. 487

Reddy E P, Reynolds R K, Santos E and Barbacid M 1982 *Nature* **300** 149–52

Registrar General's Statistical Review of England and Wales for the three years 1968–70: Supplement on Cancer 1975 (London: HMSO)

Rogers H G, Thomas J F and Bingham G E 1983 *Science* **220** 428–9

Royal College of Physicians 1983 *Health or Smoking* (London: Pitman)

Royal Commission on Environmental Pollution 1976 *Sixth Report* (London: HMSO)

——1981 *Eighth Report* (London: HMSO)

Rycroft M J 1982 *Nature* **295** 191–2

Salter S H 1974 *Nature* **249** 720–4

——1980 *IEE Proc.* **127** 308–19

Schwartz C 1975 *Science for the People* (May) (SESPA, 9 Walden Street, Jamaica Plain, Mass., USA)

Science 1976 **194** 172 *et seq.*

Scientific World 1986 no 3

Siddall E 1982 *Report* AECL 7540 (Ottawa: Atomic Energy of Canada Ltd)

Slovic, P, Fischhoff B and Lichtenstein S 1979 *Environment* **21** 14–39

Smith P G and Doll R 1981 *Br. J. Radiol.* **54** 187–94

Spiers F W 1968 *Radioisotopes in the Human Body: Physical and Biological Aspects* (London: Academic) p. 234

Spradbury J P 1973 *Wasps* (London: Sidgwick and Jackson)

Starr C 1969 *Science* **165** 1232–8

——1982 *Atom* no 314 250–5

Taylor P J 1984 *Report* RR-11 (Oxford: Political Ecology Research Group)

Tombs Sir Francis 1981 *Electron. Power* **27**(4) 284

Tuyns A J, Pequignot G and Jansen O M 1977 *Bull. Cancer (Paris)* **64** 45–60

UNSCEAR 1983 *IAEA Bull.* **24** 33–9

Warner F 1981 *Proc. R. Soc.* A **376** 205

Wash 1400 1975 *The Reactor Safety Study* NUREG 74/014 (Washington, DC: US Nuclear Regulatory Commission)

Weissburger J H and Williams G M 1981 *Science* **214** 401–7

Which? 1981 (May) 280–1

——1982 (Sept) 532–9

Winter A J B 1980 *Nature* **287** 826–9

Woodhead D S and Pentreath R K 1980 *2nd Int. Dumping Symp., Woods Hole Oceanographic Institution, Mass., USA* (Chichester: John Wiley)

Index

References to tables and figures shown in bold type.